NATURE AND NURTURE IN EARLY CHILD DEVELOPMENT

For developmental scientists, the nature versus nurture debate has been settled for some time. Neither nature nor nurture alone provides the answer. It is nature *and* nurture in concert that shape developmental pathways and outcomes, from health to behavior to competence. This insight has moved far beyond the assertion that both nature and nurture matter, progressing into the fascinating terrain of *how* they interact over the course of development. In this book, students, practitioners, policy analysts, and others with a serious interest in human development will learn what is transpiring in this new paradigm from the developmental scientists working at the cutting edge, from neural mechanisms to population studies, and from basic laboratory science to clinical and community interventions. Early childhood development is the critical focus of this book, because many of the important nature-nurture interactions occur then, with significant influences on lifelong developmental trajectories.

Daniel P. Keating is Professor of Psychology, Psychiatry, and Pediatrics; Research Professor at the Center for Human Growth and Development; and Faculty Associate in the Survey Research Center at the Institute for Social Research, all at the University of Michigan. He is also a Fellow of the Canadian Institute for Advanced Research (CIFAR) and a member of CIFAR's Successful Societies Program. Keating has held positions at the University of Minnesota; the University of Maryland; the Max Planck Institute for Human Development and Education in Berlin, Germany; and the University of Toronto. He has occasionally appeared on television, including on the *Phil Donahue Show* and the *Vision* series on TV Ontario. He has also been a guest on radio talk shows featured on the Canadian Broadcasting Corporation (CBC) and the Australian Broadcasting Corporation (ABC), focused mainly on his book (with Clyde Hertzman), *Developmental Health and the Wealth of Nations* (1999). Currently, much of his work focuses on the just-launched National Children's Study, for which he serves as an investigator in a number of capacities.

Nature and Nurture in Early Child Development

Edited by

Daniel P. Keating

University of Michigan

CAMBRIDGE
UNIVERSITY PRESS

CAMBRIDGE UNIVERSITY PRESS
Cambridge, New York, Melbourne, Madrid, Cape Town, Singapore,
São Paulo, Delhi, Dubai, Tokyo, Mexico City

Cambridge University Press
32 Avenue of the Americas, New York, NY 10013-2473, USA

www.cambridge.org
Information on this title: www.cambridge.org/9780521840408

First published 2011

Printed in the United States of America

A catalog record for this publication is available from the British Library.

Library of Congress Cataloging in Publication data
Nature and nurture in early child development / edited by Daniel P. Keating.
    p.   cm.
Includes bibliographical references and index.
ISBN 978-0-521-84040-8 (hardback)
1. Nature and nurture.   2. Genetic psychology.   3. Child psychology.   I. Keating,
Daniel P., 1949–   II. Title.
BF341.N377   2010
155.42′2 – dc22   2010030830

ISBN 978-0-521-84040-8 Hardback

CONTENTS

CONTRIBUTORS

RONALD G. BARR, MA, MDCM, FRCP(C), is Professor of Pediatrics at the University of British Columbia Faculty of Medicine; Head of Developmental Sciences and Child Health at the Child and Family Research Institute of BC Children's Hospital; and a Canada Research Chair in Community Child Health Research at the University of British Columbia. He is the former Director and is currently Fellow of the Experience-Based Brain and Biological Development Program of the Canadian Institute for Advanced Research. He received his medical training (MDCM) at McGill University and specialty training in pediatrics and child development at McGill (Montreal Children's Hospital) and Harvard University (Children's Hospital Medical Center). He studies the biological and behavioral co-regulatory determinants of behavior and has applied these concepts to the development of a primary educational prevention program for shaken baby syndrome (abusive head trauma) and infant abuse called the *Period of PURPLE Crying* in collaboration with the National Center on Shaken Baby Syndrome.

ROSSANA BISCEGLIA is a doctoral candidate in the Developmental Psychology and Education program at the University of Toronto. Her research interests include understanding the complex processes that contribute to psychopathology in children and adolescents. Her dissertation focuses on the developmental processes that influence mothers' sensitive responding to their infants. The thesis will also examine the relationship between maternal sensitivity and the vasopressin V1a receptor (AVPR1A) gene, and how this gene operates in conjunction with environmental influences in influencing mothers' sensitivity.

W. THOMAS BOYCE, MD, is the Sunny Hill Health Centre/BC Leadership Chair in Child Development in the Human Early Learning Partnership and the Centre for Community Child Health Research at the University of

British Columbia. He is also Co-Director of the Experience-Based Brain and Biological Development Program of the Canadian Institute for Advanced Research. He completed his BA at Stanford University, an MD at Baylor College of Medicine, and pediatric residency training at the University of California, San Francisco (UCSF). He was named a Robert Wood Johnson Foundation Clinical Scholar at the University of North Carolina and served on the pediatrics and public health faculties of UCSF and UC Berkeley. As a social epidemiologist and a developmental–behavioral pediatrician, his research addresses how neurogenomic and psychosocial processes work together to lead to differences in childhood disease across different socio-economic groups.

MEGAN R. GUNNAR, PHD, is a Regents Professor of Child Psychology at the Institute of Child Development, University of Minnesota. Her research interests focus on the effect of adverse early life care on the development of stress and emotion reactivity and regulation. Professor Gunnar is a member of the Society for Research in Child Development, the International Society on Infant Studies, the International Society of Psychoneuroendocrinology, the American Psychological Society, and the American Psychological Association. She directs a National Institute of Mental Health–supported Interdisciplinary Developmental Science Center on Early Experience, Stress, and Neurobehavioral Development and is a member of the Canadian Institute for Advanced Research Program on Experience-Based Brain and Biological Development.

CLYDE HERTZMAN, MD, FRCP(C), is Director of the Human Early Learning Partnership (HELP), College for Interdisciplinary Studies at the University of British Columbia (UBC); Tier I Canada Research Chair in Population Health and Human Development; and Professor in the School of Population and Public Health at UBC. Nationally, he is a Fellow of the Experience-Based Brain and Biological Development Program and the Successful Societies Program of the Canadian Institute for Advanced Research. He is a Fellow of the Royal Society of Canada and of the Canadian Academy of Health Sciences. Since 2008, he has been President of the Canadian Council on Early Child Development. He holds an honorary appointment at the Institute for Child Health, University College, London.

CHANDRA GHOSH IPPEN, PHD, is Associate Research Director of the Child Trauma Research Program at the University of California, San Francisco, and the Early Trauma Treatment Network, which is a member of the National Child Traumatic Stress Network (NCTSN). She has worked on seven longitudinal studies and has conducted treatment outcome research on the effectiveness of psychosocial intervention programs with Spanish-speaking children and parents. She is coauthor of *Losing a Parent to Death: Guidelines for the Treatment of Traumatic Bereavement in Infancy and Early Childhood* (2003); director of the NCTSN Measure Review Database; co-chair of the Cultural Competence Consortium of the NCTSN; and producer and director of *Vale la Pena Recordar*, a Spanish-language video on childhood traumatic grief. As a first-generation East Indian/Japanese American, she is committed to examining how culture and context affect perception and mental health systems.

JENNIFER JENKINS, PHD, is Professor of Human Development and Applied Psychology at the University of Toronto. Previously, Dr. Jenkins was on faculty at the University of Stirling, UK, and was a Senior Clinical Psychologist at the Hospital for Sick Children, Great Ormond St., London. She obtained her PhD from the University of London. Her work focuses on the ways in which family risks shape children's socioemotional development and the effect of family stresses, both relational and socioeconomic, that influence children's developmental trajectories; she has developed, with social statisticians, methodological techniques for studying complex family structures and influences. She directs the Kids, Families, and Places longitudinal community study involving 668 newborns, their older siblings, and their parents, which looks at the moderating role of individual vulnerabilities (biological and cognitive) on the relationship between environmental stress and children's well-being. She has written on emotional development and developmental psychopathology, including coauthoring *Understanding Emotions* and *Human Emotions*.

DANIEL P. KEATING, PHD, is Professor of Psychology, Psychiatry, and Pediatrics; Research Professor at the Center for Human Growth and Development; and Faculty Associate in the Survey Research Center at the Institute for Social Research, all at the University of Michigan. He is also a Fellow of the Canadian Institute for Advanced Research (CIFAR) and a member of CIFAR's Successful Societies Program. Keating has held faculty positions at the University of Minnesota; the University of Maryland; the Max Planck Institute for Human Development and Education in Berlin, Germany; and the University of Toronto. With Clyde Hertzman, he authored and edited *Developmental Health and the Wealth of Nations*. His research focuses on the biological and developmental mediators that can explain the link from social disparities in circumstances to differential outcomes in population developmental health.

ALICIA F. LIEBERMAN, PHD, is Professor of Psychology at the UCSF Department of Psychiatry, Director of the Child Trauma Research Project, and Senior Psychologist at the Infant-Parent Program, San Francisco General Hospital, and a Clinical Consultant with the San Francisco Department of Human Services. She received her BA degree at the Hebrew University of Jerusalem and her PhD from The Johns Hopkins University. She is on the Board of Directors of Zero to Three: The National Center for Infants, Toddlers, and Their Families and on the Board of *Parents* magazine. She is the author of *The Emotional Life of the Toddler*, which has been translated into five languages. She is also a Senior Editor of *DC: 0–3 Casebook* and coauthor of *Infants in Multiproblem Families*. Her major interests include toddler development, disorders of attachment, child–parent interventions with high-risk families, and the effects of early trauma in the first years of life.

MICHELLE M. LOMAN, MA, is a graduate student in Developmental Psychopathology and Clinical Science at the Institute of Child Development, University of Minnesota. Her research interests focus on the cognitive factors and neurophysiological correlates related to how children and adolescents who experience adversity regulate their behavior. She is a member of the Society for Research in Child Development, the Society for Research in Adolescence, and the International Society for Psychoneuroendocrinology. She was a University of Minnesota Center for Cognitive Sciences trainee and is actively involved in the University of Minnesota Center for Neurobehavioral Development.

CHARLES A. NELSON III, PHD, is Professor of Pediatrics and Neuroscience and of Psychology in Psychiatry at Harvard Medical School; an affiliate faculty member in the Harvard Graduate School of Education; and a Professor in the Harvard School of Public Health. In addition, he is the Richard David Scott Professor at Children's Hospital Boston and Director of Research in the Division of Developmental Medicine. He serves on the steering committees for both the Harvard Center on the Developing Child and the Harvard Interfaculty Initiative on Mind, Brain, and Behavior. An international leader in the field of developmental cognitive neuroscience, he has achieved numerous breakthroughs in scientific understanding of brain and behavioral development during infancy and childhood. He has a particular interest in how early experience influences the course of development and in this context has studied both typically developing children and children at risk for neurodevelopmental disorders.

MICHAEL RUTTER, MD, FRS, is Professor of Developmental Psychopathology at the Institute of Psychiatry, King's College, London. After graduating from Birmingham University Medical School and postgraduate posts in neurology, pediatrics, and cardiology, he undertook training in psychiatry at the Maudsley Hospital in London, followed by a research fellowship at Albert Einstein College of Medicine in New York. He has served at the Medical Research Council (MRC) Social Psychiatry Unit, as Senior Lecturer at the Institute of Psychiatry in London, and as Professor of Child Psychiatry and Head of the Department of Child and Adolescent Psychiatry. He was Honorary Director of the MRC Child Psychiatry Research Unit and of the Social, Genetic, and Developmental Psychiatry Research Centre, both of which he established at the Institute of Psychiatry. His research has spanned an unusually wide range, from epidemiology to molecular genetics, with clinical research foci from autism to depression and antisocial behavior. Elected to the Royal Society in 1987, he was knighted in 1992.

RICHARD E. TREMBLAY received his PhD from the University of London (UK). He is Professor at the University of Montreal, at University College Dublin, and Invited Scientist at National Institute of Health Unit 669 in Paris. He was Canada Research Chair in Child Development, Interdisciplinary Professor at Utrecht University, Invited Professor at the Paris-Sud medical faculty, and International Professor at the University of Central Lancashire. He has conducted a program of longitudinal and experimental studies on the physical, cognitive, emotional, and social development of children from conception to adulthood, focusing on the development and prevention of antisocial behavior. He coordinates the Marie Curie International Network for Early Childhood Health Development. He is the founding director of the Centre of Excellence for Early Childhood Development; the Web-based Encyclopedia on Early Childhood Development; and Quebec's Inter-University Research Unit on Children's Psycho-Social Maladjustment. He has received numerous scholarly awards, including Grand Officer of Chile's Gabriela Mistral Order and a Fellow of the Royal Society of Canada.

ACKNOWLEDGMENTS

There are many individuals to thank for their help in putting this book together, beginning with the contributors, who have been willing to work diligently on various phases of this project. There is a history to this book which entails additional thanks. The idea for bringing together a group of cutting-edge scientists working the frontiers of research on early child development, but who were also able to communicate with a broad audience, originated in discussions at the University of Toronto nearly ten years ago. With the strong support of the University of Toronto, two generous Canadian funders were identified who were interested in sponsoring such an effort: the Invest in Kids Foundation (IKF), based in Toronto, Ontario; and the Lawson Foundation, based in London, Ontario. Working collaboratively with the funders, a small group at the Ontario Institute for Studies in Education at the University (OISE/UT) – notably, Jane Bertrand, Jenny Jenkins, Dona Matthews, and Anita Zijdemans, with strong support from Carol Crill Russell from IKF – identified an ideal group to participate in this initiative. We were delighted that all of our first choices for each topic agreed to participate.

A unique feature of this effort is that it was intended from the start as an educational effort with a wide variety of audiences in mind. We also wanted to make use of new media to forward this goal. The initial result was the Millennium Dialogue on Early Child Development, held at OISE/UT in 2001. A cross section of stakeholders was invited to participate in this dialogue along with the contributors, both in person at the event and internationally through a simultaneous interactive webcast. (The experimental nature of this part of the effort was described and evaluated in *ePresence Interactive Media and Webforum 2001: An Accidental Case Study on the Use of Webcasting as a VLE for Early Child Development*, by Anita Zijdemans, Gale Moore, Ron Baecker, and Daniel Keating, which appeared in the

International Handbook of Virtual Learning Environments, Springer 2005.) The untiring efforts of many people were needed to take on the technical and logistical challenges of bringing this innovative forum together, but special thanks are owed to Jane Bertrand, Dona Matthews, and Anita Zijdemans.

One consequence of this decision, embraced by the contributors, was a conscientious effort to make the material accessible to a broad audience and to engage in dialogue with the other contributors and the larger group of participants, both in person and online. The dialogue was recorded on DVD,* and both the conference papers and the multimedia production were subsequently utilized in an educational project for early childcare students carried out by Red River College in Manitoba, Canada, under the able leadership of Janet Jamieson. Contributors were very generous with their time in supplementing and clarifying this curriculum, which has been used successfully in Canada and elsewhere. The team at OISE/UT supported these and other efforts to make educational use of the Millennium Dialogue, and that support is gratefully acknowledged, especially as it was manifested in the funding of the Atkinson Centre for Society and Child Development, which was created to enhance the work of an endowed Atkinson Chair in this field.

Recently, additional funding was made available through the University of Toronto, thanks to the efforts of two recent Chairs of the Department of Human Development and Applied Psychology at OISE/UT, Janet Wilde Astington and Esther Geva. This funding enabled us to return to the contributors with an invitation to do a thorough revision and updating of their original conference papers, and supporting them as they did so, to bring this knowledge to a broader audience. Again, all the contributors agreed, and the chapters in this book, which originated in the Millennium Dialogue of 2001, retain their original spirit but are completely revised to make the book fully up-to-date.

Over the intervening period, some contributors requested permission to use part of their efforts in other publications, and limited copyright permission to do so was granted by the University of Toronto. Some parts of the original conference paper by Charles A. Nelson III appeared in *Promoting Positive Child, Adolescent, and Family Development: Handbook of Program and Policy Interventions* (R. M. Lerner, F. Jacobs, and D. Wetlieb (eds.), Thousand Oaks, CA: Sage Publications), and a reciprocal permission from the editor of that book for the substantially revised Chapter 2 is appreciated. Similarly, a version of the original conference paper presented at the

---

* Contact keatingd@umich.edu for availability of DVD recordings of the 2001 conference.

Millennium Dialogue on Early Child Development by W. Thomas Boyce appeared (with permission) in *Developmental Psychopathology, Volume 2: Developmental Neuroscience* (2nd edition, D. Cicchetti and D. J. Cohen (eds), John Wiley & Sons). The present version in Chapter 5 draws on those prior sources but represents a distinct and contemporary contribution to the topic. Again, the consent of the editors for this arrangement is appreciated. In addition, Megan R. Gunnar acknowledges that her work was supported by the National Institute of Mental Health Research Scientist Award (MH00946), and Tom Boyce acknowledges the research support from the John T. and Catherine D. MacArthur Foundation Network on Development and Psychopathology.

An additional connection among many of the contributors runs throughout this book, and is acknowledged with great gratitude. The origins of this work reach back to the establishment of the Human Development Program (HDP) of the Canadian Institute for Advanced Research (CIFAR). The founding president of CIFAR, J. Fraser Mustard, developed the idea of a Canadian-based organization with international scope that would take on complex, interdisciplinary topics across many fields, establishing scientific networks whose members would be supported to tackle them. The Human Development Program was one of the early efforts in the human sciences, and I was privileged to be asked to lead that effort. The ongoing influence of the inspirational and scientific leadership of Fraser Mustard, and his vision of a new approach to taking on complex, multifaceted research questions, cannot be overstated.

Following the successful completion of ten years of HDP in 2003, two related programs with new mandates were established under the leadership of current CIFAR president Chaviva Hosek: the Successful Societies Program (SSP) and the Experience-Based Brain and Biological Development Program (EBBD). Contributors to this volume who have CIFAR links include: Sir Michael Rutter (Advisory Board, EBBD); Charles Nelson (Advisory Board, EBBD); Ronald G. Barr (Fellow, EBBD; Fellow, HDP); W. Thomas Boyce (Fellow and Co-Director, EBBD); Megan Gunnar (Fellow, EBBD and HDP); Richard E. Tremblay (Fellow, HDP); Clyde Hertzman (Fellow, EBBD, SSP, and HDP); and Daniel Keating (Fellow, SSP; Fellow and Director, HDP). The scientific culture of interdisciplinary collaboration and dialogue fostered and sustained by the CIFAR programs is evident throughout this volume, and I am grateful to have benefited from it.

Daniel Keating
Ann Arbor
November 2009

Introduction

DANIEL P. KEATING

For developmental scientists, the nature versus nurture debate has been settled for some time. Neither nature nor nurture alone provides the answer. It is always nature *and* nurture in concert that shape developmental pathways and outcomes, from health to behavior to competence. In recent years, this insight has moved far beyond the commonplace assertion that both nature and nurture matter, progressing into the far more complicated and fascinating terrain of understanding how they interact over the course of development. This research arena has benefited from the emergence of new tools and techniques, enabling the field to gather information on the biological processes underlying child development and to analyze more complex information in more sophisticated ways that enable a clearer view of nature-nurture interactions in operation. Another significant change has been the steady accumulation of longitudinal data sets that permit the study of developmental trajectories from birth, or even prenatally, to adolescence and into adulthood.

A NEW PARADIGM FOR EARLY CHILD DEVELOPMENT

The convergence of these factors – the uptake of a nature-and-nurture model into research endeavors on many fronts, the emergence of new research tools and techniques, and the accumulation of developmental databases to which they can be applied – has set the stage for a major expansion of knowledge about human development. Much of this work has focused the field's attention on early child development as a key period within which many of these important interactions are occurring, with significant influences on the subsequent developmental trajectories for which they provide the foundation. This is to be distinguished from infant determinism, with its claims that all-important developmental pathways have

1

been set by age three or five years. There are contingencies that operate throughout the life course, affecting eventual outcomes in all areas in important ways, both positively and negatively. However, some important pathways are heavily tilted toward nature-nurture interactions in early child development, establishing foundations that will enable or constrain a host of future pathways.

There are a number of ways in which these newly dominant models and perspectives have not been taken up fully among students, practitioners, policy thinkers and analysts, or the public at large. Clearly, moving beyond the nature versus nurture polemics of an earlier era will take some time. Nevertheless, for students, practitioners, and others with a serious interest in human development, it is crucial to provide the opportunity to learn what is transpiring at the cutting edge of the new paradigm.

THE GOAL OF THIS VOLUME

The goal of this volume is to provide such an opportunity. It aims to achieve this goal in two ways. The first is to learn from the developmental scientists who are themselves working at this cutting edge, in a range of critical areas of early child development from neural mechanisms to population studies, and from basic laboratory science to clinical interventions. Owing to the fast-moving nature of work in this area, it is important to learn from those who are generating the new knowledge and confronting the complexities that arise in doing so. As the reader will discover, this volume benefits from the willingness of these scientists to communicate effectively the excitement, challenge, and implications of this new paradigm in action.

In addition, these authors accepted an additional challenge. Without watering down the key scientific progress in their fields, they have made every effort to make the science accessible to a wide audience, from students to practitioners to policy thinkers to the interested public at large. Readers will judge how well this dual challenge – conveying the cutting edge in an accessible way – has been met, and there is no doubt that some of the material requires more careful, rather than casual, reading. In any case, the significance and the timeliness of the work deserve sound reporting from the frontlines of the new paradigm, which this volume aims to provide.

THE NEW PARADIGM IN POLICY AND PRACTICE

In addition to the mandates of explaining how the new paradigm is yielding new findings in developmental science, and of doing so in an accessible way, a third mandate was undertaken by the contributors to this volume: to call attention wherever possible to the implications for both policy and practice. The full range of these implications is beyond the scope of this volume, and in any case will continue to emerge as the new paradigm takes firmer hold in the field and in public discourse. One of the barriers is that nature-and-nurture is inherently more complex than single accounts from either perspective, but the routes to overcoming this barrier are to understand better both the science and its implications. The intent of this volume is to move that dual agenda forward. As Rutter notes in Chapter 1 of this volume:

> In summary, there can be no serious doubting of the importance of genetic influences on both normal and abnormal psychological functioning. Nevertheless, knowing that there is a genetic influence does not tell one anything very useful on its own, with respect to risk or protective causal pathways. That is because genetic influences may operate in such a diversity of ways, both direct and indirect, and the implications will vary as to the details of the mode of operation.

HOW TO USE THIS VOLUME

How could this book best be used by different audiences? It is ideally suited for graduate and advanced undergraduate courses and seminars that focus on understanding how biology, behavior, and society interact in the generation of developmental pathways in infancy and early childhood. For courses with a particular focus on early child development, this volume provides a substantial overview of most of the core issues. For courses with a broader overview of human development, it can provide an important basis for understanding the earliest processes and influences on child development. As the following overview indicates in more detail, the organization of the volume is generally from more micro to more macro, and from more basic to more applied – although these levels and approaches are intermixed throughout the volume.

For practitioners working in early child development, in settings from childcare to clinical intervention, this volume provides an important view into the underlying developmental mechanisms, a view that can be essential for grounding practice in a fully realized nature-and-nurture paradigm. It also encourages a new perspective on, and creative reworking of, some long-standing practices in the field.

For policy thinkers and analysts, the new paradigm will be challenging. Entrenched arguments arising from either perspective are hard to move, and these arguments are frequently deeply embedded in policy or even institutions. Educating the public will make easier the revamping of policies in line with the newly discovered realities of early child development. Of course, for research, policy, and practice, having a clear conceptual framework to organize the complex information arising from the new paradigm will be essential, and this volume seeks to provide a framework that can be widely used.

THE ORGANIZATION OF THIS VOLUME

There were many ways that this volume could have been organized. Each chapter deals with the nature-and-nurture perspective by considering gene-environment and biology-context interactions; each provides an overview of a broad area of the field, overlapping to some extent with all other chapters; and each chapter identifies important implications for research, policy, and practice. Accordingly, instructors and the general reader can productively read the volume in any order, or sample among the chapters on topics of the greatest interest.

Nonetheless, the sequence of the chapters was ordered to lead the reader generally from molecular processes and mechanisms toward more complex behaviors and contexts, and from more basic to more applied developmental science. The opening chapter by Rutter provides a comprehensive overview of the topics that arise as one considers the joint influence of biology and experience, from which the reader will obtain a succinct and workable framework for the key issues in the rest of the volume. In Chapter 2, Nelson provides a compelling introduction to early brain development, including the implications of the nature of those developments for lifelong growth and plasticity. It can be read productively as a primer on early brain development, laying out the major stages from conception through early child development.

In the two chapters that follow, Barr, and Gunnar and Loman take up some of the most important early experiences for lifelong behavior and health. Barr focuses especially on the nature of the most important early relationship, between infant and caregiver, most often the mother. He reports on an elegant series of studies that illuminate the complex ways in which caregiving serves as support and regulation of the infant's distress, sleep, and memory. Gunnar and Loman take up the core issue of the early development and function of the stress response system, especially as they affect the formation and later function of that system.

Boyce, and Jenkins and Bisceglia, in the two chapters that follow, focus on the specific features of the interaction of biology and context. Boyce describes an elegant overall formulation of that interaction, employing to good use the metaphor of a symphony, in order to draw our attention to the promise and the challenges of building a developmental science on the complex interaction of biology and context. Jenkins and Bisceglia address an aspect of one of the longest-lived issues in the old paradigm: how to assess environmental influences on early child development. Earlier analyses looked only at the shared environment until it was recognized that the within-family environment was quite variable between siblings. This topic is taken up by Jenkins and Bisceglia in a positive research agenda, examining in some detail the effects of differential experience by siblings within families.

In the two chapters that follow, Tremblay, and Lieberman and Ippen explore how the new perspective can be used to shape developmentally guided interventions that address early vulnerabilities through an understanding of the basic developmental processes underlying social relationships, from attachment to aggression. From these chapters, it is clear that although greater complexity is introduced by taking up the new paradigm and the importance of early experience, there are also a range of new tools and approaches that can be deployed to maximize the effectiveness of early interventions, through timing that is more appropriate and a focus on more precise targets of intervention.

In the two concluding chapters, Hertzman and Keating consider the influence of the new paradigm for understanding population patterns, and the societal responses that may be most effective in light of the new knowledge. Hertzman describes a comprehensive community-level approach to these issues, ranging from biodevelopment to population studies. Keating addresses similar issues, focusing on the persistent issue of social disparities in important developmental health outcomes, and presents a model for understanding those disparities in terms of the underlying causal pathways that run through early child development.

In combination, these chapters provide the reader with the opportunity to arrive rather quickly at the cutting edge of new research on early child development across a full range of central topics, to do so with the guidance of leading active researchers in each of these fields, and to consider some of the most significant implications of this new paradigm for policy and practice. This effort is, of course, an early contribution to what will no doubt become an ongoing and lengthy dialogue as the new paradigm grows into the mainstream not only among researchers but also eventually among the public at large, and as its implications become more apparent for how we see our children, our society, and ourselves.

1

Biological and Experiential Influences on Psychological Development

MICHAEL RUTTER

INTRODUCTION

Over recent decades there have been major developments in our understanding of the various ways in which biological and experiential factors influence psychological development (Rutter & Rutter, 1993; Rutter 2000a, 2006a; Shonkoff & Phillips, 2000). In this chapter, I focus on the conceptual issues that are involved, on possible causal mechanisms, on effects on abnormal or suboptimal functioning, and on implications for intervention. Throughout, the arguments are based on empirical research findings (placing most emphasis on those that have been replicated by independent research groups) and, when dealing with causal mechanisms, reference is made to studies in biology and medicine, as well as in psychology, to draw conclusions on likely processes.

DEVELOPMENT

In much of the literature, there has been a tendency to seek to partition influences into those that are genetic (G) and those that are environmental (E), as if between them they accounted for all possibilities. This is a seriously misleading oversimplification because it focuses exclusively on individual differences without taking account of the universals of development (see Rutter, 2002). Also, it wrongly assumes that G and E are separate and involve no co-action (see Rutter, 2006a) and ignores the role of chance. Accordingly, I start with a consideration of some of the key features of developmental processes.

Neural Development

Several features of brain development warrant particular emphasis (Greenough et al., 1987; Greenough & Black, 1992; Nelson, this volume; Nelson & Bloom, 1997, Nelson et al., 2006). It is well established that the process comprises an initial proliferation of neurons and neuronal connections, accompanied by neuronal migration, and later by myelinization of nerve fibers. This phase of overproduction of nerve cells (which takes place during the first few years of life in humans) is followed by a selective pruning resulting in a loss of nerve cells. This phased development is a consequence of the fact that brain development, like development as a whole, is probabilistic. In other words, there is a genetic programming of the pattern of development but this provides a general set of instructions, as it were, rather than a specific instruction as to what should happen to each individual neuron. This means that the phase of pruning can be used to fine-tune brain organization. This process is partially driven by neuron-neuron interaction so that the organization is shaped by the ways in which neurons fire together. This leads to the establishment of neuronal networks on the general principle of neurons that "fire together wire together."

It is also partially driven by experiential input. Another way of thinking about this process of brain development is to recognize that a key feature of biological development is that it is organized to be adaptive to circumstances. From an evolutionary viewpoint, it is important that the organism be able to develop normally in a wide range of environments (Bateson & Martin, 1999; Bateson et al., 2004). It would not "work" for there to be an undue reliance on specific environmental conditions that would apply only to some individuals. Nevertheless, within a wide range of expectable environments, there is a need for sensory input if brain development is to proceed normally. This was shown most dramatically by Hubel and Wiesel (2005) with respect to the dependence of the growth and functioning of the visual cortex on appropriate visual input during the early years. This has been amply confirmed by numerous investigators since then. The practical consequence is that, if there is not coordinated binocular visual input in the first few years, normal binocular vision at a later age is not likely. That is the reason why it is important for children's squints (strabismus) to be corrected early in life.

The most rapid and most radical period of brain growth takes place during the early years of life, but it is important to recognize that brain maturation continues throughout childhood and into early adult life (Giedd, Lalonde, Celano, White et al., 2009; Gogtay et al., 2004; Huttenlocher, 2002; Toga, Thompson, & Sowell, 2006). It used to be thought that no new neurons could be produced after the first few years but it is now apparent that new nerve cells can be produced, even in adult life, at least in the hippocampus (Baringa, 2003; Gross, 2000). It is certainly the case that brain plasticity is greatest in early life but a degree of plasticity is evident even at later ages (Nelson, this volume; Nelson et al., 2006; Roder & Neville, 2003; Stiles, 2000).

The Role of Chance

One of the consequences of the probabilistic process of development is that chance, as well as genetic and environmental influences, plays a role (Molenaar, Boomsma, & Dolan, 1993; Jensen, 1997; Goodman, 1991). Thus, females inherit two X chromosomes but one is always inactivated. It appears that it is largely a matter of chance which X chromosome is suppressed. There are numerous other examples in biology where it is evident that chance is operative. In thinking about the role of chance, however, it is necessary to differentiate between group effects and individual effects. One of the consequences of probabilistic development is that it is very common for development to go slightly awry in one way or another. This is evident, for example, in the high frequency with which individuals show minor congenital anomalies. There are predictable group effects on the frequency with which such anomalies occur. For example, they are more common in babies born to older mothers, and they are more common in twins (especially monozygotic twins) than in singletons (Vogel & Motulsky, 1997). To that extent, the occurrence of anomalies is predictable at a group level. However, the particular anomalies seen in any one individual do seem to be determined by chance to a considerable extent. Because of the importance of such chance effects, it is desirable to have some measure of the extent to which any particular individual shows developmental perturbations. One way in which this matter has been approached has been to focus on dermatoglyphic asymmetry (Naugler & Ludman, 1996), but the validity of this approach remains uncertain. Such asymmetry, measured as differences between left versus right fingerprint ridge counts, has been found in one study to increase because of increased maternal stress during prenatal weeks 14–22 (King, Mancini-Marïe, Brunet, Meaney et al., 2009).

ADAPTATION TO THE ENVIRONMENT

The notion that biological development proceeds in a way that is adapted to the environments experienced at the time is often considered under the general concept of *developmental programming*. This term has also been used to refer to the experience-expectant sculpting of brain development in relation, for example, to visual input, as outlined earlier (Greenough, Black, & Wallace, 1987). However, this type of sculpting that is contingent on an average expectable environment (hence the descriptor experience-expectant) only goes awry if the environments experienced are outside a very broad range.

The experience-adaptive notion of developmental programming is different in two key respects (Bateson et al., 2004; Barker, 1999; O'Brien et al., 1999; Rutter, 2006b). First, it deals with variations within (as well as outside) a normal range and, second, it has to be viewed in relative terms. The early experiences prepare the organism, biologically speaking, to be particularly well prepared to cope with a specific type of environment, rather than to develop optimally in some absolute sense. It is unlikely that the biological processes involve a single mechanism but there are parallels across systems. For example, initially babies across the world all show much the same skills in phonological discrimination. However, during the course of the first year discriminatory abilities become increasingly shaped by the language input experienced (Maye, Werker, & Gerken, 2002; Kuhl et al., 1997; Werker, 2003). Later on, children show phonological discriminations that are adapted to the language with which they have been brought up, and they find it quite difficult to pick up discriminations that are important in other languages, but not in their own. In somewhat comparable fashion, the early vocalizations of completely deaf children are normal, but as auditory input becomes increasingly important during the middle and latter part of the first year, sound productions become progressively distorted.

Before turning to a more detailed consideration of biological programming, it is necessary to note that there is a third variety of neural effects of experiences – what have been termed experience-dependent brain effects (Greenough, Black, & Wallace, 1987) – meaning that there are effects of experience on brain structure and function that can occur at any age and that do not involve effects on ordinary brain growth. Thus, animal studies have shown that environmental enrichment/deprivation has brain effects in adult as well as juvenile animals. Human brain imaging studies have shown the effects of psychological as well as pharmacological intervention (Goldapple et al., 2004). Other brain imaging studies have also shown variations in brain structure associated with intensive experiences in adult life (Rutter, 2009). This should not be surprising because learning has to involve the brain in some way. What is important, however, is to recognize that there are different types of brain effects resulting from experiences, and these types carry different implications. This may be considered further in relation to experience-adaptive programming.

Thus, the development of the immune system is similarly shaped by a pattern of early exposure to both infections and to antigenic substances (Bock & Whelan, 1991). The neuroendocrine system is also influenced by stress experiences in early life (Gunnar & Vasquez, 2006). There has been a particular interest lately in the possibility that dietary exposure functions in this way. Thus, numerous studies have shown that very small babies have a much-increased risk of hypertension, occlusive coronary artery disease, and diabetes in later life (Barker, 1999, 2007; O'Brien et al., 1999). The finding is striking because in midlife it is being overweight, rather than being underweight, that puts people at an increased risk of these diseases. It has been suggested that subnutrition in early life leads to developmental programming that is adapted to poor diets but, because of this, the organism is maladapted to the more affluent diets that the same individuals experience in later life.

One important consequence of this adaptational concept is that it may not be helpful to provide individuals who were deprived in early life with abundance later on. Thus, it has been suggested that feeding in middle or later childhood to compensate for subnutrition in infancy may be damaging as far as adult outcomes are concerned. The language example is a reminder that the early input provided adaptation to a particular language environment and not something that is better in an absolute sense. To what extent there may be psychological parallels to the dangers of overcompensation through rich experiences in middle childhood is completely unknown but knowing that biological adaptation often does work in this sort of way is a reminder that it would be unwise to discount the possibility totally.

The neural effects of severe experiential deprivation in humans have been most strikingly evident in the sequelae of profound institutional deprivation in children reared in institutions, who have been adopted into generally well-functioning adoptive families (Rutter, Beckett, Castle, Kreppner et al., 2007). Although no sequelae were detectable in children who left institutional care under the age of 6 months, deficits were found in about two-fifths of the children adopted after that age (Beckett et al., 2006; Kreppner, Rutter, Beckett, Colvert et al., 2007). Six aspects of the findings were particularly striking. First, the effects of profound institutional deprivation in the early preschool years proved to be remarkably persistent – so that the effects at 11 years were just about as marked as they had been at age 6. This was so despite having spent at least seven years in the environment of generally well-functioning adoptive families. Second, the rate of deficits showed a marked step-wise increase for adoptions occurring during the second half of the first year of life, with no further increase thereafter in spite of continuing institutional deprivation. The implication appears to be that profound institutional deprivation takes some months to bring about lasting effects, but after about six to twelve months, some form of biological change takes place that is associated with lasting psychological deficits.

Third, these findings applied even in those children who did not show evidence of subnutrition (Sonuga-Barke, Beckett, Kreppner, Colvert et al., 2008). In this subgroup, psychological deprivation was associated with a marked impairment in head growth and therefore, almost certainly, with brain growth. Other findings have shown that brain growth is the main driver of head growth. In any case, it is implausible that there could be normal brain growth if the skull is not expanding to accommodate the larger brain.

Fourth, despite the persistence of early experience effects, changes in children's functioning did take place between ages 6 and 11. The persistence of effects has a high degree of probability but it is not fixed and immutable. Fifth, even in the subgroup of children experiencing the most prolonged deprivation, there was still considerable heterogeneity in the outcome, with a few individuals coming through relatively unscathed. Although much remains to be learned about the phenomenon of resilience (Rutter, 2006c), it is real. Sixth, the persisting sequelae mainly concerned unusual patterns such as disinhibited attachment (Rutter, Colvert, et al., 2007) or quasi-autistic features (Rutter, Kreppner, et al., 2007), rather than more ordinary forms of emotional disturbance or disruptive behavior.

Sensitive Periods

At one time, based on the evidence from early studies of imprinting, much attention was paid to so-called critical periods. It soon became evident that these periods were broader than originally envisaged, were not entirely fixed, and were influenced by a variety of conditions (Bateson, 1966). The recognition that this was so led, for a while, to an excessive rejection of the proposition that environmental effects could be strongly influenced by the period of development at which they operated. The concept has now returned in a modified form within the concept of sensitive periods (Rutter, 2006b). The examples given earlier in relation to developmental programming illustrate the importance of their operation. With respect to language development, not just phonological discriminations are affected but also grammar to some extent. People who learn a second language in later life have no particular difficulty acquiring vocabulary, but the way in which they speak is recognizably different from that of a native first-language speaker. Both their sound production and their use of grammar lack the colloquial quality of first language acquisition. One functional imaging study has also suggested that later second language learning uses a somewhat different part of the brain to that which subserves first-language learning in early childhood (Kim et al., 1997).

The effects of sensitive periods are also evident in responses to unilateral brain injury. Such injuries in adults and older children lead to marked language and verbal cognitive deficits if the lesion is in the dominant hemisphere and to visuospatial deficits if it is in the other hemisphere. However, this differentiated pattern of effects is not seen when unilateral brain lesions occur in very early life (Bates & Roe, 2001; Vargha-Khadem et al., 1992; Vargha-Khadem & Mishkin, 1997). It used to be said that the effects of early brain injuries were much less than those of later brain injuries (the so-called Kennard principle). That is actually not an appropriate way of portraying the findings. In some respects, the effects of early lesions are greater but the main difference is that they do not show the lateralization effects seen in older people (Rutter, 1993). The precise brain processes involved are not fully understood but it is evident that there is the possibility of interhemispheric transfer or take-up of functions in early life that is not available later.

Mental Functioning

Three main features of mental development are especially important. First, humans are social animals and from infancy onward, individuals are socially interactive. This is apparent in the development of selective social attachments but it is equally obvious in the extent to which babies and children seek and get pleasure from socially interactive play, conversation, and experiences. To a very considerable extent, learning takes place in a social context. Second, again from an early age, individuals process their experiences cognitively and emotionally. They seek to make sense of what is happening to them and they conceptualize what it is all about. Obviously, they do so in ways that become increasingly complex and subtle but elementary processing is evident from the outset. The idea that babies were passive organisms on whom experiences impinged in a uniform way was clearly wrong and has been abandoned. Nevertheless, there are important ways in which children's ability to process their experiences goes through important transitions.

For example, it is evident that toddlers' ability to appreciate standards and expectations does not develop until the end of the second year of life (Kagan, 1981). Similarly, children's mentalizing abilities (i.e., their capacity for understanding what another person is likely to be thinking given their situation and circumstances) becomes well developed only during the 2- to 4-year period (Baron-Cohen, 2000). Toward the end of the second year, children also develop much greater powers of imagination and make-believe. Because of the crucial importance of the cognitive and emotional processing of experiences, and because such processing skills are quite limited during the first two years of life, it has been argued that this means that most early experiences do not have enduring effects (Kagan, 1980). In other words, persistence of effects is to a very considerable extent dependent on people's thought processes about the experiences.

The ways in which memory functions also change with age. Thus, most people have very little, if any, memory of discrete events and experiences during the first few years of their life – so-called infantile amnesia (Rutter, Maughan, et al., 1998). On the other hand, implicit memory (as reflected, for example, in children's learning of language) does persist from the infancy period onward. There seems to be an important distinction between implicit and explicit memory (Baddeley, 1990; Tulving, 1983). Probably, what is limiting early memories is not the laying down of memories but rather their retrieval – perhaps because the context and meaning in later childhood are so different from that in infancy.

Together with the growth of skills and the increasing capacity to process experiences, there develops a set of self-concepts. Individuals increasingly come to have a view of what they are like as individuals and of what they can expect in terms of interactions with other people and of experiences that they encounter. Such self-concepts are multifaceted and include features such as self-esteem, self-efficacy, and internal working models of relationships. How children respond to later experiences is likely to be shaped, to an important extent, by the cognitive set that they bring to the experiences.

Continuities and Discontinuities

At one time, many people seemed to assume that the norm in development was consistency and continuity in functioning, as if it was only change and discontinuity that required explanation. That was a very odd notion because it is obvious that the whole process of development involves change. Of course, it also involves very important continuities and stabilization of functioning. Nevertheless, both continuity and discontinuity are to be expected and both require explanation (Rutter & Rutter, 1993).

As part of this idea that it was only change that required accounting for, people tended to assume that stress experiences were likely to be one of the key ways in which this came about. However, the evidence is rather against that notion. Stress experiences overall tend to accentuate, rather than diminish, preexisting characteristics. These are what have been termed *accentuation effects* (Caspi & Moffitt, 1993). Of course, experiences can bring about change, but they tend to do so only when later experiences are markedly different from earlier ones. These have been sometimes termed *turning point effects*, the reality of which has been demonstrated in relation to a range of positive and negative experiences such as a harmonious marriage, army experiences, incarceration, or a geographical move (Rutter, 1996). Continuities and discontinuities will also be influenced by biological changes. This is most obvious in relation to the changes of puberty as they affect not only hormonal functioning but also body shape, with consequent effects on people's positive and negative feelings about such bodily changes. Although the mechanisms remain ill understood, it is also apparent that there are important age-related changes in the sex ratio of some disorders. This is evident, for example, in relation to depression, which has an approximately equal sex ratio before puberty but a marked female preponderance thereafter (Bebbington, 1996, 1998). There are also age-related differences in drug response. For example, children and adolescents with major depression do not show the beneficial response to tricyclic medication that is more typical in adulthood (Brent & Weersing, 2008). It is not that they are resistant to all antidepressant medication, because the same age effect is not found with selective serotonin uptake inhibitors (Brent & Weersing, 2008). Although not well studied, it is probably the case that the euphoriant response to stimulant medication that is typical in adult life is not seen in childhood. To the contrary, the effect may be dysphoric rather than euphoric.

As people grow older, they show a continuing responsivity to the environment. The fact that this is so has played a role in the emergence of a life span concept of development. Although physical growth ends in early adult life, many important experiences do not usually take place until the postchildhood years. This is evident, for example, in the emergence of sexuality, in the establishment of marital and other cohabiting relationships, in having children, in the establishment of work and professional careers, and in the various changes that take place in the peer group. As well as these social transitions, it is necessary to appreciate that continuities and discontinuities in psychological functioning may be affected by the regular, and especially the excessive, use of substances that give rise to dependence, either psychological or physical.

A further feature that requires mention is that of kindling effects, in which prior episodes of depression alter the magnitude of life stressors that can produce a subsequent episode (Kendler et al., 2000, 2001; Post, 1992). Research findings suggest that the experience of major depression somehow brings about bodily changes that play a role in recurrence and that are associated with a changed responsivity to environmental stressors.

GENETIC FACTORS

In considering genetic effects, a distinction must be drawn between segregating and nonsegregating genes. Nonsegregating genes are those that, through the process of evolution, have come to be present in the same way in all members of a species. Thus, universally present genes determine the programming of development and are responsible for the fact that all human beings (apart from those who are diseased) have the potential for language, have two eyes, and have one dominant brain hemisphere. Segregating genes, by contrast, are those that lead to differences among individuals (Rutter, 2006a; Rutter, Moffitt, & Caspi, 2006). Of course, in their origins, these are directly comparable with the nonsegregating genes but the difference is that the allelic variations in segregating genes mean that each person is genetically different from other people. Over the course of time, different allelic variations develop – the process of mutation. All allelic variations are, in an evolutionary sense, mutations. However, there is a difference between those that give rise to some overt pathology (as would be the case with Mendelizing medical conditions such as hemophilia, cystic fibrosis, or retinitis pigmentosa) and those that lead to variations within the normal range. By and large, the pathological mutations that are directly responsible for wholly genetic conditions account for rare disorders rather than common ones. There are a very large number of such single-gene disorders but, even in total, they are far outnumbered by multifactorial disorders and multifactorial traits. Multifactorial characteristics are those that result from the interplay among several, often a large number, of genes and a similarly large number of environmental influences. No one of these (either genetic or environmental) constitutes a necessary and sufficient cause, and the particular outcome is the result of a very complex set of influences working together. This is the probabilistic way in which genetic factors operate, with all the common psychological characteristics, both normal and abnormal.

Multifactorial Genetic Influences

At one time, behavioral geneticists thought it useful to provide quantitative estimates of the extent to which individual variation in a population was genetically determined (heritability) or environmentally determined. Although it was recognized from the outset that such figures applied only to populations and not to individuals, and although it was appreciated that heritability estimates were population-specific, it was nevertheless considered that the partitioning of variance into genetic and nongenetic components was both useful and meaningful.

It has become increasingly clear that the idea that genetic and environmental influences are separate, and therefore separately quantifiable, is a rather misleading oversimplification (Rutter, 2006a; Rutter et al., 1999a, 1999b; Rutter, Pickles, Murray, & Eaves, 2001). As already noted, the effects of chance as well as specific environments will play a role in individual variation. In addition, however, it has become apparent that many processes involve interplay among genes and environments, such that it is not meaningful to seek to estimate the separate effects of each. Gene-environment correlations mean that genetic factors play a role in bringing about individual differences in environmental risk exposure (Kendler & Baker, 2007). Gene-environment interactions mean that genetic factors also operate through effects on individual differences in susceptibility to environmental risks. In traditional quantitative genetic analyses, these effects are all attributed to genes although the reality is that the risks derive from the combination of genetic and environmental influences. It is also the case that effects may be reliant on the interplay among genes or among environmental risk factors. In other words, instead of each genetic factor having an independent effect, the influence of one gene may be dependent on the presence of other genes. These have been termed *epistatic* and *dominance effects* (the former referring to synergistic effects between different genes and the latter to synergistic effects between different allelic variations of the same gene).

Quantitative Genetic Findings

One of the extremely important findings from quantitative genetics (meaning twin, adoptee, and family studies) is that all behavior is genetically influenced to a substantial degree (Rutter, 2006a; Rutter et al., 1999b; Plomin et al., 2001). Thus, not only are there genetic effects on mental disorders such as autism or schizophrenia and on basic psychological traits such as intelligence or emotional reactivity, but so also there are genetic influences on individual differences in religious commitment, the likelihood of being divorced, or engagement in antisocial behavior. Many people have been deeply resistant to accepting that genes could influence social behaviors and attitudes because it is quite implausible that there could be a gene for crime or a gene for divorce. Of course, they are correct that there cannot be such genes, but that is to misunderstand the point. The expectation that there are behaviors that are not subject to genetic influences has to presuppose that they have no biological underpinnings. In short, one would have to suppose that, for example, the working of the mind had no connection with the functioning of the brain. That is an even more implausible assumption. As far as there is any somatic involvement, individual differences in the soma will be genetically influenced to some degree. We need to get away from the old-fashioned thinking that genes only affect disease to an appreciation that all human behavior is subject to genetic influences (along with the major effects of environmental influences).

Over the years, all sorts of doubts have been raised about the validity of twin, adoptee, and family designs and the assumptions on which they are based. Attention has been drawn, for example, to the extent to which adoptee samples seriously underrepresent high-risk rearing environments, the unrepresentativeness of many twin samples, and the likely frequent violation of the equal environments assumption that is so fundamental for twin studies (see Rutter, 2006a; Rutter et al., 1999a, 1999b, 2001). These concerns have validity and a consequence may be that, to some extent, heritability findings exaggerate genetic effects. On the other hand, other considerations (the need to take into account persistence of traits and measurement error effects) are also likely to mean that heritability figures are something of an underestimate of genetic effects. As the precise heritability figure has no theoretical or practical implications, and because none of these concerns seriously undermine the main finding of the pervasiveness of substantial genetic effects, the overall conclusion that there are genetic effects on all human behaviors remains solid.

It is necessary, however, to understand the ways in which genes operate indirectly as well as directly. The background to this issue is provided by two very well-established observations. First, there are huge individual differences in environmental risk exposure (Rutter et al., 1995). Some people go through life suffering an endless stream of seriously adverse psychosocial experiences while others have scarcely any. It is necessary to understand the origins of these individual differences. Second, with all manner of environmental hazards (physical as well as psychosocial) there are immense individual differences in people's response to them (Rutter, 1999a, 2000b, 2006c). The question then is how genetic factors might play a contributory role in both sets of individual differences (namely environmental risk exposure and environmental risk vulnerability).

Gene-environment correlations can be considered under three broad headings (Rutter et al., 1997). First, there are so-called *passive* correlations. This simply means that parents not only pass genes on to their children but also act in ways that influence their rearing experiences. A key consideration here is that the personal qualities that are influenced by genes are often the ones that are also crucial with respect to the provision of rearing experiences. Thus, for example, much research has shown that mentally ill parents have a substantially increased likelihood of providing a family environment that is characterized by discord, conflict, and scapegoating. The practical consequence is that genetically influenced mental disorders involve the creation of risk environments as well as the passing on of genes that involve an increased susceptibility to mental disorder. The point has been made here in terms of psychopathology, but the same applies to individual differences in nonpathological psychological traits. This indirect effect reflects the important role of parents in shaping the upbringing of their children.

By contrast, active and evocative gene-environment correlations reflect the importance of people's own behavior in shaping and selecting their environments. Active gene-environment correlations concern the role of people's behavior in influencing their nonsocial experiences. Thus, children's own genetically influenced behavior will play a role in whether or not they spend their leisure time on the football field, in the library, in front of the TV set, or in music lessons. Evocative gene-environment correlations refer to the role of genetically influenced behaviors on interpersonal interactions. How one person behaves will influence how other people respond. Thus, longitudinal studies have been consistent in showing that, for example, antisocial children are much more likely than other children to have high-risk psychosocial experiences in adult life (Robins, 1966; Champion et al., 1995). Their rates of unemployment, rebuffs from friends, broken love affairs and marriages, and being without friends or social support are much increased. Obviously, these effects do not stem directly from genes. The effects concern interplay between people's behavior and their environments, and genes come into the equation only because they have effects that influence individual differences in behavior.

The terms shaping and selecting environments are perhaps unfortunate in that they carry the misleading implication that active choice is involved and that therefore individuals can be "blamed" for their own predicaments. Although a degree of personal choice may be involved, quite a few of the effects do not involve active choice at all. Thus, antisocial individuals do not actively seek disadvantageous social situations. Rather, they act in ways that simply make it more likely that those will come about. Kendler and Baker (2007) have provided a systematic review of the evidence with respect to genetic influences on individual differences in environmental risk exposure.

Gene-environment interactions (G×E) refer to the role of genetics in influencing people's vulnerability to particular environmental risk factors. Thus, when people are exposed to disease-inducing bacteria or viruses, only some people come down with the infection. Similarly, although we are all exposed to pollens in the spring, only some of us develop hay fever. These individual differences in vulnerability are genetically influenced to an important extent. For a long time, behavior geneticists were skeptical about the importance of gene-environment interactions because their "black box" analyses of the interplay between "anonymous" genetic factors and "anonymous" environmental factors rarely showed statistically significant interactions (Plomin et al., 1988). However, that was a reflection of the misunderstanding of how biology works. There is absolutely no reason to suppose that there will be a general interaction between genetic and environmental risks. Rather, there are specific interactions between particular genetic risk factors and particular risk environments (see Rutter & Silberg, 2002). When these have been studied appropriately, gene-environment interactions have been found (Kendler et al., 1995; Silberg et al., 2001). Thus, they have been evident in relation to negative life events and depression and between family discord and antisocial behavior (Cadoret et al., 1986, 1995). The important implication is that genetic effects may be indirect as well as direct.

Up to 2002, the human evidence on G×E in the field of psychopathology was largely confined to quantitative genetic findings from twin and adoptee studies. Because the G×E applied to 'black box' genetic effects without reference to actual individual identified genes, the evidence was necessarily somewhat circumstantial. The situation was totally transformed by the new wave of studies using molecular genetic identification of actual specific genes and detailed measurement of environments shown to carry an environmentally mediated risk for psychopathology (see Rutter, 2007a; Rutter, Moffitt, & Caspi, 2006; Caspi & Moffitt, 2006). The research leaders have been Caspi and Moffitt with their demonstrations of G×E with respect to a variant of a gene influencing monamine oxidase activity that interacted with child maltreatment in relation to the outcome of antisocial behavior (Caspi et al., 2002); with respect to a 5-hydroxytryptamine transporter (5HTT) gene allelic variation that interacted with early child maltreatment and later life stresses in relation to the outcome of depression (Caspi et al., 2003); and with respect to the role of the variant of the catechol-o-methyltransferase (COMT) gene predisposing to schizophrenia spectrum disorders only through an interaction with heavy early use of cannabis (Caspi et al., 2005). The reality of the G×E has been shown by the rigorous methodological checks, the numerous human replications, the confirmatory studies in other animal species, and the human imaging studies that showed the effects of the G×E on neural structure and function (for reviews see Rutter, 2007a, 2009; Caspi & Moffitt, 2006). A meta-analysis purporting to challenge the G×E findings has been published (Risch, Herrell, Lehner, Liang, et al., 2009), but it has many problems and focuses just on G×E as a statistical matter rather than as a biological feature likely to throw light on biological mediating mechanisms (Rutter, Thapar, & Pickles, 2009; Rutter, 2010). The important message remains that many effects on psychopathology do not concern genes or environment operating separately, but rather from their synergistic co-action, with the consequences of each dependent on the other.

Molecular Genetic Findings

The same point derives even more strikingly from molecular genetic studies – meaning the investigation of the causal role of individual genes (Plomin et al., 2001; Rutter et al., 1999a, 1999b; Thapar & Rutter, 2008). In the field of internal medicine, individual genes have been found to play a role in individual variations in response to prescribed medication, alcohol, the effects of smoking in creating a risk for occlusive coronary artery disease, in the role of head injuries as a predisposing factor for Alzheimer's disease, and vulnerability to infections of various kinds (Rutter, 2000a). Even in the field of internal medicine, our knowledge of these effects is quite limited and in the fields of psychopathology and of normal psychological development, research of this kind is only just beginning. Accordingly, there are very few specific conclusions that can be drawn about particular genetic effects. On the other hand, what the findings do show very clearly is a range of general features about how genes work.

First, with respect to multifactorial traits and multifactorial disorders (and that means almost all of those with which we are likely to be concerned in relation to common psychological features), multiple genes are involved. There is not nor is there likely to be a single gene that is responsible for disorders such as autism or schizophrenia that are strongly genetically influenced. Rather, several different genes will be involved. No one of them causes the mental disorder but each plays a part in susceptibility to it (Kendler, 2005). Whether or not the disorder actually develops will depend on whether the person has other susceptibility genes and whether the person encounters relevant risk environments. A corollary of this feature is that the risk associated with any one gene is likely to be very small indeed. The cumulative effect of different susceptibility genes will often be quite large, but no one gene on its own has a big effect. It should be added that genes have protective as well as risk effects. Whether or not a person develops a particular disorder may derive from their lack of protective genes as much as from the possession of risk ones.

A second major feature is that genetic effects are often pleiotropic (meaning that they affect many different functions) and the same gene may have both good effects and bad effects, depending on which outcome is being considered. The best-known example of this kind is the protective effect in relation to malaria of being a heterozygote for sickle cell anemia (Weatherall & Clegg, 2001). In the behavioral arena, it is also apparent that traits that bring risk for one sort of outcome may be protective against others. For example, high emotional reactivity constitutes a risk factor for anxiety disorders but a protective factor against antisocial behavior (Rutter, Giller, & Hagell, 1998).

Third, genetic effects are not the same all over the world. Because genes are an intrinsic biological feature it is sometimes supposed that the effects must be invariant, but they are not (Rutter, 2001). To begin with, there are some quite surprisingly large geographical variations in susceptibility genes. The best-documented example of this is the findings on thalassemia where the allelic variant responsible for this Mendelian disorder varies across different parts of the world. In addition, the gene responsible for the unpleasant flushing response to alcohol is found only in Asiatic groups. Environmental circumstances can also affect gene frequencies. For example, the thalassemia genes in the relevant ethnic groups in North America are at a much lower frequency than in Africa, presumably because there is a genetic advantage to having the gene in areas where malaria is endemic whereas there is no such advantage in North America. For reasons that are not understood, the apoE-4 gene, which carries a substantially increased susceptibility to Alzheimer's disease in most populations, shows a weaker risk effect in African and Hispanic populations (Farrer et al., 1997).

A fourth consideration with respect to genetic effects is that the susceptibility genes associated with particular mental disorders may not operate directly on the disorder as such. Thus, the genetically influenced liability to both depressive and anxiety disorders is probably mediated largely through the temperamental feature of neuroticism or emotional reactivity (Kendler, 1996). Similarly, part of the genetic liability to hyperactivity disorders is probably mediated through the temperamental feature of sensation-seeking (Levy & Hay, 2001). One of the consequences of this form of mediation of risk effect is that it cannot be assumed that the genetic effects are necessarily harmful. For example, presumably sensation-seeking may predispose to rock climbing, playing the stock market, or scientific creativity as well as to various psychopathological outcomes. Which one of these is the result will depend, it is thought, on the interplay with environmental circumstances of various kinds.

As previously noted, genetic effects may also operate through creating an increased (or decreased) liability to experience environmental risks or to be susceptible to them. As far as there are effects that are more directly associated with disorder, the risks may apply to part-functions or part-elements in the disorder rather than the disorder as such. For example, it has been suggested, with some supporting evidence, that this may be the case in relation to reading disorders (Grigorenko et al., 2000).

Three other aspects of genetic influences on psychological functions and on mental disorder require emphasis. First, almost all of the known specific susceptibility genes concern common normal allelic variations and not rare pathological mutations. Thus, all of us are likely to carry genes that are associated with risk outcomes; whether or not we actually develop the risk outcome will depend on the presence of other genes, and on whether or not we encounter environmental risks. The effects are probabilistic and not deterministic. Second, it cannot be assumed that confirmed susceptibility genes have their effect on the disorder as such. Instead, they may carry risk because they influence sensitivity to the environment or because they have effects on general bodily functions such as immune responses or neuroendocrine functioning. Third, because genetic effects are crucially dependent on gene expression, psychological/psychopathological effects may derive from influences on such gene expression. In years gone by, there was the general assumption that the only genes that mattered were those that (via ribonucleic acid – RNA) had effects on proteins. It is now clear that that assumption was in error. Several of the susceptibility genes identified so far are ones that influence gene expression and do not code for proteins. Because multiple deoxyribonucleic acid (DNA) elements influence

expression, this means that the concept of a single gene is an oversimplification. Each gene that codes for proteins (and hence that involves the capacity
for direct biological effects) is enabled to operate by virtue of other genes
whose action is on gene expression. It should be added that environmental
influences could also affect gene expression, which is discussed in more
detail in the following section.

In summary, there can be no serious doubting of the importance of
genetic influences on both normal and abnormal psychological functioning. Nevertheless, knowing that there is a genetic influence does not tell one
anything very useful on its own, with respect to risk or protective causal
pathways. That is because genetic influences may operate in such a diversity
of ways, both direct and indirect, and the implications will vary as to the
details of the mode of operation.

ENVIRONMENTAL EFFECTS

During the last few decades, behavioral geneticists have raised queries as to
whether there are any important environmental effects other than in cases
of extreme environmental circumstances (Rowe, 1994; Scarr, 1992). The
skepticism derives from three main sources. First, the findings on the pervasiveness of gene-environment correlations have underlined the reality
of the possibility that some effects that have been labeled environmental may, in reality, reflect genetic mediation (Plomin & Bergeman, 1991).
Behavioral geneticists have pointed out that the great majority of studies
of psychosocial risk factors have failed to use research designs that could
differentiate between genetic and environmental mediation. Second, as far
as correlations between a family risk feature and child psychopathology
represent environmental mediation, the question is whether the effects are
those of the child's behavior on the family, or the family's influence on
the child (Bell, 1968). Again, critics have noted how few designs are able
to sort out the direction of the causal arrow. Third, behavioral geneticists have pointed to the evidence that, as far as there are environmental
effects, they are mainly nonshared rather than shared (Plomin & Daniels,
1987). The effects of family influences are to make children within the
same family different rather than the same. On this basis, doubts have been
expressed over whether or not familywide risk factors have any meaningful
importance.

The methodological challenges are real and important, and they have
to be met. Nevertheless, the general dismissal of environmental, and especially family, influences is not warranted (Rutter et al., 1999a, 1999b, 2001;

Rutter, 2006a, 2000a). The genetic effects on individual differences in environmental risk exposure are modest rather than overwhelming. Typically, genetic factors account for some 20 to 40 percent of the variance. This is substantial, but it leaves plenty of room for environmentally mediated risks. In addition, quantitative genetic findings are entirely consistent in showing the importance of nongenetic factors. For most traits and most disorders, about half the population variance is attributable to genetic factors and about half to nongenetic factors. Even with strongly genetically influenced mental disorders such as autism, schizophrenia, hyperkinetic disorders, and bipolar disorders, some 10 to 30 percent of the variance is attributable to nongenetic factors.

Second, although the reality of children's effects on parents (and other people) has been well demonstrated, it is clear that the causal arrow runs in both directions. Third, the arguments with respect to the supposed overwhelming importance of nonshared environmental effects have been misleadingly expressed and the claims are exaggerated. Thus, there are some forms of psychopathology (antisocial behavior is the most obvious example) where it is common to find that several children within the same family are affected and the genetic findings point to the importance of shared as well as nonshared effects. However, what has also become apparent is that nonshared effects are markedly reduced once measurement error and persistence of behavior over time have been taken into account (see also Jenkins & Bisceglia, this volume). Finally, the question of whether effects are shared or nonshared has nothing to do with whether the risks are familywide or not. Rather, they refer to the extent to which the effects make children within the same family similar or different. In other words, a familywide influence (such as family discord and conflict) that affected some children much more than others (perhaps through scapegoating) would appear as a nonshared effect although it derives from a familywide risk factor.

So, rather than rely on soapbox evangelism, we need to turn to the studies that have provided the necessary methodological controls and ask to what extent they show environmentally mediated risks. A review of the evidence (Rutter, 2007b; Rutter et al., 2001) has made clear that there are many different research strategies that provide such controls. These include a wide range of natural experiments that serve to pull apart variables that ordinarily go together. The study of environmental effects within monozygotic twin pairs would be one such natural experiment; the study of the consequences of radical change in rearing circumstances such as brought about by adoption would be a second; the study of the effects of disasters (such as famine) affecting whole communities would be a third; the use of migration designs would be a fourth; and the study of nonfamilial environments (such as schools) would be a fifth. The evidence from their use is consistent in showing the reality of environmentally mediated risks with respect to a range of risk factors (Rutter, 2000a). We now turn to some of the specifics of the findings.

Intrauterine and Neonatal Risk Factors

There are several routes by which intrauterine/neonatal influences may affect psychological development. First, substances in the mother's bloodstream may cross the placental barrier and cause abnormalities in fetal development. This is well established with respect to high levels of maternal alcohol early in the pregnancy (Stratton, Howe, & Battaglia, 1996; Streissguth & Kanter, 1997), with certain epileptic drugs (especially valproate), and probably also with cocaine and benzodiazepines (Mayes, 1999; Schuetze, Eiden, & Edwards, 2009). What is less clear in all these instances is the extent of risk for neuropsychological development when there are no obvious congenital physical abnormalities. One of the complications, of course, is that substance misuse during a pregnancy is likely to be followed by substance misuse as the children are growing up, and it has not proved easy to determine the relative importance of postnatal-rearing risks as against prenatal damage. Genetic mediation also needs to be considered. Nevertheless, the reality of substance-induced abnormalities in fetal development is well established, as are the effects on behavior.

Second, very low birthweight and serious obstetric complications can lead to brain damage and subsequent handicap (Cooke & Abernethy, 1999; Stewart et al., 1999). If severe damage can lead to severe handicap, it is likely that milder degrees of damage can have more subtle effects. Pasamanick and Knobloch (1966) put forward the concept of a continuum of reproductive casualty. Systematic studies have indicated that the risks from minor obstetric complications are usually negligible. However, associations have been found to both the development of schizophrenia (McDonald, Fearon, & Murray, 2000) and of suicidal behavior (Jacobson et al., 1987; Salk et al., 1985). The mechanisms remain quite unclear. The difficulties of determining the links between brain structure and mind function are underlined by the structural imaging findings from follow-up studies of very low birthweight infants. As compared with control populations, there is a substantial increase in structural brain abnormalities but, rather surprisingly, these have not been found to correlate substantially with the cognitive deficits found in the same children (Cooke & Abernethy, 1999; Stewart et al., 1999). Accordingly, the mediating mechanisms remain unclear. Third, it has been argued that prenatal exposure to toxins, such as lead, can impair later psychological development, presumably through adverse effects on the brain. Overall, it seems that the risks are very small within the range of lead levels found in most current-day populations (Rutter & Russell Jones, 1983). On the other hand, both animal and human studies do make it clear that high levels of lead do cause damage. Fourth, there is evidence that high levels of maternal anxiety or stress during the pregnancy have measurable effects on the infant's later behavior (Coe & Lubach, 2000; Schneider & Moore, 2000). Once more, the mechanisms are poorly understood, but effects via biological programming constitute one possibility.

Psychosocial Risk Factors

In testing for environmentally mediated risks from psychosocial experiences, it has been crucial to use designs that can take account of possible genetic mediation, possible effects of the child on the environment, and of third variable influences of one kind or another. As noted earlier, a substantial range of research strategies are available to meet these needs. The findings have been consistent in showing major effects on psychological development when there have been major changes in the environment. The effects are most strikingly shown in the case of various forms of extreme environments, but there are a growing number of studies showing significant effects of environmental variations within the normal range. Psychopathological risks have been found to be increased, using multiple research strategies, for serious family discord and conflict; for a lack of continuity in personalized caregiving; for a lack of reciprocal conversation, play, and social experiences; and for severely stressful life events that carry long-term threat, particularly associated with loss or humiliation (Rutter, 2000a). Effects have been found to apply to antisocial behavior, depression, substance abuse, and cognitive development. Although there have been forceful arguments put forward that peer group influences are much more important than family ones (Harris, 1998), the evidence does not support this conclusion. Rather, the effects have been demonstrated from risk experiences in the family, the peer group, the school, and the community. There are both shared and nonshared environmental effects but it probably is crucial that the risk experiences directly impinge on the individual. There is also a growing body of evidence to suggest that there are important gene-environment interactions and that the adverse effects of psychosocial risks are slight in the absence of genetic risk, but may be quite marked in its presence (see earlier section on G×E).

Although there are age-related variations in susceptibility to different forms of psychosocial risk, the effects apply across the whole period of infancy, childhood, adolescence, and adult life. All studies have emphasized the huge individual differences in response to even the most severe psychosocial risks (Rutter, 1999a, 2000b). The findings also suggest that adverse experiences can be protective as well as leading to an increased vulnerability. The precise reasons why apparently similar experiences can sometimes have steeling effects, and at other times have sensitizing effects, are not well understood. However, it seems likely that one of the crucial considerations is how the individual coped with the negative experience at the time (Rutter, 2000b; Sandberg & Rutter, 2008). Successful coping and successful adaptation can promote resilience, whereas a failure to manage the stress experience successfully can be damaging. It has also become clear that a distinction needs to be drawn between distal and proximal risk processes (Rutter, Giller, & Hagell, 1998). Proximal risk processes are those that are directly implicated in the mechanisms leading to psychopathology or other adverse outcomes. Severely depriving institutional care and the experience of scapegoating and focused hostility are both examples of this kind. By contrast, distal risk processes are those that are influential only because they predispose to proximal risk factors, not being themselves part of the direct risk process. Poverty and early parental loss probably fall into that category for many outcomes. They are associated with an increase in risk but that increase is contingent on whether or not the experience is associated with impaired parenting. The findings underlie the importance of considering risk and protective processes in terms of multilink chain effects rather than a once-and-for-all single causal effect that stems from one variable at one point in time. Overall, single psychosocial risks have negligible effects; the cumulative effect of multiple risk experiences, however, can be quite great.

Effects on the Organism

Until recently, surprisingly little attention was paid to the effects of psychosocial experiences on the organism. Nevertheless, it is crucial to determine such effects because only by understanding them can the mechanisms by which psychosocial risks are carried forward, after the negative experience ends, be understood. Although persistence of adverse sequelae is dependent on the persistence or occurrence of the psychosocial risks that started off the negative causal process, there is also good evidence that, in certain circumstances, the adverse sequelae continue even after there has been a change to a good, even superior, rearing environment (Rutter, 1989). The question then is what changes in the organism are responsible for this carry-forward of effects, despite changes in the environment.

Many possibilities have to be considered and have some evidential support. Thus, it may come about through learning processes in which what is learned has negative implications. That there are brain changes that accompany learning has been demonstrated, most especially in relation to the phenomenon of imprinting (Horn, 1990). The question arises, of course, as to why new learning in association with good environments later does not undo the negative effects of early learning that took place in quite different environmental circumstances. Sometimes this may be because of sensitive period effects, as with imprinting. Sometimes it may be because the new experiences have not been of a kind to undo the earlier effects. That would be the case with the persistence of phobias when the fear has led to avoidance of the stimulus for the phobia. Two particular forms of learning have received special attention in the psychological literature. First, there are the effects of interpersonal interactions (Robins, 1966; Champion et al., 1995). This is most obvious in the case of antisocial behavior leading on to negative interpersonal interactions of many different kinds. Persistence here comes about because children have learned to behave in ways that generate or predispose to further negative experiences. Second, there may be effects on children's cognitive set as it applies to their view of themselves and their view of the world with which they have to interact. The possible importance of this mechanism has been suggested in relation to the effects of abuse in leading to attributional biases in social encounters (Lansford, Malone, Dodge, Pettit et al., 2006; Lansford, Malone, Stevens, Bates et al., 2006), and the effects of stress experiences in leading to patterns of negative thinking that predispose to depression (see Brent & Weersing, 2008) and internal working models of relationships that lead people to form insecure later relationships because adverse early experiences created negative expectations about what social encounters and social relationships mean and what they can bring (Bretherton, 1999). In each case, it has to be said that, although the processes are clearly plausible (and their existence has evidential support), there is so far a lack of supporting evidence that these do actually constitute the mediating mechanism for the carry-forward of adverse effects of psychosocial risk experiences.

A related mechanism concerns effects on coping styles (Compas et al., 2001). The issue is whether the initial response to the adverse experience shaped a style of coping that is likely to bring with it adverse consequences.

Thus, for example, studies of institution-reared children have shown the high prevalence of a lack of planning in their approach to new challenges or new situations (Quinton & Rutter, 1988; Rutter, Quinton, & Hill, 1990). They seem to have acquired the attitude that there is nothing that they can do about their situation (which is likely to have been the case while they were at the institution) and hence that there was no point in taking positive decisions about such key decisions as careers, choice of marriage partner, and so on. Conversely, numerous studies have shown the protective value of a conviction of self-efficacy and a style of exercising that efficacy. Conversely, if the coping has involved recourse to drugs or alcohol, their use is likely to bring with it a further set of problems.

In discussing developmental programming, attention was drawn to the evidence that, at least for some somatic functions, experiences in early life lead to adaptive changes in the organism that prepare it for dealing with circumstances at that time, but that may prove maladaptive if later environmental conditions are very different. It is quite possible that there could be a psychological parallel. Perhaps the strongest contender is the effect of institutional rearing on later social relationships (Rutter, 2006b; Rutter, O'Connor et al., 2004; Rutter et al., 2001). The findings indicate that intimate peer relations tend to be somewhat different in children reared in institutions with a large roster of ever-changing caregivers. The patterns persist long after adoption into well-functioning nuclear families. The mediating mechanism for the persistence is not known but it could be some kind of developmental programming effect of this kind.

In recent years, much has been made of the possibility that early experiences so sculpt brain development that subsequent development is permanently altered. Such effects, as already noted, mainly apply to the consequences of lack of experiential input that is outside the normal expectable range. It is not likely that this applies to most psychosocial adversities in humans. Perhaps the one possible exception concerns the effects of severely depriving institutional care that lacks any appreciable personal interaction, as was evident in the studies of adoptees from Romanian institutions (see earlier). Even here, the degree of cognitive recovery was remarkable but the persisting effects associated with prolonged institutional care were equally striking. Where there were persistent sequelae, they may have been mediated by some kind of developmental programming effect of this kind.

Yet another possibility concerns the effects of stressful experiences on neuroendocrine function (Hennessey & Levine, 1979; Gunnar & Loman, this volume; Gunnar & Vásquez, 2006). That there are effects is well demonstrated but what is much more dubious is whether the consequences for later psychological functioning are mediated by these neuroendocrine changes. It remains a possibility but it seems doubtful that it constitutes a main mechanism.

Animal studies have shown, too, that severe stress can lead to hippocampal damage (O'Brien et al., 1999). It is possible that this also applies to humans. Most of the supporting evidence comes from animal studies but there are some indications that there may be human parallels.

In the past few years, the importance of environmental effects on gene expression and thereby on psychological functioning (Cameron et al., 2005; Champagne et al., 2004; Meaney & Szyf, 2005; Weaver et al., 2004) has become increasingly evident. The studies by Meaney and his colleagues were crucial in opening up this area of research. They showed that a particular form of archback nursing in rats during the first week of life (but not later) was associated with lasting individual differences in the offspring's behavior and response to later stress. Cross-fostering experiments showed that this intergenerational effect was environmentally, rather than genetically, mediated. Neurochemical research went on to demonstrate that the effect was mediated by DNA methylation of a specific glucocorticoid receptor gene promoter in the hippocampus. The causal inference was further substantiated by showing the reversal effects following use of a particular drug that affected methylation. The findings are hugely important because they showed the major impact of environmental influences on gene expression, and hence demonstrated a mechanism by which environments "get under the skin" by virtue of the biological changes on the organism that they bring about.

The main conclusion that comes out of these considerations is that there is a pressing need to learn more about the processes that mediate the long-term effects of early adverse experiences. As noted, the possibilities are many and various and range from processes that primarily concern styles of mental functioning to processes that involve lasting biological changes.

It should be added that it is necessary to appreciate that several somewhat different kinds of effects need to be considered. Thus, most of the research has focused on individual differences. These are important but so too are the effects on the overall level of a trait or disorder. There are many examples of very large changes over time of this kind. Thus, for example, they are evident in terms of the increase in IQ (Flynn, 1987, 2000), the increase in average height both during childhood and adult life (Tizard, 1975), the fall in the age of menarche (Tanner, 1962), and, more recently, in a substantial rise in many forms of psychosocial disorder in young people, including antisocial behavior, suicidal acts, substance use and abuse, and depression

(Rutter & Smith, 1995). It is also relevant that experiences can make a major difference to the level without having much effect on individual differences.

IMPLICATIONS FOR INTERVENTION

Research has greatly increased our understanding of the risk and protective mechanisms that are involved in both normal and abnormal psychological development. This knowledge provides many leads that should shape policies of prevention and intervention. Nevertheless, they do not provide a simple blueprint for what will be effective, and many crucial areas of ignorance remain. It is important to make the translation from research to policy and practice now, but it is equally crucial that we move ahead in ways that will inform us on which things work (for whom and under what circumstances) and which things do not. A two-way iterative process is needed by which research informs policy and practice but practice can also shape and inform research (Rutter, 1998, 1999b). Attention will now be drawn to some of the main principles that derive from this overview of biological and experiential influences on psychological development.

Types of Effects

When considering interventions, it is crucial to recognize the need to be concerned with several different sorts of effects. Thus, the focus may be on raising the overall level of functioning in the population or on the reduction of rates of disorder. Over this last century, society has been hugely successful in increasing life expectancy, reducing infantile mortality, improving the outcome for very low birthweight babies, and increasing height and physical fitness. The benefits to society have been enormous but most of these beneficial effects have not been accompanied by any substantial impact on individual differences or on social inequities. In the psychosocial arena, the same is likely to apply. For example, it is clear that gun control in the United States would be likely to have a major effect in markedly reducing homicide rates to levels comparable with the rest of the industrialized world (Rutter, Giller, & Hagell, 1998), but it would not make much difference to individual differences. Indeed, it could well be that, in many circumstances, improving the overall well-being of the population may even increase individual differences because the overall improvement could give more scope for biological differences to exert effects (Ceci & Papierno, 2005).

Knowledge about the causal factors involved with individual differences in psychopathology or in levels of psychological functioning does have important implications for prevention and intervention. There are many examples in medicine showing how a focus on key risk factors can make an impact. Thus, the reduction of smoking in the population has had measurable benefits with respect to lung cancer; the greater use (and availability) of condoms has reduced the spread of sexually transmitted disease; and probably a change in sleeping posture of infants has led to a reduction in sudden infant deaths (in cots or cribs). It should be appreciated, however, that most of the successful interventions in medicine have not come about through any direct focus on a risk factor. Rather, an understanding of causal mechanisms, and of the pathophysiology involved, has led to the development of successful remedies of one kind or another. Again, it is likely that there are parallels in the psychological arena. In that connection, too, it is important to appreciate that it may be as valuable to have interventions that affect the course, recurrence, or outcome of a disorder as it is to affect its origin or first onset.

Interplay among Risk and Protective Factors

It is very striking from all research that, on their own, very few risk factors have a major effect. That applies as much to single genes as it does to single psychosocial risk factors (Rutter, 1999a, 2000b). The main effects come from the cumulative impact of multiple risk factors, together with a paucity of protective influences. Moreover, many of the effects do not operate in an additive fashion. Rather, there are synergistic or catalytic effects resulting from the combination of genetic risk with environmental risk exposure, and with the interplay among genes and among psychosocial influences. The implication is that it may often be advantageous to focus on genetic risk groups when considering environmental interventions and to target individuals experiencing a multiplicity of environmental risks. A further implication is that it may well make a substantial difference to reduce exposure to risks, even if major risks cannot be eliminated. That is because it may be advantageous to diminish the possibility of adverse synergistic effects.

Multiplicity of Causal Pathways

Four main points derive from the empirical findings on causal pathways (Rutter, 1995). First, many causal processes involve multistage indirect chains of effects that cannot sensibly be reduced to a single point at which it can be said that one basic cause was operating. The implication is that there are many points on that indirect causal pathway that provide the opportunity for intervention. It also means that, in deciding at which point to intervene, it is appropriate to consider the chances of making a real difference (resulting from practical considerations) as well as the strength of the causal effect. Second, attention needs to be paid to the origins of risk factors as well as their mode of risk mediation. Interventions may sometimes be most usefully targeted on the processes involved in origins rather than on the risk mechanism as such. Third, with almost all outcomes, several rather different causal pathways may be involved. Once more, this closely parallels the situation in internal medicine more generally (Rutter, 1997). Fourth, many of the risk and protective factors operate across the whole of the range of normal variation, and not just at some supposed pathological extreme. Accordingly, interventions within that normal range may be worthwhile. Again, there are many medical parallels.

Genetic Determinism

The biological revolution, and especially the huge advances in genetics, has led some people to assume a degree of genetic or biological determinism that is out of keeping with the evidence. There is no doubt at all that genetic factors are hugely important in individual differences with respect to all psychological functions. Nevertheless, very few of the effects are direct, and scarcely any deterministic. The scenario of using DNA profiling to sort out individuals who are and are not at risk is fantasy because most of the risks involve a complicated interplay among genetic and environmental risk factors. Of course, there are circumstances in which it may be positively beneficial to know about specific genetic risks because it will allow the possibility of avoiding particular environmental risks that impinge largely because of the genetic susceptibility. However, the main value of advances in genetics will be in their contributory role in our gaining an understanding of causal pathophysiology, and in delineating the interplay among risk and protective factors, rather than because there is likely to be much future in gene manipulation in relation to multifactorial traits or disorders (Thapar & Rutter, 2008; Plomin & Rutter, 1998; Rutter, 2006a; Rutter & Plomin, 1997).

Timing of Risks

Although there is considerable value in gaining a better understanding of the ways in which age-related differences influence risk and protective processes, there is no single age period that has a monopoly on risks and there is no age beyond which it is generally too late to intervene (Rutter & Sroufe, 2000). In some cases, risks (and therefore prevention opportunities) are greatest in very early life but in other cases, they are greater at a later age. Of course, it makes sense to seek to provide an early alleviation of risks but it would certainly be a mistake to focus all preventive efforts or all therapeutic interventions on a single age period. A good beginning in life helps in relation to later adversities, but it certainly provides no guarantee of good development if serious later hazards are encountered. Conversely, many adversities in the early years provide a disadvantaged beginning but, if these are followed by good rearing conditions, the early damage can be countered and its effects reduced or eliminated. There are exceptions in which more enduring effects derive from particular sorts of experiences during sensitive periods, and we need to understand how these operate and how best to deal with such effects. However, we should not assume that that is how most effects operate.

Resilience

As an extrapolation of the same point, research findings are consistent in showing a marked degree of resilience in the face of serious risk experiences (Luthar, Cicchetti, & Becker, 2000; Rutter, 1999a, 2000b, 2006c). That does not mean that people necessarily escape from ill effects but it does mean that the powers of recuperation from adverse experiences and restoration of normal functioning are great. Our understanding of the mechanisms involved in resilience remains rather rudimentary. There are many valuable leads as to what factors might be important, and the evidence is certainly clear-cut that a multiplicity of features are operative. The findings also make it clear that there is not a general quality of resilience; rather, people vary quantitatively in their degree of resilience, and such resilience needs to be thought of in relation to specific sorts of circumstances rather than a general biological quality. Policies and practice on intervention are, however, likely to be helped by research designed to gain a better appreciation of the causal processes involved in the phenomenon of resilience.

Greater efforts are needed now to implement actions based on existing knowledge. It is at least as important, however, to accept and meet the immense research challenges that remain. If our efforts are to lead to improvements in policy and practice, it will be crucial to ensure an effective, iterative interplay between research and policy/practice, with each informing and developing the other.

REFERENCES

Baddeley, A. D. (1990). *Human Memory: Theory and Practice.* Hove & London: Lawrence Erlbaum Associates.

Baringa, M. (2003). Newborn neurons search for meaning. *Science,* 299, 32–4.

Barker, D. J. P. (1999). Fetal programming and public health. In P. M. S. O'Brien, T. Wheeler, and D. J. P. Barker (Eds.), *Fetal Programming: Influences on Developmental and Disease in Later Life* (pp. 3–11). London: RCOG Press.

Barker, D. J. P. (2007). The origins of the developmental origins theory. *Journal of Internal Medicine,* 261, 412–17.

Baron-Cohen, S. (2000). *Understanding Other Minds: Perspectives from Developmental Cognitive Neuroscience.* Oxford: Oxford University Press.

Bates, E., & Roe, K. (2001). Language development in children with unilateral brain injury. In C. A. Nelson & M. Luciana (Eds.), *Handbook of Developmental Cognitive Neuroscience* (pp. 281–307). Cambridge, MA: MIT Press.

Bateson, P. P. (1966). The characteristics and context of imprinting. *Biological Review,* 41, 177–211.

Bateson, P., Barker, D., Clutton-Brock, T., Deb, D., D'Udine, B., Foley, R. A. et al. (2004). Developmental plasticity and human health. *Nature,* 430, 419–21.

Bateson, P., & Martin, P. (1999). *Design for a Life: How Behaviour Develops.* London: Jonathan Cape.

Bebbington, P. (1996). The origins of sex differences in depressive disorder: Bridging the gap. *International Review of Psychiatry,* 8, 295–332.

Bebbington, P. (1998). Editorial: Sex and depression. *Psychological Medicine,* 28, 1–8.

Beckett, C., Maughan, B., Rutter, M., Castle, J., Colvert, E., Groothues, C., et al. (2006). Do the effects of early severe deprivation on cognition persist into early adolescence? Findings from the English and Romanian Adoptees study. *Child Development,* 77, 696–711.

Bell, R. Q. (1968). A reinterpretation of the direction of effects in studies of socialization. *Psychological Review,* 75, 81–95.

Bock, G. R., & Whelan, J. (1991). *The Childhood Environment and Adult Disease.* Chichester: Wiley.

Brent, D., & Weersing, V. (2008). Affective disorder. In M. Rutter, D. Bishop, D. Pine, S. Scott, J. Stevenson, E. Taylor, A. Thapar (Eds.), *Child and Adolescent Psychiatry. 5th edition.* Oxford: Blackwell Publishing.

Bretherton, I. (1999). Updating the "internal working model" construct: Some reflections. *Attachment and Human Development,* 1, 343–57.

Cadoret, R. J., Troughton, E., O'Gorman, T. W., & Heywood, E. (1986). An adoption study of genetic and environmental factors in drug abuse. *Archives of General Psychiatry,* 43, 1131–6.

Cadoret, R. J., Yates, W. R., Troughton, E., Woodworth, G., & Stewart, M. A. (1995). Adoption study demonstrating two genetic pathways to drug abuse. *Archives of General Psychiatry,* 52, 42–52.

Cameron, N. M., Parent, C., Champagne, F. A., Fish, E. W., Ozaki-Kuroda, K., & Meaney, M. J. (2005). The programming of individual differences in defensive responses and reproductive strategies in the rat through variations in maternal care. *Neuroscience & Biobehavioral Reviews*, 29, 843–65.

Caspi, A., McClay, J., Moffitt, T. E., Mill, J., Martin, J., Craig, I. W., et al. (2002). Role of genotype in the cycle of violence in maltreated children. *Science*, 297, 851–4.

Caspi, A., & Moffitt, T. (1993). When do individual differences matter? A paradoxical theory of personality coherence. *Psychological Inquiry*, 4, 247–71.

Caspi, A., & Moffitt, T. E. (2006). Gene-environment interactions in psychiatry: Joining forces with neuroscience. *Nature Reviews Neuroscience*, 7, 583–90.

Caspi, A., Moffitt, T. E., Cannon, M., McClay, J., Murray, R., Harrington, H., et al. (2005). Moderation of the effect of adolescent-onset cannabis use on adult psychosis by a functional polymorphism in the COMT gene: Longitudinal evidence of a gene-environment interaction. *Biological Psychiatry*, 57, 1117–27.

Caspi, A., Sugden, K., Moffitt, T. E., et al. (2003). Influence of life stress on depression: Moderation by a polymorphism in the 5-HTT gene. *Science*, 301, 386–9.

Ceci, S. J., & Papierno, P. B. (2005). The rhetoric and reality of gap closing: When the "have-nots" gain but the "haves" gain even more. *American Psychology*, 60, 149–60.

Champagne, F., Chretien, P., Stevenson, C. W., Zhang, T. Y., Gratton, A., & Meaney, M. J. (2004). Variations in nucleus accumbens dopamine associated with individual differences in maternal behaviour in the rat. *Journal of Neuroscience*, 24, 4113–23.

Champion, L. A., Goodall, G. M., & Rutter, M. (1995). Behavioural problems in childhood and stressors in early adult life: A 20-year follow-up of London schoolchildren. *Psychological Medicine*, 25, 231–46.

Coe, S. L., & Lubach, G. R. (2000). Prenatal influences on neuroimmune set points in infancy. *Annals of New York Academy of Sciences*, 917, 468–77.

Compas, B. E., Connor-Smith, J. K., Saltzman, H., Thomsen, A. K., & Wadsworth, M. E. (2001) Coping with stress during childhood and adolescence: Problems, progress, and potential in theory and research. *Psychological Bulletin*, 127, 87–127.

Cooke, R. W. I., & Abernethy L. J. (1999). Cranial magnetic resonance imaging and school performance in very low birthweight infants in adolescence. *Archives of Disease in Childhood: Fetal Neonatal Edition*, 81, F116–21.

Farrer, L. A., Cupples, L. A., Haines, J. L., Hyman, B., Kukull, W. A., Mayeux, R., Myers, R., Pericak-Vance, M. A., Risch, N., van Duijn, C. M., for the APOE and Alzheimer Disease Meta Analysis Consortium (1997). Effects of age, sex, and ethnicity on the association between apolipoprotein E genotype and Alzheimer disease. *Journal of the American Medical Association*, 278, 1349–56.

Flynn, J. R. (1987). Massive IQ gains in 14 nations: What IQ tests really measure. *Psychological Bulletin*, 101, 171–91.

Flynn, J. R. (2000). IQ gains, WISC subtests, and fluid g: g theory and the relevance of Spearman's hypothesis to race. In G. R. Bock, J. A. Goode, & K. Webb (Eds.), *The Nature of Intelligence*. London: Wiley.

Giedd, J. N., Lalonde, F. M., Celano, M. J., White, S. L., Wallace, G. L., Lee, N. R., & Lenroot, R. K. (2009). Anatomical brain magnetic resonance imaging of typically developing children and adolescents. *Journal of the American Academy of Child & Adolescent Psychiatry*, 48(5), 465–70.

Gogtay, N., Giedd, J. N., Lusk, L., Hayashi, K. M., Greenstein, D., Vaituzis, A. C. et al. (2004). Dynamic mapping of human cortical development during childhood through early adulthood. *Proceedings of the National Academy of Sciences USA*, 101, 8174–9.

Goldapple, K., Segal, Z., Garson, C., Lau, M., Bieling, P., Kennedy, S., & Mayberg, H. (2004). Modulation of cortical-limbic pathways in major depression: Treatment-specific effects of cognitive behavior therapy. *Archives of General Psychiatry*, 61, 34–41.

Goodman, R. (1991). Growing together and growing apart: The nongenetic forces on children in the same family. In P. McGuffin & R. Murray (Eds.), *The New Genetics of Mental Illness* (pp. 212–24). Oxford: Heinemann Medical.

Greenough, W. T., & Black, J. E. (1992). Induction of brain structure by experience: Substrates for cognitive development. In M. R. Gunnar & C. A. Nelson (Eds.), *Developmental Behavior Neuroscience* (pp. 155–200). Hillsdale, NJ: Erlbaum.

Greenough, W. T., Black, J. E., & Wallace, C. S. (1987). Experience and brain development. *Child Development*, 58, 539–59.

Grigorenko, E. L., Wood, F. B., Meyer, M. S., & Pauls, D. L. (2000). Chromosome 6p influences on different dyslexia-related cognitive processes: Further confirmation. *American Journal of Human Genetics*, 66, 715–23.

Gross, C. G. (2000). Neurogenesis in the adult brain: Death of a dogma. *Nature Reviews – Neuroscience*, 1, 67–73.

Gunnar, M., & Vázquez, D. M. (2006). Stress, neurobiology, and developmental psychopathology. In D. Cicchetti & D. Cohen (Eds.), *Developmental Psychopathology, Vol. 2* (pp. 533–77). New York: Wiley.

Harris, J. R. (1998). *The Nurture Assumption: Why Children Turn Out the Way They Do*. London: Bloomsbury.

Hennessey, J. W., & Levine, S. (1979). Stress, arousal, and the pituitary-adrenal system: a psychoendocrine hypothesis. In J. M. Sprague & A. N. Epstein (Eds.), *Progress in Psychobiology and Physiological Psychology* (pp. 79–111). New York: Academic Press.

Horn, G. (1990). Neural bases of recognition memory investigated through an analysis of imprinting. *Philosophical Transactions of the Royal Society*, 329, 133–42.

Hubel, D. H., & Wiesel, T. N. (2005). *Brain and Visual Perception*. Oxford: Oxford University Press.

Huttenlocher, P. R. (2002). *Neural Plasticity: The Effects of Environment on the Development of the Cerebral Cortex*. Cambridge, MA: Harvard University Press.

Jacobson, B., Eklund, G., Hamberger, L., Linnarsson, D., Sedvall, G., & Valverius, M. (1987). Perinatal origin of adult self-destructive behavior. *Acta Psychiatrica Scandinavica*, 76, 364–71.

Jensen, A. R. (1997). The puzzle of nongenetic variance. In R. J. Sternberg & E. L. Grigorenko (Eds.), *Intelligence, Heredity, and Environment* (pp. 42–88). Cambridge: Cambridge University Press.

Kagan, J. (1980). Perspectives on continuity. In O. G. Brim & J. Kagan (Eds.), *Constancy and Change in Human Development* (pp. 26–74). Cambridge, MA: Harvard University Press.

Kagan, J. (1981). *The Second Year: The Emergence of Self-Awareness.* Cambridge, MA: Harvard University Press.

Kendler, K. S. (1996). Major depression and generalised anxiety disorder. Same genes, (partly) different environments – revisited. *British Journal of Psychiatry,* 168 (suppl. 30), 68–75.

Kendler, K. S. (2005). "A gene for . . . " The nature of gene action in psychiatric disorders. *American Journal of Psychiatry,* 162, 1243–52.

Kendler, K. S., & Baker, J. H. (2007). Genetic influences on measures of the environment: A systematic review. *Psychological Medicine,* 37, 615–26.

Kendler, K. S., Kessler, R. C., Walters, E. E., MacLean, C., Neale, M. C., Heath, A. C., & Eaves, L. J. (1995). Stressful life events, genetic liability, and onset of an episode of major depression in women. *American Journal of Psychiatry,* 152, 833–42.

Kendler, K. S., Thornton, L. M., & Gardner, C. O. (2000). Stressful life events and previous episodes in the etiology of major depression in women: An evaluation of the "Kindling" hypothesis. *American Journal of Psychiatry* 157, 1243–51.

Kendler, K. S., Thornton, L. M., & Gardner, C. O. (2001). Genetic risk, number of previous depressive episodes, and stressful life events in predicting onset of major depression. *American Journal of Psychiatry,* 158, 582–6.

Kim, K. H. S., Relkin, N. R., Lee, K.-M., & Hirsch, J. (1997). Distinct cortical areas associated with native and second languages. *Nature,* 388, 171–4.

King, S., Mancini-Marïe, A., Brunet, A., Meaney, M., Laplante, D., & Walker, E. (2009). Prenatal maternal stress from a natural disaster predicts dermatoglyphic asymmetry in humans. *Development and Psychopathology,* 21(2), 343–53.

Kreppner, J., Rutter, M., Beckett, C., Colvert, E., Hawkins, A., Stevens, S., et al. (2007). Normality and impairment following profound early institutional deprivation: A longitudinal follow-up into early adolescence. *Developmental Psychology,* 43(4), 931–46.

Kuhl, P. K., Andruski, J. E., Chistovich, I. A., Chistovich, L. A., Kozhevnikova, E. V., Ryskina, V. L., Stolyarova, E. I., Sundberg, U., & Lacerda, F. (1997). Cross-language analysis of phonetic units in language addressed to infants. *Science,* 277, 684–6.

Lansford, J., Malone, P., Dodge, K., Pettit, G., Bates, J., & Crozier, J. (2006). A 12-year prospective study of patterns of social information processing problems and externalizing behaviors. *Journal of Abnormal Child Psychology,* 34(5), 715–24.

Lansford, J., Malone, P., Stevens, K., Bates, J., Pettit, G., & Dodge, K. (2006). Developmental trajectories of externalizing and internalizing behaviors: Factors underlying resilience in physically abused children. *Development and Psychopathology,* 18(1), 35–55.

Levy, F., & Hay, D. (Eds.) (2001). *Attention, Genes, and ADHD.* Hove: Erlbaum.

Luthar, S. S., Cicchetti, D., & Becker, B. (2000). The construct of resilience: A critical evaluation and guidelines for future work. *Child Development,* 71, 543–62.

Maye, J., Werker, J. F., & Gerken, L. (2002). Infant sensitivity to distributional information can affect phonetic discrimination. *Cognition,* 82, B101–11.

Mayes, L. C. (1999). Developing brain and in-utero cocaine exposure: Effects on neural ontogeny. *Development and Psychopathology*, 11, 685–714.

McDonald, C., Fearon, P., & Murray, R. M. (2000). Neurodevelopmental hypothesis of schizophrenia 12 years on: Data and doubts. In J. L. Rapoport (Ed.), *Child Onset of 'Adult' Psychopathology: Clinical and Research Advances* (pp. 193–220). Washington, DC: American Psychiatric Press.

Meaney, M. J., & Szyf, M. (2005). Environmental programming of stress responses through DNA methylation: Life at the interface between a dynamic environment and a fixed genome. *Dialogues in Clinical Neuroscience*, 7, 103–23. Review.

Molenaar, P. C. M., Boomsma, D. I., & Dolan, C. V. (1993). A third source of developmental differences. *Behavior Genetics*, 23, 519–24.

Naugler, C. T., & Ludman, M. D. (1996). Fluctuating asymmetry and disorders of developmental origin. *American Journal of Medical Genetics*, 66, 15–20.

Nelson, C. A., & Bloom, F. E. (1997). Child development and neuroscience. *Child Development*, 68, 970–87.

Nelson, C. A., de Haan, M., & Thomas, K. M. (2006). *Neuroscience and Cognitive Development: The Role of Experience and the Developing Brain.* New York: Wiley.

O'Brien, P. M. S., Wheeler, T., & Barker, D. J. P. (Eds.) (1999). *Fetal Programming: Influences on Developmental and Disease in Later Life.* London: RCOG Press.

Pasamanick, R., & Knobloch, H. (1966). Retrospective studies on the epidemiology of reproductive casualty: Old and new. *Merrill-Palmer Quarterly*, 12, 7–26.

Plomin, R., & Bergeman, C. S. (1991). The nature of nurture: Genetic influences on "environmental" measures. *Behavioral and Brain Sciences*, 10, 373–427.

Plomin, R., & Daniels, D. (1987). Why are children in the same family so different from one another? *Behavioral and Brain Sciences*, 10, 1–15.

Plomin, R., DeFries, J. C., & Fulker, D. W. (1988). *Nature and Nurture during Infancy and Early Childhood.* Cambridge: Cambridge University Press.

Plomin, R., DeFries, J., McClearn, G. E., & McGuffin, P. (2001). *Behavioral Genetics, 4th Edition.* New York: Worth.

Plomin, R., & Rutter, M. (1998). Child development, molecular genetics, and what to do with genes once they are found. *Child Development*, 69, 1223–42.

Post, R. (1992). Transduction of psychosocial stress into the neurobiology of recurrent affective disorder. *American Journal of Psychiatry*, 149, 999–1010.

Quinton, D., & Rutter, M. (1988). *Parenting Breakdown: The Making and Breaking of Intergenerational Links.* Aldershot: Avebury.

Risch, N., Herrell, R., Lehner, T., Liang, K., Eaves, L., Hoh, J., Griem, A., Kovacs, M., Ott, J., & Merikangas, K. (2009). Interaction between the serotonin transporter gene (5-HTTLPR), stressful life events, and risk of depression: A meta-analysis. *JAMA: Journal of the American Medical Association*, 301(23), 2462–71.

Robins, L. (1966). *Deviant Children Grown Up.* Baltimore: Williams & Wilkins.

Roder, B., & Neville, H. (2003). Developmental functional plasticity. In J. Grafman & I. H. Robertson (Eds.), *Handbook of Neuropsychology, 2nd Edition, Vol 9* (pp. 231–70). New York: Elsevier Science.

Rowe, D. C. (1994). *The Limits of Family Influence: Genes, Experience, and Behaviour.* New York: Guilford Press.

Rutter, M. (1989). Pathways from childhood to adult life. *Journal of Child Psychology and Psychiatry*, 30, 23–51.

Rutter, M. (1993). An overview of paediatric neuropsychiatry. In F. Besag & R. Williams (Eds.), *The Brain and Behaviour: Organic Influences on the Behaviour of Children. Special Supplement to Educational and Child Psychology,* 10, 4–11.

Rutter, M. (1995). Causal concepts and their testing. In M. Rutter & D. Smith (Eds.), *Psychosocial Disorders in Young People: Time Trends and Their Causes.* Chichester: Wiley.

Rutter, M. (1996). Transitions and turning points in developmental psychopathology: As applied to the age span between childhood and mid-adulthood. *International Journal of Behavioral Development,* 19, 603–26.

Rutter, M. (1997). Comorbidity: Concepts: claims and choices. *Criminal Behavior and Mental Health,* 7, 265–86.

Rutter, M. (1998). Practitioner review: Routes from research to clinical practice paper in child psychiatry: Retrospect and prospect. *Journal of Child Psychology and Psychiatry,* 39, 805–16.

Rutter, M. (1999a). Resilience concepts and findings: Implications for family therapy. *Journal of Family Therapy,* 21, 119–44.

Rutter, M. (1999b). The Emanuel Miller Memorial Lecture. Autism: Two-way interplay between research and clinical work. *Journal of Child and Adolescent Psychiatry,* 40, 169–88.

Rutter, M. (2000a). Psychosocial influences: Critiques, findings, and research needs. *Development and Psychopathology,* 12, 375–405.

Rutter, M. (2000b). Resilience reconsidered: Conceptual considerations, empirical findings, and policy implications. In J. P. Shonkoff & S. J. Meisels (Eds.), *Handbook of Early Childhood Intervention* (pp. 651–82). New York: Cambridge University Press.

Rutter, M. (2001). Child psychiatry in the era following sequencing the genome. In F. Levy & D. Hay (Eds), *Attention, Genes, and ADHD* (pp. 225–48). Hove: Erlbaum.

Rutter, M. (2002). Nature, nurture, and development: From evangelism through science towards policy and practice. *Child Development,* 73, 1–21.

Rutter, M. (2006a). *Genes and Behavior: Nature-Nurture Interplay Explained.* Oxford: Blackwell Publishing.

Rutter, M. (2006b). The psychological effects of early institutional rearing. In P. J. Marshall & N. A. Fox (Eds.), *The Development of Social Engagement: Neurobiological Perspectives* (pp. 355–91). Oxford: Oxford University Press.

Rutter, M. (2006c). Implications of resilience concepts for scientific understanding. *Annals of the New York Academy of Sciences,* 1094, 1–12.

Rutter, M. (2007a). Gene-environment interdependence. *Developmental Science,* 10, 12–18.

Rutter, M. (2007b). Proceeding from observed correlation to causal inference: The use of natural experiments. *Perspectives on Psychological Science,* 2(4), 377–95.

Rutter, M. (2009). Developmental perspectives on psychopathology. In D. J. Charney & E. S. Nestler (Eds.), *The Neurobiology of Mental Illness, 3rd edition* (pp. 1239–50) New York: Oxford University Press.

Rutter, M. (2010). Commentary: Gene-environment interplay: Depression and anxiety. *Cutting Edge Series.*

Rutter, M., Beckett, C., Castle, J., Kreppner, J., Stevens, S., Sonuga-Barke, E., O'Connor, T. G., Groothues, C., & Castle, J. (2007a). Effects of profound early institutional deprivation: An overview of findings from a UK longitudinal study of Romanian adoptees. *European Journal of Developmental Psychology*, 4(3), 332–50.

Rutter, M., Champion, L., Quinton, D., Maughan, B., & Pickles, A. (1995). Understanding individual differences in environmental risk exposure. In P. Moen, G. H. Elder Jr., & K. Lüscher (Eds.), *Examining Lives in Context: Perspectives on the Ecology of Human Development* (pp. 61–93). Washington, DC: American Psychological Association.

Rutter, M., Colvert, E., Kreppner, J., Beckett, C., Castle, J., Groothues, C., et al. (2007b). Early adolescent outcomes for institutionally deprived and nondeprived adoptees. I. Disinhibited attachment. *Journal of Child Psychology and Psychiatry*, 48, 17–30.

Rutter, M., Dunn, J., Plomin, R., Simonoff, E., Pickles, A., Maughan, B., Ormel, J., Meyer, J., & Eaves, L. (1997). Integrating nature and nurture: implications of person-environment correlations and interactions for developmental psychopathology. *Development & Psychopathology (Special Issue)*, 9, 335–66.

Rutter, M., Giller, H., & Hagell, A. (1998). *Antisocial Behavior by Young People*. New York: Cambridge University Press.

Rutter, M., Kreppner, J., Croft, C., Colvert, E., Castle, J., Sonuga-Barke, E., Beckett, C., & Murin, M. (2007). Early adolescent outcomes of institutionally deprived and nondeprived adoptees. III. Quasi-autism. *Journal of Child Psychology and Psychiatry*, 48(12), 1200–07.

Rutter, M., Kreppner, J., O'Connor, T., & the English and Romanian Adoptees Study Team (2001). Specificity and heterogeneity in children's responses to profound institutional privation. *British Journal of Psychiatry*, 179, 97–103.

Rutter, M., Maughan, B., Pickles, A., & Simonoff, E. (1998). Retrospective recall recalled. In R. B. Cairns, L. R. Bergman, & J. Kagan (Eds.), *Methods and Models for Studying the Individual. Essays in Honor of Marian Radke-Yarrow* (pp. 219–42). Thousand Oaks, CA: Sage Publications.

Rutter, M., Moffitt, T. E., & Caspi, A. (2006). Gene-environment interplay and psychopathology: Multiple varieties but real effects. *Journal of Child Psychology and Psychiatry*, 47, 226–61.

Rutter, M. O'Connor, T., & the English and Romanian Adoptees Research Team. (2004). Are there biological programming effects for psychological development? Findings from a study of Romanian adoptees. *Developmental Psychology*, 40, 81–94.

Rutter, M., Pickles, A., Murray, R., & Eaves, L. (2001). Testing hypotheses on specific environmental causal effects on behavior. *Psychological Bulletin*, 127, 291–324.

Rutter, M., & Plomin, R. (1997). Opportunities for psychiatry from genetic findings. *British Journal of Psychiatry*, 171, 209–19.

Rutter, M., Quinton, D., & Hill, J. (1990) Adult outcome of institution-reared children: Males and females compared. In L. Robins & M. Rutter (Eds.), *Straight & Devious Pathways from Childhood to Adulthood* (pp. 135–57). Cambridge: Cambridge University Press.

Rutter, M., & Russell Jones, R. (Eds). (1983). *Lead versus Health: Sources and Effects of Low-Level Lead Exposure*. Chichester: Wiley.

Rutter, M., & Rutter, M. (1993). *Developing Minds: Challenge and Continuity across the Lifespan*. Harmondsworth, Middx. & New York: Penguin & Basic Books.

Rutter, M., & Silberg, J. (2002). Gene-environment interplay in relation to emotional and behavioral disturbance. *Annual Review of Psychology*, 53, 463–90.

Rutter, M., Silberg, J., O'Connor, T., & Simonoff, E. (1999a) Genetics and child psychiatry: I. Advances in quantitative and molecular genetics. *Journal of Child Psychology and Psychiatry*, 40, 3–18.

Rutter, M., Silberg, J., O'Connor, T., & Simonoff, E. (1999b) Genetics and child psychiatry: II. Empirical research findings. *Journal of Child Psychology and Psychiatry*, 40, 19–55.

Rutter, M., & Smith, D. (1995). *Psychosocial Disorders in Young People: Time Trends and Their Causes*. Chichester: Wiley.

Rutter, M., & Sroufe, L. A. (2000). Developmental psychopathology: Concepts and challenges. *Development and Psychopathology*, 12, 265–96.

Rutter, M., Thapar, A., & Pickles, A. (2009). Gene-environment interactions: Biologically valid pathway or artifact? *Archives of General Psychiatry*, 66(12), 1287–9.

Salk, L., Lipsitt, L. P., Sturner, W. Q., Reilly, B. M., & Levat, R. H. (1985). Relationship of maternal and perinatal conditions to eventual adolescent suicide. *Lancet*, 1, 624–7.

Sandberg, S., & Rutter, M. (2008) The role of acute life stresses. In M. Rutter, E. Taylor, S. Scott, D. Bishop, A. Thapar, J. Stevenson, & D. Pine (Eds.), *Child and Adolescent Psychiatry, 5th Edition*. Oxford: Blackwell Publishing.

Scarr, S. (1992). Developmental theories for the 1990s: Development and individual differences. *Child Development*, 63, 1–19.

Schneider, M. L., & Moore, C. F. (2000). Effect of prenatal stress on development: A nonhuman primate model. In C. Nelson (Ed.), *Minnesota Symposium on Child Psychology* (pp. 201–43). New Jersey: Erlbaum.

Schuetze, P., Eiden, R., & Edwards, E. (2009). A longitudinal examination of physiological regulation in cocaine-exposed infants across the first 7 months of life. *Infancy*, 14(1), 19–43.

Shonkoff, J. P., & Phillips, D. A. (2000). *From Neurons to Neighborhoods: The Science of Early Childhood Development*. Washington, DC: National Academy Press.

Silberg, J., Rutter, M., Neale, M., & Eaves, L. (2001). Genetic moderation of environmental risk for depression and anxiety in girls. *British Journal of Psychiatry*, 179, 116–21.

Sonuga-Barke, E., Beckett, C., Kreppner, J., Colvert, E., Hawkins, A., Rutter, M., Stevens, S., & Castle, J. (2008). Is subnutrition necessary for a poor outcome following early institutional deprivation? *Developmental Medicine & Child Neurology*, 50(9), 664–71.

Stewart, A. L., Rifkin, L., Amess, P. N., Kirkbride, V., Townsend, J. P., Miller, D. H., Lewis, S. W., Kingsley, D. P. E., Moseley, I. F., Foster, O., & Murray, R. M. (1999). Brain structure and neurocognitive and behavioural function in adolescents who were born very preterm. *Lancet*, 353, 1653–7.

Stiles, J. (2000). Neural plasticity and cognitive development. *Developmental Neuropsychology*, 18, 237–72.

Stratton, K., Howe, C., & Battaglia, F. (1996). *Fetal Alcohol Syndrome: Diagnosis, Epidemiology, Prevention, and Treatment*. Washington, DC: National Academy Press.

Streissguth, A. P., & Kanter, J. (1997). *The Challenge of Fetal Alcohol Syndrome: Overcoming Secondary Disabilities*. Seattle: University of Washington Press.

Tanner, J. M. (1962). *Growth at Adolescence*. Oxford: Blackwell Scientific.

Thapar, A., & Rutter, M. (2008). *Genetics*. In M. Rutter, E. Taylor, S. Scott, D. Bishop, A. Thapar, J. Stevenson, & D. Pine (Eds.), *Child and Adolescent Psychiatry, 5th Edition*. Oxford: Blackwell Publishing.

Tizard, J. (1975). Race and IQ: The limits of probability. *New Behaviour*, 1, 6–9.

Toga, A. W., Thompson, P. M., & Sowell, E. R. (2006). Mapping brain maturation. *Trends in Neuroscience*, 29, 148–59.

Tulving, E. (1983). *Elements of Episodic Memory*. London and New York: Oxford University Press.

Vargha-Khadem, F., Isaacs, E., Van Der Werf, S., Robb, S., & Wilson, J. (1992). Development of intelligence and memory in children with hemiplegic cerebral palsy: The deleterious consequences of early seizures. *Brain*, 115, 315–29.

Vargha-Khadem, F., & Mishkin, M. (1997). Speech and language outcome after hemispherectomy in childhood. In I. Tuxhorn, H. Holthausen, & H. E. Boenigk (Eds.), *Paediatric Epilepsy Syndromes and Their Surgical Treatment* (pp. 774–84). Sydney, Australia: Libbey.

Vogel, F., & Motulsky, A. G. (1997). *Human Genetics: Problems and Approaches, Third Edition*. Berlin: Springer-Verlag.

Weatherall, D. J., & Clegg, J. B. (2001). *The Thalassemia Syndromes, Fourth Edition*. Oxford: Blackwell.

Weaver, I. C., Diorio, J., Seckl, J. R., Szyf, M., & Meaney, M. J. (2004). Early environmental regulation of hippocampal glucocorticoid receptor gene expression: Characterization of intracellular mediators and potential genomic target. *Annals of the New York Academy of Science*, 1024, 182–212.

Werker, J. F. (2003). Baby steps to learning language. *Journal of Pediatrics*, 143, S62–9.

2

Neural Development and Lifelong Plasticity

CHARLES A. NELSON III

The formation and growth of the human brain are undoubtedly two of the most remarkable feats of human construction. Although the 1990s were declared the "decade of the brain" in the United States, it is clear as we enter the early twenty-first century that our knowledge of brain function and development is far from complete. Knowledge of brain development is critical to understanding child development, a point made throughout this chapter. In particular, although it is commonly believed that brains develop on their own accord, largely under the direction of genes and hormones, I will make clear in this chapter that brains desperately need both endogenous and exogenous experiences to grow properly. In the sections that follow, I will describe the major events that give rise to the human brain. Once this blueprint is established, I will then discuss the role of experience in influencing the brain. I will do so by drawing on the role of both early and late experience to demonstrate that although brain development is largely limited to the first two decades of life, *brain reorganization* continues to occur through much of the life span.

BRAIN DEVELOPMENT – A PRÉCIS

Shortly after conception, rapid cell division in the zygote results in the formation of the blastocyst. By the end of the first week, the blastocyst itself has separated into an inner and outer layer. The latter will become support structures such as the amniotic sac, umbilical cord, and placenta whereas the former (inner) layer will become the embryo itself. During the second week, the embryo begins to subdivide into layers, and it is from this outer, ectodermal layer of the embryo that the nervous system will form. How this miraculous transformation from a thin layer of unspecified tissue into the highly complex organ known as the brain occurs is the subject of intense study. In the following section, the major prenatal and postnatal events that give rise to the human brain are discussed. The major prenatal events consist of neural induction and neurulation, cell proliferation and migration, followed by differentiation, apoptosis (cell death), and axonal outgrowth. Myelination and synaptogenesis begin prenatally (subsequent to the formation of processes – axons and dendrites), with both processes continuing well into the second decade of life.

PRENATAL DEVELOPMENT

Neural Induction

Neural induction is the process whereby the undifferentiated cells that comprise a portion of the ectodermal layer of the embryo go on to become neural tissue itself. In the human, this event occurs at 16 days gestation (O'Rahilly & Gardner, 1979). The mechanisms that permit this transformation are still not clear. The traditional view is that a chemical agent is secreted from the mesoderm, which induces the dorsal side ("toward the rear") of the ectoderm to develop into the nervous system (Spemann & Mangold, 1924). More recent discoveries in developmental neurobiology have revealed that members of the transforming growth factor beta (TGF-β) superfamily (e.g., activin) play an important role in induction, whereas several proteins (e.g., follistatin) permit neuralization by inhibiting these TGF-βs (Hemmati-Brivanlou, Kelly, & Melton, 1994).

Neurulation

Neurulation involves converting the neural plate into a neural tube. The plate itself emerges as a thickening along the midline of the dorsal ectoderm during induction. Once the neural plate appears, it becomes elongated along the rostrocaudal (top to bottom) axis (Smith & Schoenwolf, 1997). The neural plate is gradually transformed into a tube, which will later go on to form the brain and spinal cord. The widest section of the neural fold represents the future forebrain, and the presumptive midbrain is identified by a bend in the neural axis called the cranial flexure (Sidman & Rakic, 1982).

Assuming the neural tube closes correctly,[1] the tube itself is comprised of progenitor cells that give rise to the neurons and glia of the central nervous

---

[1] Errors in neural tube closure are generally referred to as *neural tube defects*. There is an entire class of such defects, but perhaps the most widely known is spina bifida,

system (CNS). Specifically, the rostral ("toward the front") portion of the tube will form the brain whereas the caudal ("toward the rear") portion will become the spinal cord. In addition, lying adjacent to and outside of the neural tube (i.e., sandwiched between the outer layer of the ectoderm and the neural tube) lies the neural crest. The cells that make up the neural crest will eventually give rise to the peripheral (autonomic) nervous system (which controls autonomic functions such as respiration).

Cell Proliferation

In primates and rodents, proliferation is comprised of a symmetrical and an asymmetrical stage (Rakic, 1988; Smart, 1985; Takahashi, Nowakowski, & Caviness, 1994). Chenn and McConnell (1995) have discussed how early in the proliferation period the mitosis of a progenitor cell produces two progenitor cells. Because one cell produces two identical cells, this first phase of proliferation has been described as *symmetrical*. Here the cells travel back and forth between the inner and outer sides of the ventricular zone (the first layer of the nervous system where early duplication occurs). Once duplication has occurred, the cell travels down the ventricular layer where it divides again. The two progenitor cells then independently begin the process of mitosis again. During the proliferation period, the marginal zone is formed, which contains the processes (axons and dendrites) of cells from the underlying ventricular zone (for review, see Takahashi, Nowakowski, & Caviness, 2001).

The second phase of proliferation (during which the first neurons are formed) begins at approximately 7 weeks gestation in the human, and this process continues until midgestation (Rakic, 1978). Here progenitor cells create one other progenitor cell and a postmitotic neuron (i.e., a cell no longer capable of dividing). Because two different types of cells are created, this form of proliferation is termed *asymmetrical*. Again, cells synthesize DNA and divide as they travel back and forth between the two sides of the ventricular zone. Whereas the newly formed progenitor cell goes on to generate other cells, the postmitotic neuron is believed to stop dividing and instead begins to migrate to its final destination (Rakic, 1988).

There are a multitude of subtle molecular interactions that must occur to permit and regulate cell proliferation. As a result, the embryo is very vulnerable to slight environmental perturbations. For instance, microencephaly

---

in which a portion of the tube (along the spinal cord) fails to close completely, resulting in a range of problems that can be as minor as lower motor extremity problems to serious physical handicaps.

(a heterogeneous group of disorders whose hallmark feature is that of a small brain) is due to aberrations in neural proliferation. Microencephaly can be caused by a number of exogenous experiences, including exposure to radiation, rubella, and maternal alcoholism (e.g., Warkany, Lemire, & Cohen, 1981; see Shonkoff & Phillips, 2000 for discussion). In addition, exposure to these environmental events during the proliferation phase may lead to an end of symmetrical proliferation, which in turn can cause a reduction in the final number of neurons.

Mechanisms of Migration

Once an immature neuron is formed it must migrate from the ventricular or subventricular zone to its final destination. In the human, migration begins at around 8 weeks gestation when the progenitor cells begin to produce postmitotic neurons (Rakic, 1978). Proliferation ends at approximately 4 to 5 months gestation, and the last cells begin their migration at this time.

Migration occurs in two distinct waves. In the first wave, migratory post-mitotic neurons are primarily derived from the ventricular zone, whereas in the second wave they are primarily derived from the subventricular zone (Rakic, 1972). Cortical neurons migrate in an inside-out pattern, meaning that first-born neurons migrate to lower cortical layers and the later-born neurons travel over other neurons for destinations in the outer cortex (Rakic, 1974). Consequently, neurons generated in the ventricular zone occupy the lower layers of the brain (layers 4, 5, and 6), whereas neurons derived from the subventricular zone become located in the outer regions of the brain (roughly layers 2 and 3).

There are two types of migration: radial and tangential (reviewed in Hatten, 1999; Rakic, 1995). In radial migration, neural precursors travel along radial glia from the proliferation zones to the outer areas of the CNS (Rakic, 1971, 1972, 1978). As a result, glia cells provide a path for the neurons to travel from the deep layers of the proliferation zones to their final destinations. Following the migration period, many radial glia are transformed into astrocytes, another type of glial cell (Rakic, 1990).

In contrast to radial migration, tangential migration permits neurons to travel parallel to the surface of the developing brain and, as a result, to enter and exit different brain regions (Rakic, 1990). As will be discussed in the section on differentiation in this chapter, whether cell migration follows a radial or tangential path will determine whether genetic or epigenetic influences are primarily responsible for determining the precise future location of the cell. In other words, if cells were distributed radially, then the birth date and location of the postmitotic neuron's progenitor cell will determine where the neuron will reside. However, if tangential migration were also involved, this would indicate that the cell's fate may not be completely determined by birth date and progenitor location of the progenitor, and that environmental cues could influence the cell's placement in the cortex.

Anatomical Changes due to Proliferation and Migration

As cell migration continues, the immature cortex is transformed from a single sheet composed entirely of progenitor cells to a multilayered structure with many different types of cells. By the sixth week of gestation, the marginal zone appears superficially to the ventricular zone. Between the sixth and eighth week, the intermediate zone emerges between the ventricular and marginal zones. Largely, the cells of the intermediate zone are postmitotic.

The subventricular zone (which is heavily involved in proliferation; see Rakic, 1978) emerges between the ventricular and intermediate zones between the eighth and tenth weeks of gestation (Sidman & Rakic, 1973). As the ventricular zone becomes depleted of cells from the first wave of migration, the subventricular zone provides the majority of the neurons in the second wave of migration (Rakic, 1978). The cortical plate (which later develops into the six layers of the cerebral cortex) is also formed between the intermediate and marginal zones (Rakic, 1972). Subsequently, branching neurons from the cortical plate cause the preplate to split into the marginal zone and subplate. This subplate region is involved in determining the organization of the cerebral cortex (O'Leary, Schlaggar, & Tuttle, 1994).

In the first wave of migration, the proliferation of progenitor cells in the ventricular zone leads to the formation of vesicles or bulges in the neural tube itself. By the twentieth day of human gestation, the nervous system is comprised of three primary vesicles (O'Rahilly & Müller, 1994). The three vesicles are the proencephalon (forebrain), mesencephalon (midbrain), and rhombencephalon (hindbrain). At five weeks, further proliferation gives rise to the five-vesicle stage. At this point, the proencephalon splits to become the telencephalon and the diencephalon. In addition, although the mesencephalon does not divide, the hindbrain is segmented into the metencephalon as well as the myelencephalon.

The second wave of migration occurs between the eleventh and sixteenth weeks of gestation (Rakic, 1972). During this period, the telencephalon dramatically expands and differentiates to form the cerebral cortex, the basal ganglia, the corpus callosum, and other structures, including the thalamus (Martin & Jessell, 1991). By the twentieth week of gestation, the cortical plate has divided into three layers, and by the seventh month of gestation, all six layers (lamina) of the cortex can be observed, the same number as in the adult brain (layer 6 is the deepest layer, whereas layer 1 is the most superficial, outermost layer).

Axonal Outgrowth and Synaptogenesis
Once neurons migrate to their destinations, their axons must extend to establish connections with other neurons (Tessier-Lavigne & Goodman, 1996). Growth cones (which lie at the tip of moving axons) play a key role in axonal navigation. Growth cones are comprised of two basic structures: lamellipodia and filopodia (Suter & Forscher, 1998). Whereas lamellipodia are fan shaped, filopodia protrude from the lamellipodia as small spikes. These processes expand and retract to sample the molecular surroundings and thereby determine the direction of movement for the axon.

Once the axon arrives at its target destination, synaptic connections must be formed and strengthened, a process that involves neurotrophic factors. *Neurotrophic factors* are signaling molecules that are necessary for neural survival (Henderson, 1996). Examples include neurotrophin-4/5 (NT-4/5) and brain-derived neurotrophic factor (BDNF). However, in addition to these (and other) neural growth factors (NGFs), much of synaptic modeling and remodeling depends on experience, a topic that I will return to later in this chapter. Suffice to say, as axons extend and dendrites arborize, the developing nervous system becomes more densely packed and the surface of the brain acquires convolutions (sulci and gyri) in order to accommodate this increased cortical mass. Thus, through these processes of axonal outgrowth and synaptogenesis, the brain increases in size, connections are formed, and the brain takes on a more mature appearance.

Differentiation

Cellular differentiation refers to the process whereby cells become more specialized over time. Exactly when a cell is specified to reside in a particular location is a problem of enduring interest, because different locations and cortical layers tend to exhibit distinct neural characteristics (Chenn, Braisted, McConnell, & O'Leary, 1997).

Two primary theories have been offered to account for when and how cells are specified to reside at particular destinations. Rakic's protomap hypothesis (1988, 1995) postulates that the proliferative ventricular zone contains a "blueprint" for the placement of all neurons within the adult cortex. The destination of the cell on the horizontal plane is determined by the location of the precursor cell in the ventricular zone, and the cell's position on the vertical plane is determined by the time of origin (Rakic, 1995). In contrast, in the protocortex hypothesis (Chenn et al., 1997; O'Leary et al., 1994), emphasis is placed on epigenetic influences and the role of events that occur once the neuron has reached its destination. In contrast to the protomap hypothesis, the protocortex hypothesis suggests that the spatial relationship among progenitors in the ventricular zone may not be maintained by the daughter cell in the cortex. Rather, cells from common genetic backgrounds have the capacity to travel to disparate locations of the brain via tangential migration (Walsh & Cepko, 1992, 1993). This form of migration may not follow a predetermined pattern; instead, traveling cells may be responsive to various environmental cues for determining their fate. Thus, tangential migration may provide for both positive and negative influences to alter the movement and eventual differentiation of cortical neurons.

Collectively, there is evidence to support the observation that cells travel from ventricular and subventricular zones out to the cortex in a point-to-point fashion as the protomap theory would predict. However, there is also evidence that not all cell fate is predetermined. Via tangential migration, cell movement and thus cell morphology might be somewhat plastic. This plasticity may allow for the amelioration of errors in migration and other developmental processes, but at the same time might also permit environmental perturbations to affect the developing nervous tissue adversely during the latter parts of fetal development.

Apoptosis

It has been estimated that during development, roughly half of all brain cells die, a process referred to as *apoptosis* (Raff et al., 1993). Apoptosis is a form of programmed cell death (most likely mediated by neurotrophic factors) characterized by overall shrinkage of the cell, with the organelles and plasma membrane remaining unaffected (Kerr, Wyllie, & Currie, 1972). Other cells or macrophages then quickly absorb the dead cell (Jacobson, Weil, & Raff, 1997). This rapid response prevents leakage of the cell contents as well as inflammation. Because apoptosis is involved in the degeneration of approximately *50 percent* of all neurons during development, the appropriate unfolding of this process is crucial for normal development. Indeed, perturbations in apoptosis are linked to mental retardation. For example, neurons from fetuses with Down syndrome experienced high rates of apoptosis when they were cultured relative to cells from control brains (Busciglio & Yankner, 1995). It is suggested that this neuronal defect may at least partly account for the mental retardation associated with Down syndrome.

Summary of Prenatal Development

The CNS is formed through the intricate processes of induction, neurulation, proliferation, migration, axon extension/dendritic arborization, synaptogenesis, differentiation, and apoptosis. Even the smallest perturbations during any of these periods may severely alter the developmental trajectory of the organism. On the other hand, we know that these same perturbations can have little effect on brain function (e.g., heterotopia – a pathological state in which cell bodies are found where normally only processes reside – is not uncommon among perfectly normally functioning individuals). Thus, a crucial area of future investigation lies in explaining the broad range of individual differences that lead to different functional developmental trajectories.

Having established the prenatal blueprint for the developing brain, I now turn to a discussion of crucial postnatal events.

POSTNATAL DEVELOPMENT

As stated at the outset, the major two postnatal events concern synaptogenesis and myelination, topics to which I now turn.

Formation of Axons and Dendrites

To produce a functional synapse, axons must make appropriate connections with dendrites. It is well established that axons are produced in excess numbers during perinatal life (i.e., the end of the prenatal period and the beginning of the postnatal period) and the final number may be achieved by the process of competitive elimination. For example, in the corpus callosum of the infant rhesus monkey, the number of axonal fibers is at least 3.5 times that of the adult monkey (LaMantia & Rakic, 1990). Similarly, in the visual cortex of the rhesus monkey, the peak number of axons occurs at about five postnatal months and the peak of synaptogenesis occurs around the eighth postnatal month (Michel & Garey, 1984).

During the first postnatal year, growth of dendritic trees and spines can be seen in all six layers of the cortex, although these spines are still immature. In the visual cortex, for example, there is rapid development between the second and fourth postnatal month, with maximum dendritic arborization occurring by approximately the fifth postnatal month, followed by regression to adult levels by the second postnatal year (Michel & Garey, 1984).

Synaptogenesis

Synaptic Overproduction. Although there are two types of synapses, electrical (e.g., gap junctions) and chemical, I will focus most on the latter, about which more is known. In a chemical synapse, an electrical signal from the presynaptic cell is converted into a chemical signal that can be transferred through extracellular space to the postsynaptic cell. In synaptic transmission, an electrical signal is transferred from the cell body (soma), down the axon, and signals the release of chemical messengers into extracellular space. The chemical messengers (most commonly neurotransmitters and neuropeptides) can open or close ion channels on dendritic spines, changing the electrical current in the postsynaptic cell. This process allows for intercellular communication, with most synapses occurring between axons and dendrites (but also axon to cell, dendrite to dendrite, and axon to axon).

Both spontaneous (Molliver, Kostovic, & Van der Loos, 1973) and environmentally induced neuronal activity lead to the formation and stabilization of synapses. Early developing synapses are labile. Most likely, this mechanism is preparing the system for environment input. Synapses may become stabilized through one of several mechanisms. For example, they may represent coordinated activity at pre- and postsynaptic sites (Schlaggar, Fox, & O'Leary, 1993). Formation of adult patterns of connection involves the elimination of a limited number of immature labile connections with the elaboration and addition of appropriate connections. Those synapses that make functional connections receive a larger amount of coordinated activity and are stabilized, whereas those that do not may be eliminated or reabsorbed (Changeux & Danchin, 1976).

Synapse stabilization may also occur through the local release of various neurotrophic factors (such as NGFs). It has been suggested, for example, that axons whose parent cells have recently been activated are able to respond to neurotrophic factors (e.g., Katz & Shatz, 1996; Thoenen, 1995). Glutamate receptors (the NMDA subtype) may also function in a similar manner by mediating postsynaptic activation of cortical cells (e.g., Schlaggar et al., 1993).

The first synapses may occur as early as the twenty-third week of gestation. For example, Molliver and colleagues (1973) identified the first synaptic junctions in the cortical plate at about this age. However, most synapses develop postnatally, particularly during the first year of postnatal life. Although the timing of synaptogenesis is varied, adult values and peak levels of synaptic density in the auditory cortex, visual cortex, and medial frontal gyrus show similar aggregate values, suggesting that peak densities and synaptic elimination occur to a similar degree throughout the cortex (Huttenlocher & Dabholkar, 1997). For example, Huttenlocher and colleagues (Huttenlocher, 1979a, 1979b, 1984; Huttenlocher & de Courten, 1987; Huttenlocher & Dabholkar, 1997) have carefully documented synaptogenesis in the visual cortex and prefrontal cortex using postmortem tissue. In the visual cortex, for example, the greatest increases in synaptogenesis occur between the second and eighth postnatal months, with the most rapid increases between 2 and 4 months (Huttenlocher & de Courten, 1987), even though some synapses can be seen as early as the twenty-eighth week of gestation. In contrast, in the frontal cortex, synapse formation begins at 27 weeks gestation and does not reach its maximum density until after 15 postnatal months. In the middle frontal gyrus (MFG), an area believed to be involved in higher forms of cognition, synaptic density reaches its maximum number of synapses at 3.5 years (Huttenlocher & Dabholkar, 1997).

It has been proposed that the initial overproduction of synapses in the cortex may be related to the functional property of the immature brain to allow recovery and adaptation after focal injury or malformation (Huttenlocher, 1984) and may represent a critical or vulnerable period. This overproduction may also be the mechanism by which the brain is made ready to receive specific input from the environment. Studies of synaptogenesis demonstrated important developmental increases in the postnatal period, and Goldman-Rakic (1987) proposed that the period of early overgrowth is important for the onset of cognitive function.

Synaptic Pruning. The elimination of synapses may be universal to all neuronal systems, and patterned connections within the brain may be based predominantly on large-scale regressive events (Changeux & Danchin, 1976; Huttenlocher & Dabholkar, 1997; Rakic, Bourgeois, Eckenhoff, Zecevic, & Goldman-Rakic, 1986; but see Purves, 1989). Pruning, or loss of synapses in the absence of cell death, refers to environmentally regulated changes in the density of synapses per unit of dendritic length. Synapse elimination occurs late in childhood and in adolescence (Huttenlocher & Dabholkar, 1997). Although there are topographical differences in the time course of synapse formation and elimination, quantitative measures have found a common pattern: the number of synapses seen at peak during childhood is reduced by approximately 40 percent to reach the adult value (Huttenlocher, 1979a, 1979b; Huttenlocher & de Courten, 1987).

Presynaptic neurotransmitters play a role in the stabilization of synapses and modulation of cortical neuron activity (Kostović, 1990). In particular, changes in the distribution of excitatory and inhibitory inputs may lead to pruning. For example, Diebler and colleagues (Diebler, Farkas-Bargeton, & Wehrle, 1979) propose that the amount of inhibitory neurotransmitter (GABA) at cortical synapses may drive elimination.

Second, pruning is thought to be caused by limited availability of neurotrophic factors derived from the target neuron and by trophic interactions with afferents. This may occur by way of specific neurotransmitters, NGF, NT-3, BDNF, or thyrotropin-releasing hormone (Patterson & Nawa, 1993). Thus, only collaterals that are electrically active can respond to synaptogenic factors, and synaptic contacts that are not incorporated into neuronal circuits may be gradually eliminated (Changeux & Danchin, 1976). Furthermore, it is most likely that only inappropriate synapses and their branches disappear, whereas arborization in appropriate layers may increase in size and complexity.

Summary

Overall, some synapses form as early as four months before term, although the majority of synapses are formed later in gestation, and carry forward into the postnatal period. Indeed, it is now well documented that the developing brain massively overproduces synapses, which is followed by a reduction to adult numbers. The period of overproduction and pruning varies by area, with synapses reaching adult values in the visual cortex by the fifth to sixth postnatal year, whereas the adult number of synapses in frontal cortex is not obtained until mid to late adolescence. As will be elaborated in a subsequent section, it is believed that the processes of overproduction and pruning are powerfully influenced by experience.

Myelination

Myelin is a fatty sheath that insulates axons and provides for more rapid impulse conduction. In the peripheral nervous system, myelin is comprised of Schwann cells, whereas in the CNS it is composed of oligodendrocytes.

Myelination is thought to occur in a caudal to rostral direction (front to back). Importantly, the areas of the brain to myelinate first are the same that appear to develop function first. For example, Gibson and Brammer (1981) found that primary sensory and motor projection areas of the cortex develop in advance of the association areas; layers subserving communication with the brain stem and spinal cord (layers I, IV, V, VI) myelinate prior to layers subserving communication with the cortex (II, III).

Although examining myelination before birth is still largely done on autopsy specimens, examining this same process after birth is now possible using magnetic resonance imaging (MRI). Although the use of MRI to study postnatal brain development is in its infancy, it has been used to examine brain anatomy, gray and white matter development, myelination, and the development of white matter tracts across a range of ages. For example, myelination can be inferred from the level of signal intensity. In a series of studies of typically developing newborns and young infants, Gilmore and colleagues have reported that males generally have larger intracranial volumes than females, and (in an unpublished study) that in contrast to the adult, there are no sex differences in white matter in newborns (Gilmore et al., 2004, 2006; Knickmeyer, Gouttard, Kang, Evans, et al., 2008). Interestingly, gray matter increases rapidly in the first postnatal year, both absolutely and in relation to white matter growth, although the relative growth becomes equivalent in the second year of life (Knickmeyer et al., 2008). These same authors have also reported that there is a large increase in intracranial volume in the first year, that by age 1, intracranial volume is about 72 percent of adult volume, and at 2 years, it is 83 percent (Gilmore et al., 2006; Knickmeyer et al., 2008).

From a number of longitudinal studies examining the course of myelination from early childhood through early adulthood (see Toga, Thompson, & Sowell, 2006, for recent review) comes the suggestion that there is a linear increase in white matter through age 20, and nonlinear changes in gray matter during this same time period. Moreover, Giedd et al. (1999) have shown that there is an increase in gray matter up until about the age of 12 in the frontal and parietal lobes, followed by a decrease. In the temporal lobes, the increase in gray matter occurs until age 16, followed by a decrease. Overall, the white matter increases linearly, and shows growth well into the third decade of life, whereas gray matter shows earlier rapid growth and slower growth later, depicted as a U-shaped growth curve (Giedd, Lalonde, Celano, White et al., 2009). Importantly, an ambitious project funded by the National Institutes of Health (NIH) in early 2000 is now nearing an end. This project, designed to construct an atlas of the developing brain (0–18 years) using MRI, will undoubtedly provide a rich source of information in years to come (see Almli et al., 2007)

Conclusions

The foundational elements of the developing brain are clearly laid down long before birth. They begin with the formation of the neural tube just a few weeks after conception, and conclude by the time cell migration has completed its course by about the fifth prenatal month. At this time the formation of neural circuits begins with the earliest formation of synapses, followed shortly thereafter by the myelination of axons in various sensory systems. However, the vast majority of synaptogenesis and myelination occurs postnatally. Both of these processes are under a combination of endogenous (e.g., genetic, humoral) and exogenous (e.g., experience) control. The latter occupies the next and final section of this chapter.

THE DEVELOPMENT AND MODIFICATION OF NEURAL CIRCUITS – NEURAL PLASTICITY

In the preceding section, I discussed how the brain is built from conception through the adolescent period. Clearly, some aspects of development are largely or entirely under genetic or humoral control, and thus are impervious to exogenous or endogenous experience (so-called experience-independent development). Even here, however, "experience" can exert an effect. Witness the case of prenatal exposure to teratogens or nutritional deprivation (for review and discussion, see Fuglestad, Rao, & Georgieff, 2008; Mattson et al., 2008; Shonkoff & Phillips, 2000), both of which can affect cellular differentiation (e.g., microcephaly), myelination, and possibly even neurulation. However, where the effects of experience most powerfully and compelling influence brain development pertains to the formation of synaptic circuits. Here the evidence is overwhelming that both positive and negative experiences can influence the wiring diagram of the brain. It must be kept in mind that the nature of the experience itself, coupled with the maturity of the brain at the time the experience occurs, will determine or at least influence whether the resulting neural change is beneficial or deleterious to the organism. In the sections that follow, I provide examples of both good and bad outcomes, along with examples of plasticity that are restricted to the developing organism along with those that are seemingly unconstrained by age. I shall begin, however, by discussing some general principles that account for how the structure of experience is incorporated into the structure of the brain.

Neurobiological Mechanisms underlying Neural Plasticity

As a rule, there are a number of mechanisms whereby experience induces changes in the brain. First, an *anatomical* change might reflect the ability of an existing synapse to modify its activity by forming new axons or by expanding the dendritic surface. For example, rearing rats in complex environments can lead to an increase in dendritic spines, which will ultimately lead to the formation of new synapses. Second a *neurochemical* change might be reflected in the ability of an existing synapse to modify its activity by increasing neurotransmitter synthesis and release. For example, it is now well established that N-methyl D-aspartate (NMDA), an excitotoxic amino acid used to identify a specific subset of glutamate, is known to modify pre- and postsynaptic activity and can trigger the formation of new dendritic spines (for review, see Yuste & Sur, 1999). Third, an example of a *metabolic* change might be the fluctuations in cortical and subcortical metabolic activity (e.g., glucose utilization, O_2) in response to experience. Again, as we shall see in the section on *motor plasticity*, rats taught acrobatic feats tend to increase the number of capillaries in the region of the brain involved in motor movements (see Black et al., 1998, for review). Finally, in all cases it is assumed that changes in gene expression occur with experience; for example, Rampon et al. (2000) have shown that adult mice that received a period of environmental enrichment (for as little as 3 hours and as much as 14 days) showed changes in a variety of genes associated with DNA/RNA synthesis, neuronal signaling, neuronal growth and structure, and apoptosis. Collectively, experience works its way into the brain through a number of molecular mechanisms.

Having established the mechanisms whereby experience exerts is effects on brain structure and function, I shall now turn to a discussion of specific behaviors and abilities and their neural correlates that are affected by experience.

Visual Development

Stereoscopic depth perception refers to the ability to discern depth cues based on the different visual perspective each eye receives. The development of this ability is made possible by the development of ocular dominance columns, which represent the connections between each eye and layer IV of the visual cortex. If for some reason the two eyes are not properly aligned, thereby preventing them from converging effectively on a distant target (as would occur with *strabismus*), then the ocular dominance columns that support normal stereoscopic depth perception will fail to develop normally. If this condition is not corrected by the time the number of synapses begins to reach adult values (generally in the first few years of life), the child will not develop normal stereoscopic vision. The result is not only poor stereoscopic vision but also the possibility of poor vision in one eye.

Until very recently it had been thought that the development of ocular dominance columns was largely driven by postnatal visual experience. For example, David Hubel and Torsten Wiesel reported in a series of papers in the 1960s that in monkeys and cats seeing out of both eyes was essential to the development of ocular dominance columns; thus, it was believed that such columns develop out of visual activity. Recently, however, there are reports that the initial development of such columns is not activity-dependent at all (Crowley & Katz, 1999); moreover, Crowley and Katz (2000) have suggested that the formation of these columns may depend most on innate molecules that guide growing axons to their target destination, not on visual experience per se. These findings do raise intriguing questions about the role of experience in the development of ocular dominance columns. At the very least, they suggest that such columns develop before visual experience can exert its effect (i.e., before birth), a finding consistent with an observation by Bourgeois and colleagues that being born prematurely or even removing the eyes of monkeys prior to birth has little effect on the overproduction of synapses in the visual cortex (Bourgeois, Reboff, & Rakic, 1989); however, less is known about the retraction of such synapses.

Turning this issue on its head, Maurer and colleagues (Maurer, Lewis, Brent, & Levin, 1999; for review, see Maurer et al., 2008) have reported on a longitudinal study of infants born with cataracts. An elegant aspect of this work was that the investigators were able to study infants who had the cataracts removed at different ages. The basic finding was the powerful role of experience in facilitating vision postcataract removal; thus, for example, in infants operated on within months of birth, even just a few minutes of visual experience led to a rapid change in visual acuity. In addition, as expected, the longer the infant lived with cataracts, the less experience led to favorable changes in vision. This process resembles the experience-expectant model of plasticity discussed earlier: specifically, our species has evolved to expect certain visual experiences to occur within a certain period of time, and if the organism is deprived of such experience beyond this sensitive period, development is deleteriously affected.[2]

Second Language Acquisition/The Neural Representation of Language

Dehaene and colleagues (Dehaene et al., 1997) reported a number of years ago that the neural representation of a second language was identical to that of a first language *if* the individual was truly bilingual; if, however,

---

[2] Sale et al. (2007) have recently demonstrated, in a rodent model, that under certain conditions environmental enrichment can restore visual function in adult animals suffering from amblyopia. If replicated, such work would have important implications for our current understanding of sensitive periods involved in visual development.

the mastery of the second language was not as strong as the first, then the functional neuroanatomy (based on positron emission tomography [PET] neuroimaging) was different. Because nearly all of the bilinguals studied in this work had acquired their second language at an early age, the initial conclusion drawn was that the second language needed to be learned early in life in order to share the same neural representation as for the first language. This conclusion has recently been questioned, however. These same authors (see Perani et al., 1998) wondered whether it was the *age* at which the second language was acquired that was the critical variable, or rather, the subject's *proficiency* in speaking this language. In a follow-up study, the age at which the second language had been mastered was crossed with the proficiency of speaking this language. The authors observed that it was the latter dimension that proved critical. Thus, regardless of when the second language was acquired, speaking this language with equal proficiency as the first language led to shared neural representation for both languages. Similar findings have recently been obtained with congenitally deaf individuals with mastery in sign language: the areas of the brain involved in "speaking" in sign are the same as those of hearing speakers using spoken language (see Petitto et al., 2000). Collectively, these findings call into question the critical period hypothesis for acquiring a second language.

Note, however, that the issue of shared neural representation for multiple languages should not be confused with the issue of speaking a second language without an accent. Thus, Newport (e.g., Johnson & Newport, 1989) demonstrated that individuals who acquire a second language before the age of ten are far more likely to speak that language without an accent than those who acquire that language after age 10.

Learning and Memory

In the context of the effects of experience on brain and behavior, no area has received more attention than that of learning and memory. For example, it has been known for more than twenty years that rats raised in complex laboratory environments (i.e., those containing lots of toys and social contacts) outperform rats reared in isolation on certain cognitive tasks (e.g., the former make fewer errors on tasks of spatial cognition; Greenough, Madden, & Fleischmann, 1972). At the cellular level, some of the changes observed among rats raised in such environments include: (1) several regions of the dorsal neocortex (e.g., visual areas) are heavier and thicker and have more synapses per neuron; (2) dendritic spines and branching patterns increase in number and length; and (3) there is increased capillary branching, thereby increasing blood and oxygen volume (for examples, see Black et al., 1998; Greenough et al., 1987; Greenough, Juraska, & Volkmar, 1979; Greenough et al., 1972; for recent review, see Black et al., 1998; Nelson, 2000; Nelson, de Haan, & Thomas, 2006a). Importantly, these effects are not simply due to increased motor activity; for example, Black and colleagues have demonstrated that rats engaged in repetitive motor acts that require no learning (e.g., simply running on a treadmill) show only a subset of the changes comparable to rats that were engaged in a learning task (see Black et al., 1998, for review).

What about more functional changes at the system level that is correlated with learning and memory? Erickson, Jagadeesh, and Desimone (2000) reported a study in which monkeys were presented with multicolored complex stimuli (some objects, some abstract designs). Some of the stimuli were novel (never seen before) and others were familiar. Single neurons were recorded from the perirhinal cortex, an area of the temporal lobe known to be strongly involved in episodic memory. The authors reported that after only 1 day of experience viewing the stimuli, performance of neighboring neurons became highly correlated, whereas viewing novel stimuli revealed little correlated neuronal activity. The implication of these findings is that visual experience leads to functional changes in an area of the brain known to be involved in memory. Although this finding may not be surprising on the surface, it is among the first to provide concrete evidence of how experience influences brain function.

Collectively, it is now well established that learning and memory are correlated with changes in the brain at multiple levels, from the molecular (e.g., changes in pre- and postsynaptic functioning mediated by glutamate receptors) to the molar (e.g., changes in neuronal firing). There is no sense that there is a sensitive period for learning and memory to occur (for a tutorial on the development of learning and memory, see Nelson, 1995, 2000). Indeed, there is some sense that activities that engage the learning and memory system may confer some protection on lifelong learning and memory function (see Nelson, 2000, for discussion).

Motor and Somatosensory Systems

There is now extensive work to suggest that the motor and somatosensory systems are modified by experience regardless of the age of the organism. For example, Pons et al. (1991) studied a group of monkeys that many years earlier had served in a study of deafferentation (severing the neuronal connections) of somatosensory cortex. The investigators recorded neuronal responses from a region of somatosensory cortex that would normally correspond to the deafferented portion of the limb, including the fingers, palm, and adjacent areas (area S1). Because the limb had been deafferented for twelve years, the investigators were not expecting much in the way of cortical reorganization. Much to their surprise, however, it was revealed that this region of the brain now responded to stimulation in an area of the face (this region would normally border the cortical region innervated by the deafferented limb), pointing to a reorganization of somatosensory cortex of rather massive proportions (i.e., 10–14 mm).

This report was followed by others (see Pons, 1995, for discussion), and collectively demonstrated that large-scale cortical reorganization can occur following injury even in the mature primate. Reports of the adult human followed. For example, Ramachandran, Rogers-Ramachandran, and Stewart (1992) speculated that an individual who had experienced the amputation of a limb (such as the forearm) should show sensitivity on the area of the body represented by the area of the brain adjacent to the amputated limb. Adults who had experienced various forms of amputation were examined, and were found to experience sensation in the limb that had in fact been amputated. Ramachandran examined one patient's sensitivity to tactile stimulation along the region of the face known to innervate the somatosensory cortex adjacent to the area previously innervated by the missing limb; the patient reported sensation in both the face and the missing limb. Using magnetoencephalography (MEG), Ramachandran was able to determine the degree to which the cortical surface had been reorganized to take over responsibility for the area previously occupied by the missing limb.

In both the monkey and human work, cortical reorganization was thought to be facilitated by stimulation of those parts of the body that were adjacent to the deafferentated/amputated limb, suggesting a form of "natural" intervention. To evaluate more directly whether the motor and/or somatosensory cortex can be reorganized based on experience, Nudo and Milliken (1996) mapped the motor cortex of monkeys before and after an ischemic lesion was made. As would be expected to occur in the human suffering a stroke in the same region of the brain, the infarct led to a deficit in use of that limb – in this case, the animal's inability to retrieve food pellets. The animals then received intensive training in hand use, which resulted in a return to performance comparable to preinjury levels. Cortical mapping revealed substantial rearrangement of the area of the brain that represented the hand surrounding the lesion site.

These findings, coupled with those reported on monkeys and humans, suggest that the representation of the limbs in the adult primate can be altered as a function of experience. Building on this model, Taub (2000; see also Liepert et al., 2000) has developed a form of rehabilitation that involves restraining the patient from using the unaffected limb in order to "teach" the affected (by stroke) limb to work. Relatively brief periods of restraint (on the order of weeks) coupled with massed activity of the affected limb appear to show very beneficial effects; that is, such patients gain dramatic use of the affected limb.

In summary, there is now evidence to support the thesis that cortical reorganization is possible following injury to the peripheral nervous system in the adult human and nonhuman primate. Might similar reorganization occur in the noninjured, "healthy" individual? Also using MEG, Elbert, Pantev, Wienbruch, Rockstroh, and Taub (1995) mapped the somatosensory cortex of adults with and without experience playing a stringed instrument (e.g., guitar, violin). The investigators reported that in the musicians, the area of the somatosensory cortex that represented the fingers of the left hand (used on the finger board) was larger than the area represented by the right hand (which was used to bow), and larger than the left hand area in the nonmusicians. Moreover, there was a tendency for there to be greater cortical representation in individuals who had begun their musical training before age 10. Collectively, this work suggests that the brain of the adult human can reorganize based not just on negative experience (e.g., injury) but also on positive experiences (e.g., musical training).

CONCLUSIONS

The goal of this chapter has been to explicate the unfolding of the human brain during the first years of life and to explore some of the ways experience influences this process. Space limitations have precluded an exposition of how changes in brain development lead to changes in behavioral development (for examples of such work, see Kagan & Herschkowitz, 2005; Nelson, de Haan, & Thomas, 2006b). However, it is hoped that the implications of understanding the neural underpinnings of behavioral development are self-evident. Suffice to say, only by understanding the neurobiological processes that make possible the miracle of child development will we come to understand the whole child.

REFERENCES

Almli, C. R., Rivkin, M. J., & McKinstry, R. C. Brain Development Cooperative Group. (2007). The NIH MRI study of normal brain development (Objective-2): Newborns, infants, toddlers, and preschoolers. *Neuroimage, 35*(1), 308–25.

Black, J. E., Jones, T. A., Nelson, C. A., & Greenough, W. T. (1998). Neuronal plasticityand the developing brain. In N. E. Alessi, J. T. Coyle, S. I. Harrison, & S. Eth (Eds.), *Handbook of Child and Adolescent Psychiatry. Vol 6. Basic Psychiatric Science and Treatment* (pp. 31–53). New York: John Wiley & Sons.

Bourgeois, J., Reboff, P., & Rakic, P. (1989). Synaptogenesis in visual cortex of normal and preterm monkeys: Evidence from intrinsic regulation of synaptic overproduction. *Proceedings from the National Academy of Sciences, 86,* 4297–4301.

Busciglio, J., & Yankner, B. A. (1995). Apoptosis and increased generation of reactive oxygen species in Down's syndrome neurons *in vitro. Nature, 378,* 776–9.

Changeux, J. P., & Danchin, A. (1976). Selective stabilization of developing synapses as a mechanism for the specification of neuronal networks. *Nature, 264*(5588), 705–12.

Chenn, A., Braisted, J. E., McConnell, S. K., & O'Leary, D. D. M. (1997). Development of the cerebral cortex: Mechanisms controlling cell fate, laminar and areal patterning, and axonal connectivity. In W. M. Cowan, T. M. Jessell, & S. L. Zipursky (Eds.), *Molecular and Cellular Approaches to Neural Development.* New York: Oxford University Press.

Chenn, A., & McConnell, S. K. (1995). Cleavage orientation and the asymmetric inheritance of Notch1 immunoreactivity in mammalian neurogenesis. *Cell, 82,* 631–41.

Crowley, J. C., & Katz, L. C. (1999). Development of ocular dominance columns in the absence of retinal input. *Nature Neuroscience, 2,* 1125–30.

Crowley, J. C., & Katz, L. C. (2000). Early development of ocular dominance columns. *Science, 290,* 1321–4.

Dehaene, S., Dupoux, E., Mehler, J., Cohen, L., Paulesu, E., Perani, D., et al. (1997). Anatomical variability in the cortical representation of first and second languages. *Neuroreport, 8,* 3809–15.

Diebler, M. F., Farkas-Bargeton, E., & Wehrle, R. (1979). Developmental changes of enzymes associated with energy metabolism and the synthesis of some neurotransmitters in discrete areas of human neocortex. *Journal of Neurochemistry, 32*(2): 429–35.

Elbert, T., Pantev, C., Wienbruch, C., Rockstroh, B., & Taub, E. (1995). Increased cortical representation of the fingers of the left hand in string players. *Science, 270*(5234), 305–7.

Erickson, C. A., Jagadeesh, B., & Desimone, R. (2000). Clustering of perirhinal neurons with similar properties following visual experience in adult monkeys. *Nature Neuroscience, 3,* 1066–8.

Fuglestad, A. J., Rao, R., & Georgieff, M. K. (2008). The role of nutrition in cognitive development. In C. A. Nelson & M. Luciana (Eds.), *Handbook of Developmental Cognitive Neuroscience,* 2nd edition (pp. 623–42). Cambridge, MA: MIT Press.

Gibson, A., & Brammer, M. J. (1981). The influence of divalent cations and substrate concentration on the incorporation of myo-inositol into phospholipids of isolated bovine oligodendrocytes. *Journal of Neurochemistry, 36*(3), 868–74.

Giedd, J. N., Blumenthal, J., Jeffries, N. O., Castellanos, F. X., Liu, H., Zijdenbos, A., Paus, T., Evans, A. C., & Rapoport, J. L. (1999). Brain development during childhood and adolescence: a longitudinal MRI study. *Nature Neuroscience, 2*(10), 861–3.

Giedd, J. N., Lalonde, F. M., Celano, M. J., White, S. L., Wallace, G. L., Lee, N. R., & Lenroot, R. K. (2009). Anatomical brain magnetic resonance imaging of typically developing children and adolescents. *Journal of the American Academy of Child & Adolescent Psychiatry, 48*(5), 465–70.

Gilmore, J. H., Lin, W., Knickmeyer, R., Hamer, R. M., Smith, J. K., & Gerig, G. (2006). *Imaging early childhood brain development in humans.* Presentation at the Society for Neuroscience, Fall 2006.

Gilmore, J. H., Zhai, G., Wilber, K., Smith, J. K., Lin, W., & Gerig, G. (2004). 3 Tesla magnetic resonance imaging of the brain in newborns. *Psychiatry Research, 132,* 81–5.

Goldman-Rakic, P. S. (1987). Development of cortical circuitry and cognitive function. *Child Development, 58*(3) 601–22.

Greenough, W. T., Black, J. E., & Wallace, C. S. (1987). Experience and brain development. *Child Development, 58*(3) 539–59.

Greenough, W. T., Juraska, J. M., Volkmar, F. R. (1979). Maze training effects on dendritic branching in occipital cortex of adult rats. *Behavioral & Neural Biology. 26*(3), 287–97.

Greenough, W. T., Madden, T. C., & Fleischmann, T. B. (1972). Effects of isolation, daily handling, and enriched rearing on maze learning. *Psychonomic Science, 27,* 279–80.

Hatten, M. E. (1999). Central nervous system neuronal migration. *Annual Review of Neuroscience, 22,* 511–39.

Hemmati-Brivanlou, A., Kelly, O. G., & Melton, D. A. (1994). Follistatin, an antagonist of activin, is expressed in the Spemann organizer and displays direct neuralizing activity. *Cell, 77,* 283–95.

Henderson, C. E. (1996). Role of neurotrophic factors in neuronal development. *Current Opinion in Neurobiology, 6,* 64–70.

Huttenlocher, P. R. (1979a). Synaptic and dendritic development and mental defect. In N. Buchwalk & M. Brazier (Eds.), *Brain Mechanisms in Mental Retardation.* New York: Academic.

Huttenlocher, P. R. (1979b). Synaptic density in human frontal cortex: Developmental changes and effects of aging. *Brain Research, 163,* 195–205.

Huttenlocher, P. R. (1984) Synapse elimination and plasticity in developing human cerebral cortex. *American Journal of Mental Deficiency, 88*(5), 488–96.

Huttenlocher, P. R., & Dabholkar, A. S. (1997) Regional differences in synaptogenesis in human cerebral cortex. *Journal of Comparative Neurology, 387*(2), 167–78.

Huttenlocher, P. R., & de Courten, C. (1987). The development of synapses in striate cortex of man. *Human Neurobiology, 6*(1), 1–9.

Jacobson, M. D., Weil, M., & Raff, M. C. (1997). Programmed cell death in animal development. *Cell, 88,* 347–54.

Johnson, J. S., & Newport, E. L. (1989). Critical period effects in second language learning on the production of English consonants. *Cognitive Psychology, 21,* 60–99.

Kagan, J., & Herschkowitz, N. (2005). *A Young Mind in a Growing Brain*. Hillsdale, NJ: Lawrence Erlbaum Associates.

Katz, L. C., & Shatz, C. J. (1996). Synaptic activity and the construction of cortical circuits. *Science, 274*, 1133–8.

Kerr, J. F. R., Wyllie, A. H., & Currie, A. R. (1972). Apoptosis: A basic biological phenomenon with wide-ranging implications in tissue kinetics. *British Journal of Cancer, 26*, 239–57.

Knickmeyer, R. C., Gouttard, S., Kang, C., Evans, D., Wilber, K., Smith, J. K., Hamer, R. M., Lin, W., Gerig, G., & Gilmore, J. H. (2008). A structural MRI study of human brain development from birth to 2 years. *Journal of Neuroscience, 28*, 12176–82.

Kostović, I. (1990). Structural and histochemical reorganization of the human prefrontal cortex during perinatal and postnatal life. *Progress in Brain Research, 85*, 223–39.

LaMantia, A. S., & Rakic, P. (1990). Axon overproduction and elimination in the corpus callosum of the developing rhesus monkey. *Journal of Neuroscience, 10*(7), 2156–75.

Liepert, J., Bauder, H., Miltner, W. H. R., Taub, E., & Weiller, C. (2000). Treatment-induced cortical reorganization after stroke in humans. *Stroke, 31*, 1210–16.

Martin, J. H., & Jessell, T. M. (1991). Development as a guide to the regional anatomy of the brain. In E. R. Kandel, J. H. Schwartz, & T. M. Jessell (Eds.), *Principles of Neural Science* (3rd edition). Norwalk, CT: Appleton & Lange.

Maurer, D., Lewis, T. L., Brent, H. P., & Levin, A. V. (1999). Rapid improvement in the acuity of infants after visual input. *Science, 286*, 108–10.

Maurer, D., Lewis, T. L., &. Mondloch, C. J. (2008). Plasticity of the visual system. In C. A. Nelson & M. Luciana (Eds.), *Handbook of Developmental Cognitive Neuroscience*, 2nd Edition (pp. 415–37). Cambridge, MA: MIT Press.

Mattson, S. N., Fryer, S. L., McGee, C. L., & Riley, E. P. (2008). Fetal alcohol syndrome. In C. A. Nelson & M. Luciana (Eds.), *Handbook of Developmental Cognitive Neuroscience*, 2nd Edition (pp. 643–52). Cambridge, MA: MIT Press.

Michel, A. E., & Garey, L. J. (1984). The development of dendritic spines in the human visual cortex. *Human Neurobiology, 3*(4), 223–7.

Molliver, M., Kostovic, I., & Van der Loos, H. (1973). The development of synapses in the human fetus. *Brain Research, 50*, 403–7.

Nelson, C. A. (1995). The ontogeny of human memory: A cognitive neuroscience perspective. *Developmental Psychology, 31*, 723–38.

Nelson, C. A. (2000). Neural plasticity and human development: The role of early experience in sculpting memory systems. *Developmental Science, 3*, 115–30.

Nelson, C. A., de Haan, M., & Thomas, K. M. (2006a). Neural bases of cognitive development. In W. Damon, R. Lerner, D. Kuhn, & R. Siegler (Volume Editor), *Handbook of Child Psychology, 6th Edition, Vol. 2: Cognitive, Perception and Language* (pp. 3–57). New Jersey: John Wiley & Sons, Inc.

Nelson, C. A., de Haan, M., & Thomas, K. M. (2006b). *Neuroscience and Cognitive Development: The Role of Experience and the Developing Brain*. New York: John Wiley & Sons.

Nudo, R. J., & Milliken, G. W. (1996). Reorganization of movement representations in primary motor cortex following focal ischemic infarcts in adult squirrel monkeys. *Journal of Neurophysiology, 75*(5):2144–9.

O'Leary, D. D., Schlaggar, B. L., & Tuttle, R. (1994). Specification of neocortical areas and thalamocortical connections. *Annual Review of Neuroscience, 17,* 419–39.

O'Rahilly, R., & Gardner, E. (1979). The initial development of the human brain. *Acta Anatomica, 104,* 123–33.

O'Rahilly, R., & Muller, F. (1994). Neurulation in the normal human embryo. *CIBA Foundation Symposium, 181,* 70-82,

Patterson, P. H., & Nawa, H. (1993) Neuronal differentiation factors/cytokines and synaptic plasticity. *Cell, 72,* 123–37.

Perani, D., Paulesu, E., Galles, N. S., Dupoux, E., Dehaene, S., Bettinardi, V., Cappa, S. F., Fazio, F., & Mehler, J. (1998). The bilingual brain. Proficiency and age of acquisition of the second language. *Brain, 121,* 1841–52.

Petitto, L. A., Zatorre, R. J., Gauna, K., Nikeiski, E. J., Dostie, D., & Evans, A. C. (2000). Speech-like cerebral activity in profoundly deaf people processing signed languages: Implications for the neural basis of human language. *Proceedings of the National Academy of Sciences of the United States of America. 97*(25), 13961–6.

Pons, T. (1995). Abstract: Lesion-induced cortical plasticity. In B. Julesz & I. Kovacs (Eds.), *Maturational Windows and Adult Cortical Plasticity* (pp. 175–8). Reading, MA: Addison-Wesley Publishing Company.

Pons, T. P., Garraghty, P. E., Ommaya, A. K., Kaas, J. H., Taub, E., & Mishkin, M. (1991). Massive cortical reorganization after sensory deafferentation in adult macaques. *Science, 252,* 1857–60.

Purves, D. (1989). Assessing some dynamic properties of the living nervous system. *Quarterly Journal of Experimental Physiology, 74*(7), 1089–1105.

Raff, M. C., Barres, B. A., Burne, J. F., Coles, H. S., Ishizaki, Y., & Jacobson, M. D. (1993). Programmed cell death and the control of cell survival: Lessons from the nervous system. *Science, 262,* 695–9.

Rakic, P. (1971). Guidance of neurons migrating to the fetal monkey neocortex. *Brain Research, 33,* 471–6.

Rakic, P. (1972). Mode of cell migration to the superficial layers of fetal monkey neocortex. *Journal of Comparative Neurology, 145,* 61–83.

Rakic, P. (1974). Neurons in rhesus monkey visual cortex: Systematic relation between time of origin and eventual disposition. *Science, 183,* 425–7.

Rakic, P. (1978). Neuronal migration and contact guidance in the primate telencephalon. *Postgraduate Medical Journal, 54,* 25–40.

Rakic, P. (1988). Specification of cerebral cortical areas. *Science, 241,* 170–6.

Rakic, P. (1990). Principles of neural cell migration. *Experientia, 46,* 882–91.

Rakic, P. (1995). Radial versus tangential migration of neuronal clones in the developing cerebral cortex. *Proceedings of the National Academy of Sciences, 92,* 11323–7.

Rakic, P., Bourgeois, J. P., Eckenhoff, M. F., Zecevic, N., & Goldman-Rakic, P. S. (1986). Concurrent overproduction of synapses in diverse regions of the primate cerebral cortex. *Science, 232*(4747) 232–5.

Ramachandran, V. S., Rogers-Ramachandran, D., & Stewart, M. (1992). Perceptual correlates of massive cortical reorganization. *Science, 258,* 1159–60.

Rampon, C., Jiang, C. H., Dong, H., Tang, Y-P., Lockhart, D. J., Schultz, P. G., Tsien, J. Z., & Hu, Y. (2000). Effects of environmental enrichment on gene expression in the brain. *Proceedings of the National Academy of Sciences, 97,* 12880–4.

Sale, A., Vetencourt, J. F. M., Medini, P., Cenni, M. C., Baroncelli, L., De Pasquale, R., & Maffei, L. (2007). Environmental enrichment in adulthood promotes amblyopia recovery through a reduction of intracortical inhibition. *Nature Neuroscience, 10,* 679–81.

Schlaggar, B. L., Fox, K., O'Leary, D. D. (1993) Postsynaptic control of plasticity in developing somatosensory cortex. *Nature, 364*(6438), 623–6.

Sidman, R. L., & Rakic, P. (1973). Neuronal migration, with special reference to developing human brain: A review. *Brain Research, 62,* 1–35.

Sidman, R., & Rakic, P. (1982). Development of the human central nervous system. In W. Haymaker & R. D. Adams (Eds.), *Histology and Histopathology of the Nervous System.* Springfield, IL: Charles C. Thomas.

Shonkoff, J. P., & Phillips, D. A. (2000). *From Neurons to Neighborhoods: The Science of Early Childhood Development.* Washington, DC: National Academy of Sciences Press.

Smart, I. H. M. (1985). A localized growth zone in the wall of the developing mouse telencephalon. *Journal of Anatomy, 140,* 397–402.

Smith, J. L., & Schoenwolf, G. C. (1997). Neurulation: Coming to closure. *Trends in Neurosciences, 20,* 510–17.

Spemann, H., & Mangold, H. (1924). Uber induktion von embryonalanlagen dürch implantation artfremder organisatoren. *Archiv fuer Mikroskopische Anatomie Entwicklungsmechanik, 100,* 599–638.

Suter, D. M., & Forscher, P. (1998). An emerging link between cytoskeletal dynamics and cell adhesion molecules in growth cone guidance. *Current Opinion in Neurobiology, 8,* 106–16.

Takahashi, T., Nowakowski, R. S., & Caviness, V. S., Jr. (1994). Mode of cell proliferationin the developing mouse neocortex. *Proceedings of the National Academy of Sciences, 91,* 375–9.

Takahashi, T., Nowakowski, R. S., & Caviness, V. S., Jr. (2001). Neocortical neurogenesis: regulation, control points, and a strategy of structural variation. In C. A. Nelson & M. Luciana (Eds.), *Handbook of Developmental Cognitive Neuroscience.* Cambridge, MA: MIT Press.

Taub, E. (2000). Constraint-induced movement therapy and massed practice. *Stroke 31*(4), 986–8.

Tessier-Lavigne, M., & Goodman, C. S. (1996). The molecular biology of axon guidance. *Science, 274,* 1123–33.

Thoenen, H. (1995) Neurotrophins and neuronal plasticity. *Science, 270*(5236), 593–8.

Toga, A. W., Thompson, P. M., & Sowell, E. R. (2006). Mapping brain maturation. *Trends in Neuroscience, 29,* 148–59.

Walsh, C., & Cepko, C. L. (1992). Widespread dispersion of neuronal clones across functional regions of the cerebral cortex. *Science, 255,* 434–40.

Walsh, C., & Cepko, C. L. (1993). Clonal dispersion in proliferative layers of developing cerebral cortex. *Nature, 362,* 632–5.

Warkany, J., Lemire, R. J., & Cohen, M. M. (1981). *Mental Retardation and Congenital Malformations of the Central Nervous System.* Chicago: Year Book Medical Publishers.

Yuste, R., & Sur, M. (1999). Development and plasticity of the cerebral cortex: From molecules to maps. *Journal of Neurobiology, 41,* 1–6.

3

Mother and Child: Preparing for a Life

RONALD G. BARR

INTRODUCTION

In trying to understand what helps the lives of human infants get off to a good start, we have concentrated on the quite remarkable and particular relationship that occurs between mothers and their infants in the first few days and weeks of life. Of course, everyone is interested in mother-infant relationships, but the word "relationship" can take on many meanings. For most, the term conjures up a picture of a mother lovingly holding her infant in her arms, perhaps in an *en face* position in which they are looking directly into each other's eyes. The assumption is that this is a manifestation of a maternal nature of loving and supportive caring that the mother is hardwired to provide. However, in Sarah Blaffer Hrdy's thoughtful critique (Hrdy, 1999), she challenges the assumption that there is any such thing as an "essential" mothering nature. She argues that mothering, as with many other aspects of animal behavior, is subject to choices (or trade-offs) that may differ depending on the experiences and contexts in which mothering is taking place.

If Hrdy's position is correct, then the implications for understanding what gets the lives of human infants off to a good start are enormous. Among other things, the argument implies that good mothering is not guaranteed, but rather that it is contingent on a variety of factors, only some of which we know very much about. In our work, we have tried to understand what the effects of various components of possible caregiving strategies might be. The idea is that understanding these component effects would provide the basis for making informed choices in the face of several contingencies. As a way of deciding what might be most important to study, we have been guided by two concepts that provide what we think of as useful heuristic frameworks for trying to understand and investigate

70

some of the components that are likely to be relevant to a good start in different, contingent, caregiving contexts.

The first concept focuses on biological pathways by which mothers and infants are related to each other in their common, daily interactions. The particular biological pathways in which we are interested are those that are activated by and depend for their effects on *caregiving behaviors* on the part of the mother (or other caregivers). As one way of talking about this, we have borrowed from Myron Hofer's concept of "hidden regulators" in mother-infant interaction (Hofer, 1994b). The second concept focuses on the apparent dissociation in historical time between the relatively rapid recent cultural changes in caregiving strategies within which infants grow up currently compared to the relatively stable biological constitutions with which infants have come into the world for centuries. As a way of thinking about this dissociation, we will consider what is typically different between the caregiving strategies of Western caregivers (that we will generally call a "separation" paradigm) and those of the !Kung San hunter-gatherers (that we will generally call a "closeness" paradigm).

Concept I: "Hidden Regulators" in Mother-Infant Interaction

Our approach to understanding biological pathways implicated in mother-infant interactions applied to human infants and their caregivers borrows significantly from the work of Myron Hofer and colleagues in their studies of caregiving behaviors in infant and mother rat dyads. Hofer coined a term that nicely describes the biological interrelatedness of the mother-infant dyad within caregiving behavior, referring to these biological pathways as "hidden regulators" of early infant behavior (Hofer, 1984, 1994a, 1994b). *Hidden regulators* capture the idea that, within the observable interactions of mothers and infants, there are a number of unobserved sensory, motor, nutrient, and thermal events that exert relatively discrete (or specific) regulatory control by way of central nervous system neuro-modulator subsystems over particular behavioral (such as distress vocalizations, or crying) or physiological (such as activation of the hypothalamic-pituitary-adrenal [or HPA] axis) response systems of the infant (Hofer, 1994b).

Hofer has helpfully summarized some of the cardinal features of the hidden regulators system in a diagram (Fig. 3.1) (Hofer, 1996). We can use Figure 3.1 to get an overview of how caregiving might help to regulate the infant's physiological and behavioral functioning. It illustrates a number of general principles gleaned from Hofer's experimental paradigm.

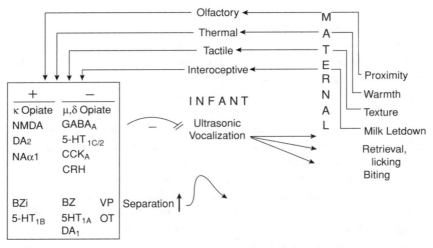

Figure 3.1. Hidden regulators system. Reprinted from Hofer, M. A. (1996). Multiple regulators of ultrasonic vocalization in the infant rat. *Psychoneuroendocrinology*, 21(2), 203–17: Figure 1, p. 205, with permission from Elsevier.

First, the pathways have been elicited using a now classic maternal separation experimental paradigm. When caregiving is withdrawn acutely (by removing the mother from the infant), the complete set of regulatory interactions that are part of typical caregiving are all withdrawn from the infant together. Then, when one (or more) caregiving components are reintroduced (for example, reinstitution of suckling alone, or delivery of nutrients alone), then one (or more) of the components of the separation response (for example, distress vocalizations) are reduced or prevented, without affecting the other components. Second, when combinations of caregiving components (for example, texture plus odor plus warmth) are reintroduced, they generally restore more of the typical response to caregiving than single components, and do so in a cumulative fashion. Third, the closer these combinations of caregiving components are to social companion-like caregiving (as opposed to inanimate caregiving) contact, the more effective they are. Fourth, the sensory systems that are activated by the caregiving components seem to be organized in parallel. Activation of one sensory component of the system can have the same effect as another sensory component on an outcome response (like crying). Fifth, within the complex central circuitry to which the sensory systems project, at least some of the neuromodulatory systems also seem to be organized in parallel. This is indicated by the fact that selective pharmacologic blockade of specific systems (e.g., opioidergic, serotonergic, or benzodiazepinergic receptors) affects specific output responses, but not output responses to agonists of other types. Sixth, because of this, some sensory pathways are preferentially linked to particular central neuromodulatory systems. An example that we will come back to later is that orogustatory (taste) effects on distress vocalizations are mediated, at least in part, by central opioid-dependent systems whereas orotactile (contact) effects are not. Seventh, some stimuli such as milk nutrients may engage not just one, but a "cascade" of parallel regulatory systems. For example, milk may first engage a faster-acting preingestive taste pathway that is opioid-dependent, and then a slower-acting postingestive pathway that is cholecystokinin-mediated (and opioid-independent).

In general, the hidden regulators operate as a quite exquisite feedback system in which the more molar act of caregiving engages a number of parallel, and probably redundant, systems of regulation operating via physiological pathways that are reflected in infant behavior. The infant behavioral outputs (such as distress vocalizations) also act in turn to engage maternal caregiving, thereby closing the loop in the regulatory system, as it were. In the studies that we will mention soon, some examples of these pathways will be shown to be present and active in human infants.

Concept II: Dissociation between Cultural Changes in Caregiving Strategies and the Biological Constitution of Infants

The second heuristic concept that helps to provide us with some insight about the effects of early experience and setting is the apparent dissociation between behavioral and biological levels of functioning because of recent cultural changes in caregiving behavior in the face of the relative stability of our biological constitutional makeup over generations. The cultural changes can be illustrated by considering some of the quite impressive differences between caregiving features typical of the !Kung San hunter-gatherers and those more typical of Western industrialized societies, and then noting their historical roots. Elsewhere (Barr, 1990a), I have described in more detail five of these caregiving dimensions that seem to distinguish these two approaches to early infant caregiving. !Kung San and Western caregiving can be characterized according to:

1. **Direct physical contact.** !Kung San caregiving includes almost constant physical contact during the day and during the night. Konner reports that daytime contact with a caregiver (not just mother) in the first three months occurred more than 80 percent of the time,

compared to less than 25 percent in Western observations (Konner, 1976). At night, contact is maintained by co-sleeping.

2. **Carrying/holding.** In the !Kung San, infants are constantly carried in a sling (or *kaross*) that permits infants to accompany their mothers while gathering food as well as in the camp settings. In Western settings, by comparison, holding and carrying (not including feeding) occurred for 2.7 hours/day at 5 weeks of age in 1986 in Montreal mothers (Hunziker & Barr, 1986). In 2006, St. James-Roberts and colleagues reported that London mothers held their infants for 3.5 hours/day, and Copenhagen mothers held theirs for 6.4 hours/day (St.James-Roberts et al., 2006). Despite the wide range and cultural differences, this is still far from the constant holding typical of the !Kung San.

3. **Feeding frequency/interfeed interval.** !Kung San infant feeding has been described as continuous, occurring during the daytime approximately four times an hour for 1–2 minutes per feed for the first two to three years of life, with the longest interval between feedings averaging less than one hour (Konner & Worthman, 1980). In Western societies, feeding frequencies follow a more "pulse"-like pattern, occurring in the range of eight to fourteen times per day, with the higher end frequencies being more typical of committed breastfeeding advocates (Barr & Elias, 1988; St.James-Roberts et al., 2006).

4. **Posture.** Both because of the constant use of slings as well as because of a cultural belief that upright, posture promotes development, !Kung San infants are usually upright, whereas Western infants spend more time in supine positions.

5. **Responsiveness to infant distress.** Among the !Kung San, responsiveness to infant distress signals, even to small frets, is virtually universal and immediate. A fret will elicit a response within 10 seconds 90 percent of the time (Barr et al., 1987). Deliberate nonresponse rates of 40 to 50 percent are reported in Western samples (Bell & Ainsworth, 1972; Hubbard & van IJzendoorn, 1987).

The term "indulgent" has been used to describe !Kung San caregiving, but this term carries unnecessary value connotations as to whether or not this is appropriate caregiving. It is perhaps less susceptible to value judgments to contrast !Kung San and Western caregiving packages as strategies emphasizing "closeness" and "separation," respectively. Note, however, that the "closeness" or "separation" includes both spatial (in terms of contact and proximity) and temporal (in terms of responsivity and feeding frequency) dimensions.

Although it remains controversial, there is considerable interest in the likelihood that something like the contemporary differences in !Kung San and Western caregiving "packages" are reflective of differences in historical time, with !Kung San-like caregiving being more characteristic of preindustrial modes of social organization, and Western caregiving more typical of postindustrialized Western societies. It has been argued that such Western caregiving strategies are historically very recent, and have been normative for less than 1 percent of human history (Kennell, 1980; Lozoff & Brittenham, 1979). Furthermore, cross-species comparisons of caregiving dimensions, breast milk composition, and suckling rates suggest that, at least in mammalian species and possibly in evolutionary time, there has been a predictable relationship between breast milk composition (low protein and fat), low sucking rates, and frequent feeding at short intervals associated with close and continuous proximity ("carrying" species; as in the "closeness" caregiving of the !Kung San). High protein and fat content and high sucking rates at longer intervals have been associated with intermittent proximity ("caching" species), more analogous to a "separateness" caregiving strategy in Western societies (Ben Shaul, 1962; Blurton-Jones, 1972; Ewer, 1968). Because human breast milk is low in protein and fat content, one might expect that humans would coevolve a carrying strategy (as have the !Kung San). Further, from this point of view, the caching or "separateness" strategy more typical in Western societies appears to represent a relatively recent abandonment of this strategy.

Importantly, this is not to say that human infants cannot be successfully raised using a separateness strategy. Indeed, the evidence is quite compelling that they can, and have, been so raised. What it does do is provide us with some guidelines for investigating what the trade-offs are, or might be, if different strategies are used, especially in early infancy. Let us turn now to some of the experimental observations that we and others have made that might be able to indicate the kinds of trade-offs implicated in choosing "closeness" or "separateness" caregiving strategies. To illustrate this, I will offer examples of how one or more components (e.g., contact, nutrients, etc.) of caregiving that would be more or less available depending on caregiving strategy can be shown to affect one or more infant responses. The three infant responses we have chosen to study are infant distress (or crying) behavior, infant responses to pain experiences, and infant cognition (specifically, memory). Arguably, each of these is essential to early survival, and each appears to be contingent on early caregiving strategy in some way.

Contact and Nutrient Effects on Early Distress Behavior

An early indication of the potential of contact to affect early distress behavior came from a randomized controlled trial of increased carrying and holding behavior on the part of caregivers (Hunziker & Barr, 1986). As background, it is important to know that distress (crying and fretting behavior) in the first three months follows an age-related curve, sometimes referred to as the "normal crying curve" (Barr, 1990b) in which the overall duration of crying and fussing increases in the first two months, peaks, and then decreases subsequently. What we wanted to see was whether the amount of crying could be reduced if carrying and holding were increased. The caregivers assigned to the "supplemental carrying" group carried their infants an average of 4.4 hours/day, compared to 2.7 hours/day in the control group, beginning during the fourth week of life. The amounts of carrying (as well as other potential confounding variables such as the number of feedings) were documented by validated parental diary records (Barr et al., 1988, 1989; St. James-Roberts et al., 1993). On the basis of previous anecdotal observations among the !Kung San and other groups in which carrying was common, we had expected some diminution in distress behaviors, but the effect was even greater than we expected. In Figure 3.2, the results of the individual distress patterns are depicted. As can be seen, the infants with increased carrying (Supplemented infants) cried and fussed considerably less than those with typical amounts of carrying (Control infants).Overall, this amounted to a reduction from 2.2 to 1.2 hours of distress behavior (or a 43-percent reduction) during the sixth week of life, when distress behavior is usually at its peak (Hunziker & Barr, 1986). The decreased distress was replaced by increased awake alert behavior, but there was no change in sleep, suggesting some specificity of the effect of increased carrying and holding. Although this amount of carrying and holding does not approach the amount practiced by the !Kung San, it nevertheless confirmed that this form of contact could have substantial effects on infant distress behavior.

In another study, we investigated the effects of different nutrient meals provided to newborn infants on the second or third day of life (Oberlander et al., 1992). This too was a randomized controlled trial in which infants were assigned to receive one of three types of feeds as an "extra" feed three hours after their last feed. The three feeds were (1) a balanced formula (Enfalac® with iron), (2) lactose in water (6.9 percent), and (3) water only. The infants were then observed for 40 minutes alone in a bassinet for fretting/crying, noncry wakefulness, and sleep. The results of this study

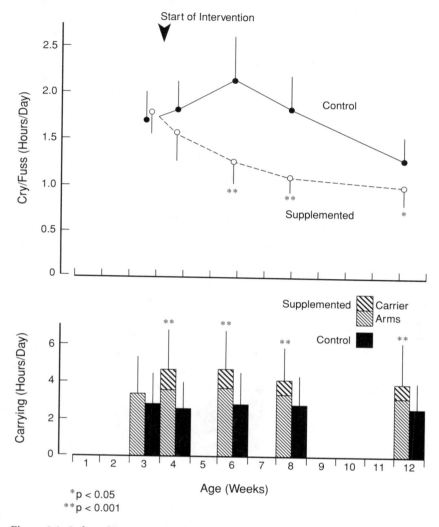

Figure 3.2. Infant distress patterns. Reproduced with permission from Hunziker, U., & Barr, R. (1986). Increased carrying reduces infant crying: A randomized control trial. *Pediatrics*, 77, 641–8: Figure 1, p. 644, Copyright © 1986 by the AAP.

suggested that nutrient composition of meals had significant effects on infant behavioral states. More intriguingly, there seemed to be a degree of *specificity* to these effects. As shown in Figure 3.3, both the lactose and balanced formula feeds reduced crying and fretting relative to the water feed. Figure 3.4 shows that when sleep was the outcome measure, balanced

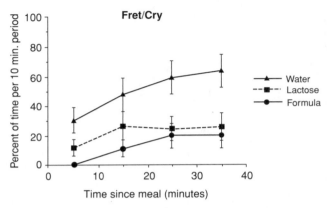

Figure 3.3. Crying differences by feeding. Reproduced with permission from Oberlander, T., Barr, R., et al. (1992). Short-term effects of feed composition on sleeping and crying in newborn infants. *Pediatrics*, 90(5), 733–40: Figure 2, p. 735, Copyright © 1992 by the AAP.

formula increased sleep relative to water throughout the observation time. However, lactose formula had effects *like water* early (no increased sleep), but *like formula* later. In Figure 3.5, we see the effects on noncrying wakefulness. Here there was no difference among any of the feeds. Rather, it appears that the recency of *the feeding act itself* facilitated more wakefulness that clustered in the first 10 minutes after the feed.

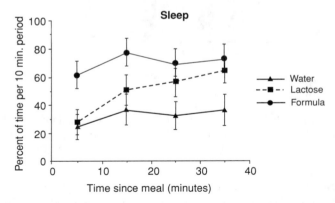

Figure 3.4. Sleeping differences by feeding. Reproduced with permission from Oberlander, T., Barr, R., et al. (1992). Short-term effects of feed composition on sleeping and crying in newborn infants. *Pediatrics*, 90(5), 733–40: Figure 1, p. 735, Copyright © 1992 by the AAP.

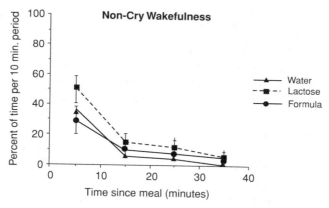

Figure 3.5. Noncry wakefulness differences by feeding. Reproduced with permission from Oberlander, T., Barr, R., et al. (1992). Short-term effects of feed composition on sleeping and crying in newborn infants. *Pediatrics*, 90(5), 733–40: Figure 3, p. 735, Copyright © 1992 by the AAP.

These studies imply that both contact and nutrient caregiving behaviors – carrying/holding in the first and nutrient composition in the second – had substantial effects on infant behavioral states, and sometimes quite specific effects. Furthermore, the size of each intervention was somewhere between what might be expected when comparing a !Kung San-like "closeness" and a Western "separation" caregiving strategy. Nevertheless, the size of the effects was quite impressive, despite the modest dose change in contact and nutrient interventions. However, in both studies, what constituted contact and what constituted nutrient stimuli were still quite complex, each being constituted by a number of components, and each component of which might have been determining to different extents. In subsequent studies, we began to look more closely at less complex and more *discrete* contact and nutrient stimuli for effects on behavioral, pain, and cognitive outcomes.

More Specific "Hidden Regulators" in Humans: The Sucrose Story

A number of studies related to the effect of sucrose tastes compared to pacifier contact have provided a possible strategy for probing more specifically some of the pathways by which nutrient and contact stimuli may be acting, at least in regard to some behavioral outcomes. Elliott Blass and his colleagues were the first to report what has come to be called "the sucrose effect," that is, the response of crying infants to tastes of sucrose (Blass et al., 1989; Blass & Ciaramitaro, 1994; Blass & Smith, 1992). The response consists of three components: first, in response to a taste of sucrose, a crying newborn will cease crying for two minutes or more following application of the sucrose taste; second, there will be an increase in mouthing activity for the first minute or so; and, third, there will be an increase in hand-mouth activity (Blass et al., 1989; Blass & Ciaramitaro, 1994; Blass & Smith, 1992). We replicated this finding using two drops (about 250 ul) of 24 percent sucrose solution (Barr et al., 1994), and the results are shown in Figure 3.6. The upper panel shows the time course of the crying reduction over five minutes. The middle panel shows the briefer time course for mouthing that reflects the presence of the sucrose taste in the mouth. Note that the mouthing was essentially gone by 2 minutes, whereas the quieting effect persisted after that. The lower panel shows the increase in hand-mouth activity. As it turns out, what seems to be the operative property in this quieting effect is the fact that the sucrose is sweet. We were able to demonstrate this by comparing the sucrose response to other sweet but noncarbohydrate tastes (such as aspartame), and other nonsweet carbohydrate, fat, and bitter tastes (Barr et al., 1999a; Graillon et al., 1997).

Another important observation is that this *taste* (or orogustatory) calming effect is functionally different from that which occurs with a nontaste *contact* (or orotactile) stimulus, such as a pacifier. This was demonstrated in a study of Blass and his colleagues (Blass & Ciaramitaro, 1994) in which the calming effect of sucrose was shown to persist following the sucrose stimulus whereas the calming effect of a pacifier was only operative when the pacifier was in place, but not after. Furthermore, this functional difference in behavioral calming appears to reflect different central mechanisms that mediate these behavioral changes. Evidence from rat pups (and other species) indicates that the sucrose taste effect requires a functioning opioid-dependent system (Blass et al., 1990; Kehoe & Blass, 1986a, 1986b, 1986c; Panksepp, 1986), whereas many forms of contact quieting do not (Blass et al., 1990, 1995; Kehoe & Blass, 1986a). Although similar pharmacologic manipulations cannot ethically be done in infant humans, a number of observations provide convergent evidence that the sucrose (or, more accurately, sweet) taste effect is opioid-mediated in human infants, whereas pacifier calming is not. For example, newborn infants of methadone-dependent mothers do not respond with quieting to sucrose taste as do normal infants, but nevertheless have the same response to a pacifier stimulus (Blass & Ciaramitaro, 1994). This suggests that it is not that these infants generally are less soothable, but rather that it is the relatively specific opioid-dependent calming that is compromised.

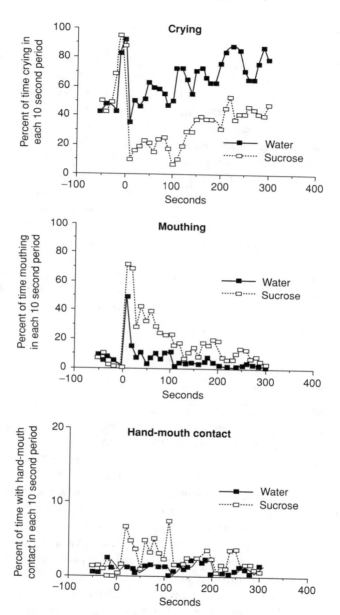

Figure 3.6. Infant response to sucrose. From Barr, R. et al. (1994). Effects of intraoral sucrose on crying, mouthing, and hand-mouth contact in newborn and six-week-old infants. *Developmental Medicine and Child Neurology, 36,* 608–18. Published by John Wiley & Sons. Copyright © 2008, 1994 Mac Keith Press.

In summary, these and other observations lead to the conclusion that there are at least two systems available to the infant through sensory pathways by which caregiving differences might contribute to calming, and therefore more generally to state regulation. One of these is stimulated by sweet taste (and possibly, hedonically positive experiences; Barr & Young, 1999) and appears to be dependent, at least in part, on an intact functioning opioid-dependent system, whereas another one is stimulated by nontaste contact that does not require central opioid-dependent systems. Thus, very much like Hofer's hidden regulators model, there appear to be parallel pathways via different peripheral sensory systems preferentially mediated by different central nervous system pathways to the common outcome of behavioral calming. In this sense, the pathways seem to be redundant.

Having two, relatively independent, systems for calming and soothing provides us with a potentially valuable set of tools – we call them bio-behavioral probes – for trying to find out what is going on in early infancy when these pathways are (or are not) available to the infant in the face of typical, everyday challenges. Let's consider now how they may help us to understand two of the challenges of early infancy. One is regulation of the distress state, and the other is regulation of response to painful medical procedures.

Regulation of Distress Behavior in the Early Weeks of Life

Recall from the earlier discussion that distress behavior in infants tends to follow an *n*-shaped curve over the first three months of life. Furthermore, there is considerable between-subject variability in the amount of crying and fussing behavior each infant does, ranging from quite extreme amounts (sometimes referred to as colic) to quite minimal amounts. So, one question is what mechanisms might be implicated in this developmentally normal, but from the caregiver's point of view, very distressing behavior.

A number of observations implicate the sucrose-accessible, distress-reduction system as perhaps particularly relevant to these early distress phenomena. For example, both we (Barr et al., 1994) and Elliott Blass and his colleagues (Zeifman et al., 1996) have shown that sucrose effects are substantial in the newborn period, and detectable but much weaker by about 6 weeks of age. In other words, crying amounts increase while the sucrose-accessible system for distress reduction appears to become less efficacious. Second, if infants with colic are on the upper end of the spectrum of distress behavior in otherwise normal infants, then one might expect that the sucrose-accessible calming system would be less effective in them than in infants without colic, at least at 6 weeks of age, when the crying behavior is greatest. This turned out to be true in one assessment of that prediction. Already crying infants who met clinical criteria for colic had an initial response to sucrose (while the sucrose was in their mouths) that was equivalent to infants without colic, but the *persistent* calming effects after the first minute were not detectable in infants with colic; only in those without (Barr et al., 1999b). Although encouraging, an attempted replication of colic/noncolic differences in sucrose-mediated soothing was not as successful (Ghosh, Barr, Paterson, Huot, & Brody, 2007). A third observation is that this difference between infants with and without colic does not appear to extend to pacifier (or orotactile contact) calming. Already crying infants with colic are just as likely to calm to the insertion of a pacifier in their mouths as are infants without colic at 6 weeks of age (and at 4 months of age, after colic has resolved as well). Finally, the increased carrying that was so effective in infants without colic (Hunziker & Barr, 1986) was not effective when tried as a treatment for infants with colic (Barr et al., 1991), again consistent with the idea that contact-accessible calming may not be an operative system in infants with the excessive crying characteristic of colic. These observations lend credibility to the idea that the sucrose-accessible taste system may be relevant to this early crying increase and possibly also to differences among infants with and without colic, although more work needs to be done to confirm this suggestion. However, understanding how these calming systems may or may not be *differentially* implicated in regard to manifestations of early distress behavior is providing some new insights into possible mechanisms of what have been fairly mysterious behaviors to date. As well, they illustrate how different components of caregiving might differentially contribute to early behavioral state regulation.

Regulation of Pain Response by Caregiving Components

Analogously, it makes sense to ask whether caregiving components might recruit "hidden regulators" that help the infant deal with pain-stress procedures. In one study (Gormally et al., 2001), we asked the question as to whether sucrose taste, holding contact, or both might be able to recruit physiological systems that modified the infant pain response to a heel-stick procedure done to obtain a blood sample in normal newborns. In addition we were interested in whether, if the component interventions both worked, did they do so additively (the effect of one was added to the effect of the other) or interactively (the effect of both was greater than the effect of either alone or both together). The results confirmed a number of interesting things about how caregiving components might function. One is that the effects of either intervention or both depended on the outcome being considered. For example, *crying behavior* was reduced by taste and holding, and the effects were additive. However, *facial grimacing* was only reduced by holding and not by taste. Furthermore, the effects on physiological responses like heart rate and vagal tone depended on preintervention levels (they worked better when preintervention levels were high), and were interactive with each other. The picture that emerges from this study is that the effects of caregiving components can be complex, and depend on the behavioral or physiological response system being studied. Overall, however, it seems that in one way or another, the more such calming pathways that were able to be recruited, the better.

If this is true, then we reasoned that the most efficacious caregiver intervention that we could imagine to buffer pain experiences would be if the infant was breastfeeding during the painful procedure. At least in principle, and extrapolating from a large literature in the field of psychobiology, breastfeeding has the capacity to recruit taste-accessible opioid-dependent systems (as indicated by the previous sucrose studies), peripheral motor-activity accessible serotonergic-dependent systems, nonopioid (possibly oxytonergic) systems by contact, and possibly others. To test this, we compared the behavioral responses of infants who were being breastfed *before and during* the time they were receiving their 2-month DPT immunizations to infants who were being held but only breastfed *after* the immunization procedure (Barr et al., 2005). We measured crying behavior and flushing response as well as a number of facial indices typical of pain and stress in both groups. The results were quite dramatic in infants who were breastfeeding during the immunization. There were a few infants in whom it was virtually impossible to know when the needle had entered the skin. However, the more typical response to needle entry was that the infant paused in its sucking, emitted two to three cries, and then returned (after about 10 to 15 seconds) to breastfeeding comfortably. Compared to the responses of those not breastfeeding, there was an approximately 50-percent reduction in responses across the board in those who were breastfeeding. This was similar whether one looked at the whole minute of pain and recovery, or just at the 10 seconds during which the injection itself took place. These findings certainly imply that painful and stressful challenges in early infancy appear to be buffered to a significant extent when experienced in a context in which caregiving components, preferably many of them, are accessible to the infant. Furthermore, they seem to act very much as might be predicted by Hofer's "hidden regulators" concepts.

Nutrient Effects on Memory in Newborn Infants

Encouraged by the relatively substantial effects that caregiving components seem to have on infant behavioral state regulation and infant pain response, we have recently been exploring another domain of possible influence; namely, whether there are nutrient effects on infant cognition and, more specifically, memory. We were stimulated by findings from research in adult humans and in other species that elevations in levels of blood glucose appeared to improve memory performance, at least for certain kinds of memory. Most of the research in humans has been reported in the elderly (in whom questions of regulation of glucose metabolism are paramount) and young adults (in whom they are not). Previously, the youngest humans in whom glucose enhancement of any cognitive function had been assessed had been 6–7 year olds, in whom glucose ingestion improved reaction time and reduced frustration in a computer game task (Benton et al., 1987).

Despite the challenging question of how newborn memory does or does not map onto adult memory, understanding potential nutrient enhancement of memory in infants seemed to us a compelling question. We know little or nothing about how glucose levels might affect cognition of any kind in infants, never mind memory. Given the wide range in feeding frequency and even postfeeding glucose levels in infants, the possibility that glucose might affect memory seemed of some importance. Added to this is the fact that, in the human experience, feeding patterns have ranged from one feed every four hours in postindustrial societies to four feeds per hour in hunter-gatherers (Konner & Worthman, 1980). Consequently, the possible relationship among feeding, glucose levels, and memory in infants seemed worthy of study.

To do this, we utilized a technique usually referred to as an orientation-habituation-recovery paradigm that allows us to measure a form of developmentally appropriate memory that is germane to the way newborn infants process their experience. In this study, we used the form of this paradigm in which infants turn their heads toward a sound (the stimulus words "tinder" or "beagle") played to them through speakers to the right and left (Horne et al., 2006).

The procedure has four phases. In three of the phases, a word is presented through the speakers to the infants every 2 seconds in trial blocks of 30 seconds each.

1. In the *Orientation-Habituation phase*, the infant initially responds to the sound with a head turn of more than 45 degrees (Orientation),

but on repeated trial presentations, ceases to do so (Habituation). When the infant stops turning to the sound, this represents behavioral evidence that the infant no longer treats the sound stimulus as "new."

In our procedure, the criterion for Orientation is three turns toward the stimulus sound within four successive trials. The criterion for Habituation is three successive trials with no turns toward the stimulus. Note that it may take a few or many trials for an infant to meet criteria for orientation or habituation. Consequently, the number of trials to Orientation can be used as a measure of *attention* to the stimulus words, and the number of trials to habituation can be used as a measure of *rate of learning*.

2. The second phase of the procedure is a *Pause* (or delay period) once the criterion for habituation has been reached. Because the pause is contingent on habituation being achieved, this is an "infant-controlled" (rather than investigator-controlled) procedure.
3. The third phase is *Recovery*, at which time the same word presented in the first phase is *re*presented in six to eight trials. If the infant turns toward the sound, this implies that it is again treating the sound as "new"; that is, that it does not remember that it heard the sound before. If, however, the infant does not turn toward the sound, it implies that it is treating the sound as having been heard before, or "remembered."
4. The fourth phase is called *Novelty*, in which a new word is presented to the infant. Head turning to the new stimulus assures that any lack of head turning in the previous phase was not due to fatigue, distraction, or inability to respond to a new sound late in the procedure. Only the responses of infants who respond to the novelty word are included in the analyses.

Consequently, the critical test of memory in 2- to 3-day-old infants is whether, after having habituated to a word sound, the infant can remember that word sound when it is *re*presented following a pause; that is, whether the infant continues to treat it as old and does not turn its head toward the previously learned sound. This postdelay response is considered to reflect true information processing; that is, that the infant registers the repeated information, creates a mental representation, and subsequently responds based on the presence or absence of discrepancy between the mental representation it holds in memory and the *re*presented stimulus itself.

In previous studies, Zelazo and his colleagues had determined that new-born infants do remember word sound stimuli after a 55-second pause, but do not after a 100-second pause (Zelazo et al., 1987). In pilot trials, we determined that 100 seconds was the threshold delay, after which a spoken word sound was not remembered. Thus, our task in this study was to determine whether an increase in glucose nutrients resulted in a word sound being remembered across a 100-second delay.

In order to determine this, we randomly assigned normal 2- to 3-day-old newborns to a feeding condition or a control condition. Those in the feeding condition were given an extra feed of 2 gm/kg glucose as a 20 percent solution approximately 1.5 hours after their previous feed. Those in the control condition were given a water feed. Then the infants underwent the memory testing to see whether the word stimulus to which they were habituated was retained after the 100-second pause. If glucose levels make a difference, then those who had received a glucose feed should be able to remember the word for 100 seconds, whereas those who received the water feed should not.

As expected, there was a significant increase in the glucose levels of an average of 2.3 mmol/L in the infants who received a glucose feed. In terms of head turns toward the sound, both groups needed a similar number of trials to meet criteria for orientation (a measure of attention to the sound stimulus) and to meet criteria for habituation (a measure of rate of learning). However, as predicted, infants who received glucose turned their head toward the sound stimulus a mean of only 32 percent of the time (a low rate indicating they remembered the word stimulus), whereas those who received water turned their head a mean of 58 percent of the time. In other words, the infants receiving water were, after the pause, turning their heads toward the word sound stimulus at a higher rate that was similar to that which occurred during orientation, indicating that they did not remember the word stimulus. Finally, there were no differences in response to the new word during the Novelty phase, demonstrating that the previous low response rate in the infants who had received glucose appeared to be specific to the previously learned word.

In sum, on all four measures in the study, the only one in which a significant difference was demonstrated was in regard to whether or not infants remembered the word learned during the Orientation-Habituation phase of the procedure. This was pretty clear evidence for the following:

1. A 2-gm/kg glucose feed appeared to be sufficient to enhance the memory of 2- to 3-day-old infants for repeated spoken words;
2. The enhancement of memory for spoken words appeared to be relatively specific for (or at least preferentially affects) memory, relative to attention or rate of learning of spoken words; and

3. A difference of 2.3 mmol/L of blood glucose appeared to be sufficient to enhance the memory of 2- to 3-day-old infants for a repeated auditory stimulus.

Of course, nobody should assume from this single study that what has been demonstrated for the words "beagle" and "tinder" and a single glucose feeding applies to all forms of memory in newborns, for all forms of sensory stimuli, or for normal everyday feeding. These questions all need to be tested empirically. However, we view this as an impressive demonstration that nutrient delivery as a component of the caregiving package, in this case in the form of glucose, may have effects that have been hitherto unrecognized.

As it turns out, the mean level of glucose that occurred following the glucose feed in this study was about equivalent to the mean plus one standard deviation of glucose seen 30 minutes following a typical breast or formula feed. Consequently, there is considerable overlap between the levels typically obtained postfeeding and those that were attained in this study. This suggests that at least some of the time, postprandial blood glucose concentrations are in the range that will enhance memory for spoken words. To determine whether memory for spoken words is enhanced following a more typical feeding, we repeated the study with infants who were receiving their typical breast or formula feedings (Valiante et al., 2006). This time we compared those who were tested before a feeding (preprandial feeding group) with those in whom their blood glucose following the feeding was high or low (above or below the median for the postprandial feeding group, respectively). With that division, the blood glucose levels in the low postprandial group were no different than those in the preprandial group. Again, the three groups (preprandial, and high and low postprandial) demonstrated no difference in attention or rate of learning of the spoken words. What was interesting was that the preprandial group had higher rates of turning to the sound before feeding (a mean of 71 percent of the time) compared to both the postprandial low glucose group (51 percent of the time) and the postprandial high glucose group (41 percent of the time). Again, all groups turned with similar frequency to the Novelty word. In sum, this study confirmed the previous finding that typical feeding of newborns was associated with enhanced memory for spoken words, and that this enhancement appeared to be relatively specific to memory rather than attention or rate of learning. It raised the interesting possibility, however, that the increase in blood glucose may not be necessary for the enhanced memory to occur, because some enhancement appeared to occur without the increase in blood glucose levels in the low glucose postprandial group (Valiante et al., 2006).

These findings raise a number of interesting questions that are yet to be answered. For example, there may be some implications relating to our tendency to feed in a pulsed pattern as compared to the continuous feeding pattern of the !Kung San. It has long been assumed that the glucose levels in the brain are maintained at relatively constant levels sufficient to assure appropriate cortical functioning, except in extreme circumstances. Clearly, blood glucose is only the most indirect index of cerebral blood glucose. However, if increases in blood glucose within physiological ranges are sufficient to have significant effects on cognitive functions such as memory for spoken words, either the mechanism of these effects in the brain is only indirectly related to available blood glucose, or the assumption that brain glucose is maintained at near optimal levels may need to be reconsidered.

A second implication concerns the effects of our feeding patterns. In the human experience, feeding practices for newborn infants have varied widely, including feeding at frequencies of six to seven times/day in North American settings up to four times/hour among hunter-gatherers such as the !Kung San. Although we know something about the time course of blood glucose following widely separated single feeds, we know almost nothing about the time course in closely spaced continuous feeding practices typical of the !Kung San. At least one possible consequence of a !Kung-like feeding system is that it keeps blood glucose at a more constant and less variable level relative to widely spaced pulse feeding regimens. Another is that mean blood glucose levels may remain higher. If either of these assumptions is true, then it is possible that this range of feeding regimens has different effects on the memory experience of newborn infants. The debate as to what is the optimal feeding regimen has been joined on a number of levels, but arguments concerning possible differential effects on early cognition have not been one of them. If this phenomenon turns out to be robust and generalizable to typical feeding conditions, it is conceivable that the question of interfeed intervals may deserve reconsideration for another reason.

SUMMARY AND CONCLUSIONS

These observations suggest a number of things about the importance of early caregiving for the infant that could be important for assuring a good start in life. First, they all provide rather striking examples of quite substantial effects that caregiving behaviors in general, and components of caregiving packages in particular, have on the behavior of infants. Second, these effects are present in at least three domains of activity: (1) behavioral state regulation, (2) responses to stressful, painful experiences, and (3) cognition (or memory). It is likely, therefore, that early caregiving has very wide effects, across almost all domains of infant functioning. Third, although not nearly as finely worked out in human infants as in nonhuman species, the findings are remarkably parallel to those delineated by Hofer and his colleagues, in terms of the general concept of hidden regulators. As he described, behavioral and physiological systems, often operating in parallel, and with preferential connections to central systems, appear to be activated or recruited by particular caregiving components. Fourth, one can quite easily imagine that many more of these systems would be recruited much more often with a caregiving strategy that had the features of a closeness caregiving paradigm, as exemplified by the !Kung San, as compared to a separateness paradigm more typical in Western, industrialized societies.

If all of this is true, an important question is what this means for helping infants get off to a good start in life. One version of this is the question of whether these observations have any immediate importance for the experiences of infants early in life. It would seem that, for some specific effects such as the breastfeeding analgesia phenomenon, there are obvious implications for the performance of medical procedures that could be of immediate benefit to the infant (Blass & Barr, 2000).

Another version of this question is to ask whether any of these observations relate to long-term consequences for the infant; that is, are there any lasting benefits of a closeness versus a separateness strategy of caregiving? Here, the answer is a little less certain. For one thing, the evidence for what is sometimes referred to as early determinism, by which early experiences set up developmental trajectories, is quite controversial, and not completely clear. Furthermore, there are interesting and as yet unanswered questions about just how these early experiences might be instantiated in ways that would set up such trajectories.

One challenging hypothesis – more accurately a metaphor than a hypothesis – was suggested by the authors of *Developmental Health and the Wealth of Nations* (Keating & Hertzman, 1999). They argued that early experiences might affect life trajectories by a process of "biological embedding," such that some important biological processes (e.g., stress reactivity, or genes that are turned on or turned off) might be set or biased by early experience. This is not, as the editors point out, a strict determinist hypothesis, both because the actual outcomes are still determined by subsequent interactions with the environment, and because changes in biology are not fixed once and for all either. Consequently, the ways we have been suggesting in which caregiving behaviors (and caregiving packages) affect biological hidden regulators in the way Hofer implies would seem to be an important necessary, if not sufficient, step in understanding how this might happen.

Furthermore, there are other interesting properties of caregiving that might contribute to biological embedding. Caregiving behaviors have the properties of (1) being present very *early* in life, when the mother-infant dyad is arguably the closest, (2) occurring *frequently*, many times a day, and (3) being *iterative* (repetitive); that is, occurring in more or less the same way each time. Therefore, if caregiving behaviors are important in affecting change in biological systems, this may well happen through a process of repeated small influences over time, rather than one large effect at a single point in time (as might occur with a traumatic accident, for example). Whether or not this occurs, and whether or not it occurs at physiological levels or molecular biological levels, remains to be determined.

A third way to ask the question about the importance of these observations is to ask whether what we have learned about caregiving and early infancy implies that there is one "right" way to interact with our infants. To put it another way, do the results of these studies allow us to answer the question, "Should we carry our infants more, breastfeed more frequently, and so on?" As tempting as it is to think that the answer might be considered by most people to be "yes," it is a temptation that I think we need to be very wary of for at least three reasons.

First, the demonstration that something is or can be the case does not mean that it *should* be the case. For example, there is no logical implication for the parenting that we do that, because there is less crying with increased carrying, we should all be carrying our infants more. This is a version of what is sometimes referred to as the naturalistic fallacy (usually expressed as "what is, should be"). Although this fallacy is usually raised in the context of moral or ethical discourse, it applies equally well here.

Second, if we try to answer this question on the basis of these sorts of studies, we implicitly are making recommendations about a single behavior (e.g. carrying), independent of context. In part, this is because of the way science works, by experimentally isolating a determinant factor to the extent possible. These observations all constitute pieces of a bigger, more integrated process. What if, for example, increasing carrying decreased crying but at the same time increased sleep? Is that something we want to encourage? Is it better to sleep more and cry less? We may not like crying, but what are we getting in its place? Perhaps being awake and crying is better than being asleep. Perhaps the increased caregiver attention that crying generates has a necessary and important function. Perhaps it is the infant who cries *less* that is deprived, not the infant who cries more.

Third, these questions, and others like it, often have as an implicit starting assumption that there is an optimal form of caregiving, and that if only we did everything right, we would produce perfect infants and children. The parent advice literature is, arguably, obsessed with this optimality assumption, and it is an assumption that I have come to think is wrong. Lost in the battles over how much carrying, how frequent the feeds, or whether or how long one can co-sleep is a more important truth; namely, that evolution has produced a most remarkable organism that is flexible and resilient enough to be able to survive, and indeed thrive, in a remarkably wide range of caregiving environments. Imagine, if you will, who would ever have designed an organism that could survive and flourish under conditions of being fed once every 4 hours with milk from a different species on the one hand, to being fed breast milk four times every hour from its own mother on the other. What evolution seems to have provided is not a rigid organism for whom there is a single optimal caregiving requirement, but rather a flexible, adaptable organism that can make facultative adjustments to a variety of caregiving niches, depending on the environment in which it finds itself.

What these studies, and many others like them, *do* do is describe potential caregiving strategies rather than demonstrate optimal caregiving. They do not imply that good mothers must or should carry their infants more or feed more frequently. Rather, they imply that if one wishes to decrease crying, one may do so by increasing the amount of carrying and holding that one does. In considering the alternatives, we should introduce a different and rarely mentioned construct into decision making regarding caregiving. Following Hrdy's argument, I think of this construct as that of deciding what to do not on the basis of some ideal optimality, but on the basis of making trade-off decisions in light of the context and its demands. Increasing carrying may be feasible and easy for a stay-at-home mother with a singleton and no other competing interests, but much less so for a working mother or a single mother with many children already. Similarly, it is not necessarily the case that we would want to trade the increased caregiver attention resulting from the crying for the decreased experiential stimulation that might be associated with increased sleeping.

For many of these questions, we simply do not know what the trade-offs are. Yet this is not a reason for not being aware of the importance of trade-offs; it simply provides us with an increased challenge if we are to meet our obligations as committed caregivers to getting infants off to a good start in life.

REFERENCES

Barr, R. G. (1990a). The early crying paradox: a modest proposal. *Human Nature*, 1, 355–89.

Barr, R. G. (1990b). The normal crying curve: what do we really know? *Developmental Medicine and Child Neurology*, 32, 356–62.

Barr, R. G., Bakeman, R., Konner, M., & Adamson, L. (1987). Crying in !Kung infants: distress signals in a responsive context [Abstract]. *American Journal of Diseases of Children* 141, 386.

Barr, R. G., & Elias, M. F. (1988). Nursing interval and maternal responsivity: effect on early infant crying. *Pediatrics*, 81, 529–36.

Barr, R. G., Holsti, L., Young, S. N., Paterson, J. A., & MacMartin, L. M. (2005). Breast feeding reduces biobehavioral responses to diphtheria-pertussis-tetanus immunizations in infants at two and four months Presented at the 3rd Annual Canadian Child Health Clinician Scientist National Symposium, Oct. 14-16, 2005.

Barr, R. G., Kramer, M. S., Leduc, D. G., Boisjoly, C., McVey-White, L., & Pless, I. B. (1988). Parental diary of infant cry and fuss behaviour. *Archives of Disease in Childhood*, 63, 380–7.

Barr, R. G., Kramer, M. S., Pless, I. B., Boisjoly, C., & Leduc, D. (1989). Feeding and temperament as determinants of early infant cry/fuss behaviour. *Pediatrics*, 84, 514–21.

Barr, R. G., McMullan, S. J., Spiess, H., Leduc, D. J., Yaremko, J., Barfield, R. et al. (1991). Carrying as colic "therapy": a randomized controlled trial. *Pediatrics*, 87, 623–30.

Barr, R. G., Pantel, M. S., Young, S. N., Wright, J. H., Hendricks, L. A., & Gravel, R. G. (1999a). The response of crying newborns to sucrose: is it a "sweetness" effect? *Physiology and Behavior*, 66, 409–17.

Barr, R. G., Quek, V., Cousineau, D., Oberlander, T. F., Brian, J. A., & Young, S. N. (1994). Effects of intraoral sucrose on crying, mouthing, and hand-mouth contact in newborn and six-week-old infants. *Developmental Medicine and Child Neurology*, 36, 608–18.

Barr, R. G., & Young, S. N. (1999). A two-phase model of the soothing taste response: implications for a taste probe of temperament and emotion regulation. In M. Lewis & D. Ramsay (Eds.), *Soothing and Stress* (pp. 109–37). Mahwah, NJ: Lawrence Erlbaum Associates.

Barr, R. G., Young, S. N., Wright, J. H., Gravel, R., & Alkawaf, R. (1999b). Differential calming responses to sucrose taste in crying infants with and without colic. *Pediatrics*, 103, 1–9.

Bell, S. M. & Ainsworth, D. S. (1972). Infant crying and maternal responsiveness. *Child Development*, 43, 1171–1190.

Ben Shaul, D. M. (1962). The composition of the milk of wild animals. *International Zoo Yearbook*, 4, 333–42.

Benton, D., Brett, V., & Brain, P. F. (1987). Glucose improves attention and reaction to frustration in children. *Biological Psychology*, 24, 95–100.

Blass, E. M., & Barr, R. G. (2000). Evolutionary biology and medical practice: management of infant pain experience. *Journal of Developmental and Behavioral Pediatrics*, 21, 283–4.

Blass, E. M., & Ciaramitaro, V. (1994). A new look at some old mechanisms in human newborns: taste and tactile determinants of state, affect, and action. *Monographs of the Society for Research in Child Development*, 59(1), 1–80.

Blass, E. M., Fillion, T. J., Rochat, P., Hoffmeyer, L. B., & Metzher, M. A. (1989). Sensorimotor and motivational determinants of hand-mouth coordination in 1–3-day-old human infants. *Developmental Psychology*, 25, 963–75.

Blass, E. M., Fillion, T. J., Weller, A., & Brunson, L. (1990). Separation of opioid from nonopioid mediation of affect in neonatal rats: nonopioid mechanisms mediate maternal contact influences. *Behavioral Neuroscience*, 104, 625–36.

Blass, E. M., Shide, D. J., Zaw-Mon, C., & Sorrentino, J. (1995). Mother as shield: differential effects of contact and nursing on pain responsivity in infant rats – evidence for nonopioid mediation. *Behavioral Neuroscience*, 109, 342–53.

Blass, E. M., & Smith, B. A. (1992). Differential effects of sucrose, fructose, glucose, and lactose on crying in 1- to 3-day-old human infants: qualitative and quantitative considerations. *Developmental Psychology*, 28(5), 804–10.

Blurton-Jones, N. (1972). Comparative aspects of mother-infant contact. In N. Blurton-Jones (Ed.), *Ethological Studies of Child Behaviour* (pp. 305–28). Cambridge: Cambridge University Press.

Ewer, R. F. (1968). *Ethology of Mammals*. London: Logos Press.

Ghosh, S, Barr R., Paterson J., Huot, S., and Brody, M. (2007). *Infants with and without colic demonstrate a similar diurnal pattern of sucrose-induced soothing.* 10th International Infant Cry Research Workshop, Dragor, Denmark [July 5, 2007].

Gormally, S. M., Barr, R. G., Wertheim, L., Alkawaf, R., Calinoiu, N., & Young, S. N. (2001). Contact and nutrient caregiving effects on newborn infant pain responses. *Developmental Medicine and Child Neurology*, 43, 28–38.

Graillon, A., Barr, R. G., Young, S. N., Wright, J. H., & Hendricks, L. A. (1997). Differential response to oral sucrose, quinine and corn oil in crying human newborns. *Physiology and Behavior*, 62, 317–25.

Hofer, M. A. (1984). Relationships as regulators: a psychobiologic perspective on bereavement. *Psychosomatic Medicine*, 46(3), 183–97.

Hofer, M. A. (1994a). Early relationships as regulators of infant physiology and behavior. *Acta Paediatrica Supplement*, 397, 9–18.

Hofer, M. A. (1994b). Hidden regulators in attachment, separation, and loss. In N. A. Fox (Ed.), *The Development of Emotion Regulation: Biological and Behavioral Considerations* (pp. 192–207). Chicago: The University of Chicago Press.

Hofer, M. A. (1996). Multiple regulators of ultrasonic vocalization in the infant rat. *Psychoneuroendocrinology*, 21(2), 203–17.

Horne, P., Barr, R. G., Zelazo, P. R., Valiante, G., & Young, S. N. (2006). Glucose enhances newborn memory for spoken words. *Developmental Psychobiology*, 48, 574–82.

Hrdy, S. B. (1999). *Mother Nature: A History of Mothers, Infants, and Natural Selection*. New York: Pantheon Books.

Hubbard, F. O. A., & van IJzendoorn, M. H. (1987). Maternal unresponsiveness and infant crying. a critical replication of the Bell and Ainsworth study. In L. W. C. Tavecchio & M. H. van IJzendoorn (Eds.), *Attachment in Social Networks* (pp. 339–75). North Holland: Elsevier Science.

Hunziker, U. A., & Barr, R. G. (1986). Increased carrying reduces infant crying: a randomized controlled trial. *Pediatrics*, 77, 641–8.

Keating, D. P., & Hertzman, C. (1999). *Developmental Health and the Wealth of Nations: Social, Biological, and Educational Dynamics*. New York: The Guilford Press.

Kehoe, P., & Blass, E. M. (1986a). Behaviorally functional opioid systems in infant rats: II. Evidence for pharmacological, physiological, and psychological mediation of pain and stress. *Behavioral Neuroscience*, 100, 624–30.

Kehoe, P., & Blass, E. M. (1986b). Central nervous system mediation of positive and negative reinforcement in neonatal albino rats. *Developmental Brain Research*, 27, 69–75.

Kehoe, P., & Blass, E. M. (1986c). Opioid-mediation of separation distress in 10-day-old rats: reversal of stress with maternal stimuli. *Developmental Psychobiology*, 19, 385–98.

Kennell, J. H. (1980). Are we in the midst of a revolution? *American Journal of Diseases of Children*, 134, 303–10.

Konner, M. J. (1976). Maternal care, infant behavior, and development among the !Kung. In R. B. Lee & I. DeVore (Eds.), *Kalahari Hunter-Gatherers, Studies of the !Kung San and Their Neighbors* (pp. 218–45). Cambridge, MA: Harvard University Press.

Konner, M. J., & Worthman, C. (1980). Nursing frequency, gonadal function, and birth spacing among !Kung hunter-gatherers. *Science*, 207, 788–91.

Lozoff, B., & Brittenham, G. (1979). Infant care: cache or carry. *The Journal of Pediatrics*, 95, 478–83.

Oberlander, T. F., Barr, R. G., Young, S. N., & Brian, J. A. (1992). Short-term effects of feed composition on sleeping and crying in newborn infants. *Pediatrics*, 90(5), 733–40.

Panksepp, J. (1986). The neurochemistry of behavior. In M. R. Rosenzweig & L. W. Porter (Eds.), *Annual Review of Psychology*, Volume 37 (pp. 77–107). Palo Alto: Annual Reviews Inc.

St. James-Roberts, I., Alvarez, M., Csipke, E., Abramsky, T., Goodwin, J., & Sorgenfrei, E. (2006). Infant crying and sleeping in London, Copenhagen, and when parents adopt a 'proximal' form of care. *Pediatrics*, 117, e1146–e1155.

St. James-Roberts, I., Hurry, J., & Bowyer, J. (1993). Objective confirmation of crying durations in infants referred for excessive crying. *Archives of Disease in Childhood*, 68(1), 82–4.

Valiante, G., Barr, R. G., Zelazo, P. R., Papageorgiou, A., & Young, S. N. (2006). A typical feeding enhances memory for spoken words in healthy 2- to-3-day-old newborns. *Pediatrics*, 117, 476–86.

Zeifman, D., Delaney, S., & Blass, E. M. (1996). Sweet taste, looking, and calm in 2- and 4-week-old infants: the eyes have it. *Developmental Psychology*, 32, 1090–99.

Zelazo, P. R., Weiss, M. J., Randolph, M., Swain, I. U., & Moore, D. S. (1987). The effects of delay on neonatal retention of habituated head-turning. *Infant Behavior and Development*, 10, 417–34.

4

Early Experience and Stress Regulation
in Human Development

MEGAN R. GUNNAR AND MICHELLE M. LOMAN

Since the work of Hans Selye (1973), the idea that stress can be detrimental to health has become common knowledge. Less commonly known is the evidence that stress may have detrimental effects on development (De Bellis, 2001; Gunnar & Vazquez, 2006; Heim, Plotsky, & Nemeroff, 2004). In this chapter, we describe what is known about the physiology of stress and its potential influence on young children. We then turn to studies of human development to examine the ways that stress is regulated early in life and the evidence that the stress system is responsive to adverse conditions during infancy and early childhood.

Stress results when demands exceed immediately available resources (Lazarus & Folkman, 1984). These demands may be physical or psychological. Regardless, the imbalance in demands and resources requires that resources need to be found to meet the demands (Gunnar, 2000). These resources may be external, such as the help and support provided by parents and friends, or internal, such as a novel solution to a problem. Obtaining resources requires energy. Finding the energy needed for action and tuning the brain and body to meet the demands of the moment are the jobs of the stress system (Sapolsky, 1994). The stress system finds the energy we need to deal with immediate demands by putting future-oriented processes on hold. If there is an immediate threat to our survival, we do not need to put energy into fighting off a virus, digesting our lunch, or growing an extra inch. We need that energy to fuel the mental and physical processes that increase our chances of surviving to face tomorrow. As this example suggests, stress is not necessarily detrimental. The capacity to mount an effective stress response allows us to adapt to the changing and sometimes extreme demands of our daily existence, to stretch our abilities, and to achieve more than we might were we to avoid situations of high demand (Porges, 1995). Stress becomes detrimental, however, when it is excessive and/or when it is not followed by adequate periods of rest and repair. Stress may also be detrimental during periods of rapid growth when putting the future on hold may have long-term consequences.

The shifting of resources from the future to the here and now is accomplished through activation of two interacting systems: the sympathetic adrenomedullary system (SAM) and the pituitary-adrenocortical system (Johnson, Kamilaris, Chrousos, & Gold, 1992). These systems regulate the production of epinephrine from the medulla and cortisol from the cortex of the adrenal glands. Increases in epinephrine and cortisol have widespread effects including breaking down fat and protein stores, increasing heart rate and blood pressure, inhibiting digestion, inhibiting the growth system, and inhibiting actions of the immune system. A hierarchy of control systems in the brain regulates the SAM and pituitary-adrenocortical system (Palkovits, 1987). Nuclei in the hypothalamus and brain stem coordinate the activity of these systems in response to signals from the periphery of the body and from higher up in the brain. The involvement of higher centers in the brain allows for the activation and regulation of the stress system in anticipation of physical threats and in response to psychological and social threats (Gunnar & Vasquez, 2006; Rosen & Schulkin, 1998; Schulkin, McEwen, & Gold, 1994).

Corticotropin-releasing hormone (CRH), the neuroactive peptide that activates the pituitary-adrenocortical system, is believed to orchestrate the stress system (Nemeroff, 1998; Strand, 1999). CRH is produced in areas in the limbic system and frontal cortex that are involved in the emotions associated with stress (e.g., fear/anxiety, anger), and it is involved in brain pathways that stimulate the production of neurotransmitters that narrow attention, heighten vigilance, and promote rapid, emotion-based thought processes. Importantly, there is evidence that throughout development frequent or prolonged stress levels of cortisol increase the activity of CRH in emotion areas of the brain (de Kloet, Rots, & Cools, 1996; Heim, Owen, Plotsky, & Nemeroff, 1997). Furthermore, in several animal models, early experiences have been shown to shape the development of emotion-related CRH activity, permanently altering the threshold and magnitude of stress reactions to potentially stressful events (Caldji et al., 1998; Cirulli, Berry, & Alleva, 2003; Levine, 1994).

Most of the early experience research on stress has been conducted using rats and mice (see reviews: Cirulli et al., 2003; Gunnar & Quevedo, 2007; Meaney et al., 1994; Meaney & Szyf, 2005). For these animals it is well documented that the stress system can be altered by stress to a mother during pregnancy and/or postnatal disturbances that impair normal maternal behavior (Barbazanges, Piazza, Moal, & Maccari, 1996; Liu, Diorio, Day,

Francis, & Meaney, 2000; Liu et al., 1997). Depending on the nature and timing of these experiences, the stress system becomes either hyperreactive, responding with intensity even to mildly challenging events, or hyporesponsive, failing to respond adequately to events that threaten well-being (van Oers, de Kloet, & Levine, 1997). For example, depriving the infant rat of its mother for 24 hours when it is just a few days old produces a hyperresponsive stress system; a similar period of deprivation later in development produces a hyporesponsive stress system. Less drastic but more chronic disturbances accumulating over the equivalent of a rat's infancy and early childhood shape a hyperresponsive stress system; early weaning followed by prolonged social isolation during the equivalent of the middle childhood period produces a hyporesponsive stress system (Liu et al., 1997; Sanchez, Aguado, Sanchez-Toscano, & Saphier, 1998).

Our knowledge about early experiences and development of the stress system in nonhuman primates, although not as extensive, is growing (reviewed in Sanchez et al., 2005; Sanchez, 2006; McCormack, Newman, Higley, Maestripieri, & Sanchez, 2009). There is evidence that both pre- and postnatal disturbances can have long-term effects on behavioral and physiological responses to emotionally and/or socially challenging situations (Clarke, Wittwer, Abbott, & Schneider, 1994; Schneider, Coe, & Lubach, 1992). For example, in macaques, disrupting mother-infant interactions for what would be the equivalent of the toddler and preschool period in humans results in offspring who even as adults have higher brain levels of CRH and disturbances in the neurotransmitter systems that foster stress resilience (Rosenblum & Andrews, 1994; Rosenblum et al., 1994). In monkey species where the father plays a significant role in caregiving, similar detrimental impacts result from intermittent parental separation (Dettling, Fielding, & Pryce, 2002). Thus, the monkey studies clearly implicate parenting more generally, and not just maternal care.

The brain is a major target of stress (de Kloet & de Wied, 1980). Hormones like cortisol can only affect tissues that have receptors for that hormone. There are receptors for cortisol in many areas of the brain (Lopez, Chalmers, Little, & Watson, 1998; Sanchez, Young, Plotsky, & Insel, 2000; Vythilingam et al., 2000). The receptors involved when cortisol is at stress levels mediate energy-demanding neural processes that are required for learning (de Wied & Croiset, 1991). However, these learning-related neural events place neurons at risk by creating by-products that need to be cleaned out of the neuron (Sapolsky, Krey, & McEwen, 1986). High levels of cortisol also reduce the neuron's capacity to take up glucose. This means that the hardworking neurons have trouble fueling the *cleanup* crew in the presence of high levels of stress hormones. If this goes on too long, the by-products build up to levels that can damage neurons (McEwen, 1998). Stress also increases the production of neurotransmitters that speed information flow through emotion centers in the brain, while at the same time reducing the activity of neurotransmitters and other neurochemicals that foster the rest and repair of neurons (Rosen & Schulkin, 1998). In sum, the changes in brain biochemistry produced by stress allow the processing of information relevant to survival at the expense of activities needed to support rest, repair, and growth of the brain (Gould & Cameron, 1996). The net result, particularly during periods of brain development, is that if stress is too prolonged, the brain is compromised.

In studies of animals, the impact of chronic stress is evident in impairments in learning and memory (Born, Kern, Fehm-Wolfsdorf, & Fehm, 1987; de Wied & Croiset, 1991; Diamond & Rose, 1994; McGaugh, 1983). In studies of early experience in rats, the maternal care that produces better regulation of the stress system as the rat develops also fosters the production of brain growth factors and results in an adult animal that is better at learning and remembering (Liu et al., 2000). Similarly in studies of monkeys, those that are genetically vulnerable to disturbances in the neurotransmitter that fosters healthy neuronal functioning grow up to show those disturbances if they experience impoverished care as infants, but not if they receive good maternal care (Bennett et al., 2002). The same conditions that allow the expression of this genetic vulnerability also produce a stress-reactive monkey that tends, when given the opportunity, to drink large amounts of alcohol (Fahlke et al., 2000). Recent evidence suggests that rats exposed to early life stress because of maternal separation are more defensively aggressive, which is associated with changes in neurotransmitter systems (Veenema, Blume, Niederle, Buwalda, & Neuman, 2006).

The overall picture provided by the animal studies is that disturbances in care early in life disrupt normal brain development and produce animals that are vulnerable to stress later in life, exhibit impairments in learning and memory, are prone to self-administering drugs such as alcohol, and may be more defensively aggressive. We do not know if these animal data apply to human infants and children; however, the accumulating animal data certainly indicate that we need to understand whether similar processes are occurring during human development.

STRESS AND HUMAN DEVELOPMENT

Although often overlooked in discussions of early experience and stress, the prenatal period is likely a period of great plasticity for the human stress system (Graham, Heim, Goodman, Miller, & Nemeroff, 1999; Weinstock, 1997). The development of the stress system begins well before birth. By 18–20 weeks gestation, increases in stress hormones can be observed following invasive medical procedures (e.g., Giannakoulpoulous, Sepulveda, Kourtis, Glover, & Fisk, 1994). In animal models, a wide range of environmental (e.g., loud noises) and psychosocial (e.g., changing social groups) stresses on the mother have been shown to affect her developing offspring (see Weinstock, 1997, for review). Human studies have focused on prenatal stress ranging from that related to maternal daily hassles and symptoms of psychopathology to the effects of natural disasters and terrorism (for review see Talge et al., 2007). The activity of the mother's pituitary adrenocortical system appears to play a critical role in affecting fetal development. When the mother's stress hormone levels are controlled experimentally, psychosocial stressors in her life do not translate into alterations in fetal development (Barbazanges et al., 1996).

One pathway through which maternal stress may influence the development of her fetus is through stimulating CRH production by the placenta (Schulkin, 1999; Wadhwa, Sandman, & Garite, 2001). It is normal during pregnancy for the mother's levels of cortisol to rise. Her rising cortisol levels stimulate increased placental production of CRH that, in turn, increases the activity of the fetus' pituitary-adrenocortical system. None of this is cause for concern. In fact, the developing fetus needs cortisol to stimulate development. For example, cortisol fosters lung development, and this is why women who are about to deliver babies prematurely are given doses of dexamethasone, a synthetic form of cortisol. However, although a sufficient amount of cortisol is needed for the healthy development of the fetus, too much cortisol can be detrimental (Wadhwa, Porto, Garite, Chica-DeMet, & Sandman, 1998). Stress to the mother during pregnancy can increase her cortisol levels above the normal rise during pregnancy, stimulate more placental CRH than would be typical, and create a fetus whose pituitary-adrenal system is overly active during development. This, in turn, may produce changes in the developing brain that result in a stress-vulnerable infant.

Proving beyond a doubt that maternal stress influences fetal development in humans is hampered by our inability to perform experiments. Nonetheless, the evidence that is now accumulating indicates that, controlling statistically for numerous obstetric risk factors, mothers who feel overwhelmed by the challenges in their lives during pregnancy produce more stress hormones. In turn, the placenta produces more CRH, and babies who have higher and less variable heart rates, and those born earlier with lower birth weights (DiPietro, Hodgson, Costigan, Hilton, & Johnson, 1996; Huizink, de Medina, Mulder, Visser, & Buitelaar, 2000a, 2000b; Wadhwa et al., 2001). Perhaps importantly, lower versus higher socioeconomic status is also associated with many of these same effects (DiPietro, Costigan, Shupe, Pressman, & Johnson, 1998; however, see also DiPietro et al., 2004). The impact of maternal stress during pregnancy does not end with the child's birth. Maternal stress, anxiety, and elevated cortisol all increase the risk that the baby will have a difficult temperament (Davis et al., 2007; Wadhwa et al., 2001). In addition, there is increasing evidence that maternal stress during pregnancy also is associated with an increased risk of cognitive problems for the child (Talge et al., 2007).

At birth, healthy, full-term newborns are capable of mounting large stress responses to many kinds of stimulation. Elective surgeries, such as circumcision, produce striking elevations in cortisol, heart rate, and other physiological indicators of stress (Gunnar, Fisch, Korsvik, & Donhowe, 1981; Porter, Porges, & Marshall, 1988). Even minor manipulations like undressing, weighing, and measuring the newborn elevate heart rate and cortisol levels (e.g., Gunnar, 1992). However, the healthy newborn shows remarkable stress-regulatory capacities. For example, following a stressful experience, the newborn withdraws into periods of quiet sleep from which he or she is difficult to rouse (Gunnar, Malone, & Fisch, 1985). During this deep sleep, she rapidly clears stress hormones from circulation and returns to her normal, low baseline levels. Sleep is an important coping mechanism that fosters repair and recovery from stress throughout life (Dahl, 1996). The newborn also learns about stressful events. She shows the same or larger stress reactions to repetitions of painful experiences (e.g., blood draws); she ceases to respond to repetitions of arousing but not painful experiences (e.g., being handled during physical examinations, see Gunnar, 1992, for review). Indeed, intense pain appears to sensitize the stress system in ways that can be detected for days and perhaps even months (Gunnar, Porter, Wolf, & Rigatuso, 1995; Taddio, Katz, Ilarslch, & Koren, 1997).

Beginning in the newborn period, the things we do to care for and soothe distressed babies help to regulate and reduce stress. This has been shown quite powerfully in studies of the effects of contact comfort and feeding on the newborn's response to pain. Sweet tastes on the tongue, sucking, and being held reduce crying and heart rate reactions during painful medical procedures such as blood draws and inoculations (see for review Blass & Watt, 1999; see also Barr, this volume). All of these forms of stimulation come together in the act of nursing, and there is accumulating evidence that nursing serves to trigger opioid- and nonopioid-mediated analgesic pathways in the infant's brain. Similarly, the hormones and nutrients the baby receives through breast milk facilitate calming and enhance brain development (Hrdy, 1999). Thus, the act that sets mammals apart from other animals, that of nursing, appears to provide not only the nutrients for healthy development but also a major external regulator of the infant's stress system (Hofer, 1987). Perhaps it is not surprising that across all cultures studied, a distressed infant first elicits attention, then approach, then contact, and finally feeding from her caregivers (Barr, 1990).

Although the stress system is quite responsive at birth, within a short period of time it becomes highly buffered. Remarkably, part of this buffering is seen in an unhooking of crying and increases in stress hormones. Thus, for example, between ten and twelve weeks after birth, physical exams cease to provoke increases in cortisol, even though babies continue to cry and fuss when they are undressed and examined (Larson, White, Cochran, Donzella, & Gunnar, 1998). By the last part of the first year of life, about the time that infants learn to crawl and walk, it becomes very difficult to produce increases in cortisol when the child is in the presence of sensitive and responsive caregivers (see review by Gunnar, 2000). This is especially apparent in secure attachment relationships where even the temperamentally fearful toddler does not show increases in cortisol when confronted by events that scare him or her (Ahnert, Gunnar, Lamb, & Barthel, 2004; Nachmias, Gunnar, Mangelsdorf, Parritz, & Buss, 1996). The dissociation between distressed behavior and activation of the stress system probably allows the infant to recruit parental attention at less biological cost. That is, she can cry and gain adult protection from threat, without putting the future on hold through the activation of her stress system.

During the same developmental period when secure attachment relationships buffer the child from the potentially detrimental effects of stress system activation, long hours of out-of-home childcare produce a rise in cortisol over the day (Dettling, Gunnar, & Donzella, 1999; Dettling, Parker, Lane, Sebanc, & Gunnar, 2000; Tout, de Haan, Kipp-Campbell, & Gunnar, 1998; Watamura, Sebanc, & Gunnar, 2002; Watamura, Donzella, Alwin, & Gunnar, 2003). This increase is greatest around 2 years of age and decreases over the preschool years. By 7 years of age, the pattern of cortisol production over the day in out-of-home care is the same as it is at home (Dettling, Gunnar, & Donzella, 1999; reviewed in Gunnar & Donzella, 2002). Even at its peak, the average rise over the day at childcare is small, and currently there is no reason to suspect that it has harmful consequences for children.

Indeed, we believe that these small increases reflect the demands that group care places on the young child's emerging social skills (reviewed in Gunnar & Donzella, 2002; Gunnar & Quevedo, 2007). It is hard work playing nicely with other children who are just learning how to play nicely. It is really hard to keep this up for many hours. However, most children in childcare benefit from the challenge. Both cognitive and social skills are advanced by good childcare (Belsky et al., 2007; Peisner-Feinberg et al., 2001; see review in ch. 11, National Research Council and Institute of Medicine, 2000).

Nonetheless, for some children the rise in cortisol is fairly large and might be reason for concern. Children who are more aggressive and less socially competent tend to experience larger increases in cortisol as the day wears on (Dettling et al., 1999, 2000; Tout et al., 1998). Angry, aggressive children are also less well liked by their peers (Hartup, 1992). Being socially rejected appears to be stressful for young children, as it is for older children and adults (Gunnar, Bruce, & Donzella, 2000; Gunnar, Sebanc, Tout, Donzella, & van Dulmen, 2003). Larger increases in cortisol are also seen for children in lower quality childcare (Dettling et al., 2000). The combination of low quality childcare, angry and aggressive behavior, and young age (e.g., 2 to 3) can result in fairly large increases in cortisol over the childcare day. Low quality and long hours are also factors that predict behavior problems for children in childcare (NICHD Early Child Care Research Network, 2001). Perhaps importantly, angry and aggressive behaviors are the behavior problems these childcare factors predict.

Poor quality childcare is a risk factor experienced by many young children (see review in ch. 11, National Research Council and Institute of Medicine, 2000). Fewer children, but still a significant number, experience maltreatment during early development. Maltreating environments that neglect children's basic physical and social needs and/or are physically and emotionally threatening are detrimental to development (Cicchetti & Valentino, 2006). Several recent studies have shown alterations in brain development for maltreated children, some with particular focus on those who develop chronic posttraumatic stress disorders (PTSD) as a function of their early abuse (Bremner et al., 2003; Carrion, Weems, & Reiss, 2007; De Bellis, Keshavan et al., 1999; Pynoos, Steinberg, & Wraith, 1995; Teicher, Anderson, Polcari, Anderson, Navalta, & Kim, 2003). Disturbances in the stress system are hypothesized to contribute to these effects on the developing brain. Given this, throughout the last decade, researchers have become increasingly more interested in exploring patterns of cortisol and epinephrine production in maltreated children.

Although the precise effects are often dependent on the type of stressor (pharmacological or psychological) and personal factors such as psychiatric status or current life adversity, studies demonstrate that early neglectful and/or abusive rearing environments have a long-term influence on stress physiology (Tarullo & Gunnar, 2006). Several studies document a loss of the normal daily rhythm in cortisol production over the daytime hours (see review by Gunnar & Vazquez, 2001). Children reared in orphanages, those reared by parents who are so depressed that they cannot be sensitive or responsive to their children, children placed into foster care, and those who develop a form of stress-related failure to grow (i.e., psychosocial dwarfism) all appear to fail to produce the normal peak in cortisol in the early morning hours (see Gunnar, Fisher, & the Early Experience, Stress, and Prevention Network, 2006; Dozier et al., 2006). Instead, these children show a pattern of low and flat basal cortisol production over the daytime hours. In one study of children with PTSD, intrusive thoughts about the traumatic event also were associated with the degree of suppression in cortisol in the early morning hours (Goenjian et al., 1996). The normal daily rhythm appears to return once the child is placed in more supportive care. However, the early signature of chronic stress may then reveal itself in elevated levels of stress hormone production.

Indeed, several studies have also shown elevations in stress hormones in children who were maltreated early in childhood. Two of these studies involved children who were older (aged 7 or older), were now in stable or good living conditions, but who had been abused as infants and toddlers (De Bellis, Baum et al., 1999; Gunnar, Morison, Chisholm, & Schuder, 2001). One study showed that elevations in cortisol were particular to children who had developed deprivation dwarfism in response to their adverse care (Kertes, Gunnar, Madsen, & Long, 2008). Another study showed that only children who had experienced multiple, prolonged forms of maltreatment and, perhaps, ongoing stress in their home life produced chronic high cortisol levels following early maltreatment (Cicchetti & Rogosch, 2001). Retrospective studies of adults who reported maltreatment in their early years also have shown disturbances in stress system functioning. In one of these studies, women who reported being sexually abused as children showed larger stress reactions to a standard social stress test (i.e., giving a speech and doing mental arithmetic) (Heim et al., 2000). These responses were greater among those abused women who were clinically depressed, but were also shown to some extent by the ones who had not developed clinical depression pursuant to their abuse history.

CONCLUSIONS

Animal studies of early experience and stress document the profound and seemingly permanent effects that adverse early conditions have on development. The offspring of stressed pregnancies and compromised early care develop into individuals who are more vulnerable to stress later in life, show altered patterns of brain functioning, and often have impaired cognitive abilities. Research on the biology of early experiences and stress in human infants and children is emerging. The accumulating evidence suggests that adverse conditions early in life can shape the developing brain in our species as well. There is much, however, that we have yet to learn about the role of stress in human development. Among the unanswered questions are how genetic vulnerability may interact with early conditions to affect development and the extent to which later experiences may ameliorate the impact of early adversity. However, while we are still in search of answers to these basic questions, we cannot ignore the implications of what we already know. Clearly, the present state of our knowledge argues strongly for protecting and enhancing the quality of the pregnant woman's emotional and physical health and for protecting and enhancing the quality of care that infants and young children receive after birth. Stress is a normal part of human existence, but overwhelming stress in the face of inadequate physical and emotional support appears to pose a risk for human development throughout life.

REFERENCES

Ahnert, L., Gunnar, M., Lamb, M., & Barthel, M. (2004). Transition to child care: Associations with infant-mother attachment, infant negative emotion, and cortisol elevations. *Child Development* 75, 639–50.

Barbazanges, A., Piazza, P. V., Moal, M. L., & Maccari, S. (1996). Maternal glucocorticoid secretion mediates long-term effects of prenatal stress. *The Journal of Neuroscience*, 16(12), 3943–9.

Barr, R. G. (1990). The early crying paradox: A modest proposal. *Human Nature*, 1(4), 355–89.

Bennett, A. J., Lesch, K. P., Heils, A., Long, J. C., Lorenz, J. G., Shoal, S. E., et al. (2002). Early experience and serotonin transporter gene variation interact to influence primate CNS function. *Molecular Psychiatry*, 7, 118–22.

Belsky, J., Lowe Vandell, D., Burchinal, M., Clarke-Stewart, K. A., McCartney, K., Tresch Owen, M., et al. (2007) Are there long-term effects of early child care? *Child Development*, 78, 681–701.

Blass, E. M., & Watt, L. B. (1999). Suckling- and sucrose-induced analgesia in human newborns. *Pain*, 83, 611–12.

Born, J., Kern, W., Fehm-Wolfsdorf, G., & Fehm, H. L. (1987). Cortisol effects on attentional processes in man as indicated by event-related potentials. *Psychophysiology*, 24(3), 286–92.

Bremner, J. D., Vythilingam, N., Vermetten, E., Southwick, S. M., McGlashan, T., Nazeer, A., et al. (2003). MRI and PET study of deficits in hippocampal structure and function in women with childhood sexual abuse and posttraumatic stress disorder. *American Journal of Psychiatry*, 160, 924–32.

Caldji, C., Tannenbaum, B., Sharma, S., Francis, D., Plotsky, P. M., & Meaney, M. J. (1998). Maternal care during infancy regulates the development of neural systems mediating the expression of fearfulness in the rat. *Proceedings of the National Academy of Sciences of the United States of America*, 95(9), 5335–40.

Carrion, V. G., Weems, C. F., & Reiss, A. L. (2007). Stress predicts brain changes in children: A pilot longitudinal study on youth stress, posttraumatic stress disorder, and the hippocampus. *Pediatrics*, 119, 509–16.

Cicchetti, D., & Rogosch, F. A. (2001). Diverse patterns of neuroendocrine activity in maltreated children. *Development and Psychopathology*, 13, 677–93.

Cicchetti, D., & Valentino, K. (2006). An ecological transactional perspective on child maltreatment: Failure of the average expectable environment and its influence upon child development. In D. Cicchetti & D. J. Cohen (Eds.), *Developmental Psychopathology*, 2nd ed. (Vol. 3, pp. 129–201). New York: Wiley.

Cirulli, F., Berry, A., & Alleva, E. (2003). Early disruption of the mother-infant relationship: Effects on brain plasticity and implications for psychopathology. *Neuroscience and Biobehavioral Reviews*, 27, 73–82.

Clarke, A. S., Wittwer, D. J., Abbott, D. H., & Schneider, M. L. (1994). Long-term effects of prenatal stress on HPA axis activity in juvenile rhesus monkeys. *Developmental Psychobiology*, 27, 257–70.

Dahl, R. E. (1996). The regulation of sleep and arousal: Development and psychopathology. *Development and Psychopathology*, 8, 3–27.

Davis, E. P., Glynn, L. M., Schetter, C. D., Hobel, C., Chicz-Demet, A., & Sandman, C.A. (2007). Prenatal exposure to maternal depression and cortisol influences infant temperament. *Journal of American Academy of Child and Adolescent Psychiatry*, 46,(6), 737–46.

De Bellis, M. (2001). Developmental traumatology: the psychobiological development of maltreated children and its implications for research, treatment, and policy. *Development & Psychopathology*, 13, 539–64.

De Bellis, M. D., Baum, A. S., Birmaher, B., Keshavan, M. S., Eccard, C. H., Boring, A. M., et al. (1999). Developmental traumatology, Part 1: Biological stress systems. *Biological Psychiatry*, 45, 1259–70.

De Bellis, M. D., Keshavan, M. S., Clark, D. B., Casey, B. J., Giedd, J. N., Boring, A. M., et al. (1999). Developmental traumatology, Part 2: Brain development. *Biological Psychiatry*, 45, 1271–84.

de Kloet, E. R., & de Wied, D. (1980). The brain as target tissue for hormones of pituitary origin: Behavioral and biochemical studies. In L. Martini & W. F. Ganong (Eds.), *Frontiers in Neuroendocrinology* (Vol. 6, pp. 157–201). New York: Raven.

de Kloet, E. R., Rots, N. Y., & Cools, A. R. (1996). Brain-corticosteroid hormone dialogue: Slow and persistent. *Cellular and Molecular Neurobiology*, 16(3), 345–56.

de Wied, D., & Croiset, G. (1991). Stress modulation of learning and memory processes. In G. Jasmin & L. Proschek (Eds.), *Stress Revisited*. 2. Systemic Effects of Stress (Vol. 15, pp. 167–99). New York: Basel.

Dettling, A., Fielding, J., & Pryce, C. R. (2002). Repeated parental deprivation in the infant common marmoset (Callithrix Jacchus, Primates) and analysis of its effects on early development. *Biological Psychiatry*, 52, 1037–46.

Dettling, A., Gunnar, M. R., & Donzella, B. (1999). Cortisol levels of young children in full-day childcare centers: Relations with age and temperament. *Psychoneuroendocrinology*, 24(5), 505–18.

Dettling, A. C., Parker, S. W., Lane, S. K., Sebanc, A. M., & Gunnar, M. R. (2000). Quality of care and temperament determine whether cortisol levels rise over the day for children in full-day childcare. *Psychoneuroendocrinology*, 25, 819–36.

Diamond, D. M., & Rose, G. M. (1994). Stress impairs LTP and hippocampal-dependent memory. *Annals of the New York Academy of Sciences*, 746, 411–14.

DiPietro, J. A., Caulfield, L. E., Costigan, K. A., Merialdi, M., Nguyen, R. H., & Zavaleta, N. (2004). Fetal neurobehavioral development: A tale of two cities. *Developmental Psychology*, 40, 443–56.

DiPietro, J. A., Costigan, K. A., Shupe, A. K., Pressman, E. K., & Johnson, T. R. (1998). Fetal neurobehavioral development associated with social class and fetal sex. *Developmental Psychobiology*, 33, 79–81.

DiPietro, J. A., Hodgson, D. M., Costigan, K. A., Hilton, S. C., & Johnson, T. R. (1996). Fetal neurobehavioral development. *Child Development*, 67, 2553–67.

Dozier, M., Manni, M., Gordon, M. K., Peloso, E., Gunnar, M. R., Stovall-McClough, K. C. et al. (2006). Foster children's diurnal production of cortisol: An exploratory study. *Child Maltreatment*, 11, 189–97.

Fahlke, C., Lorenz, J. G., Long, J., Champoux, M., Suomi, S. J., & Higley, J. D. (2000). Rearing experiences and stress-induced plasma cortisol as early risk factors for excessive alcohol consumption in nonhuman primates. *Alcoholism, Clinical & Experimental Research*, 24(5), 644–50.

Giannakoulpoulous, X., Sepulveda, W., Kourtis, P., Glover, V., & Fisk, N. M. (1994). Fetal plasma and beta-endorphin response to intrauterine needling. *Lancet*, 344, 77–81.

Goenjian, A. K., Yehuda, R., Pynoos, R. S., Steinberg, A. M., Tashjian, M., Yang, R. et al. (1996). Basal cortisol, dexamethasone suppression of cortisol, and MHPG in adolescents after the 1988 earthquake in Armenia. *American Journal of Psychiatry*, 153, 929–34.

Gould, E., & Cameron, H. A. (1996). Regulation of neuronal birth, migration, and death in the rat dentate gyrus. *Developmental Neuroscience*, 18(1–2), 22–35.

Graham, Y. P., Heim, C., Goodman, S. H., Miller, A. H., & Nemeroff, C. B. (1999). The effects of neonatal stress on brain development: Implications for psychopathology. *Development and Psychopathology*, 11, 545–65.

Gunnar, M. R. (1992). Reactivity of the hypothalamic-pituitary-adrenocortical system to stressors in normal infants and children. *Pediatrics*, 90(3), 491–97.

Gunnar, M. R. (2000). Early adversity and the development of stress reactivity and regulation. In C. A. Nelson (Ed.), *The Effects of Adversity on Neurobehavioral*

Development. The Minnesota Symposia on Child Psychology (Vol. 31, pp. 163–200). Mahwah, NJ: Erlbaum.

Gunnar, M. R., Bruce, J., & Donzella, B. (2000). Stress physiology, health, and behavioral development. In A. Thornton (Ed.), *Family and Child Well-Being: Research and Data Needs.* Ann Arbor, MI: University of Michigan Press.

Gunnar, M., & Donzella, B., 2002. Social regulation of cortisol levels in early human development. *Psychoneuoendocrinology,* 27, 199–220.

Gunnar, M. R., Fisch, R., Korsvik, S., & Donhowe, J. (1981). The effect of circumcision on serum cortisol and behavior. *Psychoneuroendocrinology,* 6(3), 269–76.

Gunnar, M., Fisher, P., & the Early Experience, Stress, and Prevention Network. (2006). Bringing basic research on early experience and stress neurobiology to bear on preventive interventions for neglected and maltreated children. *Development and Psychopathology,* 18, 651–77.

Gunnar, M. R., Malone, S., & Fisch, R. O. (1985). Coping with aversive stimulation in the neonatal period: Quiet sleep and plasma cortisol levels during recovery from circumcision in newborns. *Child Development,* 56, 824–34.

Gunnar, M. R., Morison, S. J., Chisholm, K., & Schuder, M. (2001). Salivary cortisol levels in children adopted from Romanian orphanages. *Development and Psychopathology,* 13, 611–28.

Gunnar, M. R., Porter, F., Wolf, C., & Rigatuso, J. (1995). Neonatal stress reactivity: Predictions to later emotional temperament. *Child Development,* 66, 1–14.

Gunnar, M. R. & Quevedo, K. (2007). The neurobiology of stress and development. *Annual Review of Psychology,* 58, 145–73.

Gunnar, M. R., Sebanc, A. M., Tout, K., Donzella, B., & van Dulmen, M. M. H. (2003). Peer rejection, temperament, and cortisol activity in preschoolers. *Developmental Psychobiology,* 43, 346–58.

Gunnar, M. R., & Vazquez, D. M. (2001). Low cortisol and a flattening of the expected daytime rhythm: Potential indices of risk in human development. *Development and Psychopathology,* 13, 516–38.

Gunnar, M., & Vazquez, D. (2006). Stress neurobiology and developmental psychopathology. In D. Cicchetti & D. J. Cohen (Eds.), *Developmental Psychopathology,* 2nd ed., (Vol. 2, pp. 533–77). New York: Wiley.

Hartup, W. W. (1992). Peer relations in early and middle childhood. In V. B. Van Hasselt & M. Herson (Eds.), *Handbook of Social Development: A Lifespan Perspective* (pp. 257–81). New York: Plenum.

Heim, C., Newport, D. J., Heit, S., Graham, Y. P., Wilcox, M. Bonsall, R., et al. (2000). Pituitary-adrenal and autonomic responses to stress in women after sexual and physical abuse in childhood. *Journal of the American Medical Association,* 284(5), 592–97.

Heim, C., Owen, M. J., Plotsky, P. M., & Nemeroff, C. B. (1997). The role of early adverse life events in the etiology of depression and posttraumatic stress disorder: Focus on corticotropin-releasing factor. *Annals of the New York Academy of Sciences,* 821, 194–207.

Heim, C., Plotsky, P., & Nemeroff, C. B. (2004). The importance of studying the contributions of early adverse experiences to the neurobiological findings in depression. *Neuropsychopharmacology,* 29, 641–48.

Hofer, M. (1987). Shaping forces within early social relationships. In N. A. Krasnegar (Ed.), *Perinatal Development: A Psychobiological Perspective* (pp. 251–74). Orlando, FL: Academic.

Hrdy, S. B. (1999). *Mother Nature.* New York: Pantheon.

Huizink, A. C., de Medina, P. G. R., Mulder, E. J. H., Visser, G. H. A., & Buitelaar, J. K. (2000a). Prenatal psychosocial and endocrinologic predictors of infant temperament. In A. C. Huizink (Ed.), *Prenatal Stress and its Effects on Infant Development* (pp. 171–200). Hoorn, Netherlands: Drukkerij Van Vliet.

Huizink, A. C., de Medina, R., Mulder, E. J. H., Visser, G. H. A., & Buitelaar, J. K. (2000b). Psychosocial and endocrinologic measures of prenatal stress as predictors of mental and motor development in infancy. In A. Huizink (Ed.), *Prenatal Stress and Its Effects on Infant Development* (pp. 147–70). Hoorn, Netherlands: Drukkerij Van Vliet.

Johnson, E. O., Kamilaris, T. C., Chrousos, G. P., & Gold, P. W. (1992). Mechanisms of stress: A dynamic overview of hormonal and behavioral homeostasis. *Neuroscience and Biobehavioral Reviews*, 16, 115–30.

Kertes, D. A., Gunnar, M. R., Madsen, N. J., & Long, J. (2008). Early deprivation and home basal cortisol levels: A study of internationally adopted children. *Development & Psychopathology*, 20(2), 473–91.

Larson, M., White, B. P., Cochran, A., Donzella, B., & Gunnar, M. R. (1998). Dampening of the cortisol response to handling at 3 months in human infants and its relation to sleep, circadian cortisol activity, and behavioral distress. *Developmental Psychobiology*, 33(4), 327–37.

Lazarus, R. S., & Folkman, S. (1984). *Stress, Appraisal, and Coping.* New York: Springer.

Levine, S. (1994). The ontogeny of the hypothalamic-pituitary-adrenal axis: The influence of maternal factors. *Annals of the New York Academy of Sciences*, 746, 275–88.

Liu, D., Diorio, J., Day, J. C., Francis, D. D., & Meaney, M. J. (2000). Maternal care, hippocampal synaptogenesis and cognitive development in rats. *Nature Neuroscience*, 3(8), 799–806.

Liu, D., Diorio, J., Tannenbaum, B., Caldji, C., Francis, D., Freedman, A., et al. (1997). Maternal care, hippocampal glucocorticoid receptors, and hypothalamic-pituitary-adrenal responses to stress. *Science*, 227, 1659–62.

Lopez, J. F., Chalmers, D. T., Little, K. Y., & Watson, S. J. (1998). Regulation of serotonin 1A, glucocorticoid, and mineralocorticoid receptor in rat and human hippocampus: Implications for the neurobiology of depression. *Biological Psychiatry*, 43, 547–73.

McCormack, K., Newman, T., Higley, J., Maestripieri, D., & Sanchez, M. (2009). Serotonin transporter gene variation, infant abuse, and responsiveness to stress in rhesus macaque mothers and infants. *Hormones and Behavior*, 55(4), 538–47.

McEwen, B. (1998). Protective and damaging effects of stress mediators. *New England Journal of Medicine*, 338, 171–79.

McGaugh, J. L. (1983). Hormonal influences on memory. *Annual Review of Psychology*, 34, 297–323.

Meaney, M. J., Diorio, J., Francis, D., LaRocque, S., O'Donnell, D., Smythe, J. W., et al. (1994). Environmental regulation of the development of glucocorticoid receptor systems in the rat forebrain: The role of serotonin. *Annals of the New York Academy of Sciences*, 746, 260–74.

Meaney, M. & Szyf, M. (2005). Environmental programming of stress responses through DNA methylation: Life at the interface between a dynamic environment and a fixed genome. *Dialogues in Clinical Neuroscience*, 7, 103–23.

Nachmias, M., Gunnar, M. R., Mangelsdorf, S., Parritz, R., & Buss, K. (1996). Behavioral inhibition and stress reactivity: Moderating role of attachment security. *Child Development*, 67(2), 508–22.

National Research Council and Institute of Medicine (2000). *From Neurons to Neighborhoods: The Science of Early Child Development*. Committee on Integrating the Science of Early Childhood Development. J. P. Shonkoff & D. A. Phillips, (Eds.) *Board on Children, Youth, and Families, Commission on Behavioral and Social Sciences and Education*. Washington, DC: National Academy Press.

Nemeroff, C. B. (1998). The neurobiology of depression. *Scientific American*, June, 42–9.

NICHD Early Child Care Research Network (2001). Childcare and children's peer relationships at 24 and 36 months: The NICHD Study of Early Child Care. *Child Development*, 72(5), 1478–1500.

Palkovits, M. (1987). Organization of the stress response at the anatomical level. In E. R. de Kloet, V. M. Wiegant & D. de Wied (Eds.), *Progress in Brain Research* (Vol. 72, pp. 47–55). Amsterdam, Holland: Elsevier.

Peisner-Feinberg, E. S., Burchinal, M. R., Clifford, R. M., Culkin, M. L., Howes, C., Kagan, S. L. et al. (2001) The relation of preschool childcare quality to children's cognitive and social developmental trajectories through second grade. *Child Development*, 72, 1534–53.

Porges, S. W. (1995). Cardiac vagal tone: A physiological index of stress. *Neuroscience and Biobehavioral Reviews*, 19(2), 225–33.

Porter, F. L., Porges, S. W., & Marshall, R. E. (1988). Newborn pain cries and vagal tone: Parallel changes in response to circumcision. *Child Development*, 59, 495–505.

Pynoos, R. S., Steinberg, A. M., & Wraith, R. (1995). A developmental model of childhood traumatic stress. In D. Cicchetti & D. J. Cohen (Eds.), *Developmental psychopathology* (Vol. 2, pp. 72–95). New York: Wiley.

Rosen, J. B., & Schulkin, J. (1998). From normal fear to pathological anxiety. *Psychological Review*, 105(2), 325–50.

Rosenblum, L. A., & Andrews, M. W. (1994). Influences of environmental demand on maternal behavior and infant development. *Acta Paediatrica, Supplement* 397, 57–63.

Rosenblum, L. A., Coplan, J. D., Friedman, S., Bassoff, T., Gorman, J. M., & Andrews, M. W. (1994). Adverse early experiences affect noradrenergic and serotonergic functioning in adult primates. *Biological Psychiatry*, 35(4), 221–27.

Sanchez, M., 2006. The impact of early adverse care on HPA axis development: Nonhuman primate models. *Hormones and Behavior*, 50, 623–31.

Sanchez, M. M., Aguado, F., Sanchez-Toscano, F., & Saphier, D. (1998). Neuroendocrine and immunocytochemical demonstrations of decreased hypothalamo-pituitary-adrenal axis responsiveness to restraint stress after long-term social isolation. *Endocrinology*, 139(2), 579–87.

Sanchez, M., Noble, P., Lyon, C., Plotsky, P., Davis, M., Nemorff, C., et al. (2005). Alterations of diurnal cortisol rhythm and acoustic startle response in nonhuman primates with adverse rearing. *Biological Psychiatry*, 57, 373–81.

Sanchez, M. M., Young, L. J., Plotsky, P. M., & Insel, T. R. (2000). Distribution of corticosteroid receptors in the Rhesus brain: Relative absence of glucocorticoid receptors in the hippocampal formation. *The Journal of Neuroscience*, 20, 4657–68.

Sapolsky, R. (1994). *Why Zebras Don't Get Ulcers: A Guide to Stress, Stress-Related Diseases, and Coping.* New York: Freeman.

Sapolsky, R. M., Krey, L. C., & McEwen, B. S. (1986). The neuroendocrinology of stress and aging: The glucocorticoid cascade hypothesis. *Endocrine Reviews*, 7(3), 284–301.

Schneider, M. L., Coe, C. L., & Lubach, G. R. (1992). Endocrine activation mimics the adverse effects of prenatal stress on the neuromotor development of the infant primate. *Developmental Psychobiology*, 25(6), 427–39.

Schulkin, J. (1999). Corticotropin-releasing hormone signals adversity in both the placenta and the brain: Regulation by glucocorticoids and allostatic overload. *Journal of Endocrinology*, 161(3), 349–56.

Schulkin, J., McEwen, B. S., & Gold, P. S. (1994). Allostasis, amygdala, and anticipatory angst. *Neuroscience and Behavioral Reviews*, 18(3), 385–96.

Selye, H. (1973). The evolution of the stress concept. *American Scientist*, 61(6), 692–99.

Strand, F. L. (1999). Chapter 10: Hypophysiotropic neuropeptides: TRH, CRH, GnRH, GHRH, SS, PACAP, DSIP. In *Neuropeptides: Regulators of Physiological Processes* (pp. 179–228). Cambridge, MA: MIT Press.

Taddio, A., Katz, J., Ilarslch, A. L., & Koren, G. (1997). Effects of neonatal circumcision on pain response during subsequent routine vaccination. *The Lancet*, 340(March 1), 599–603.

Talge, N. M., Neal, C., Glover, V., & the Early Stress, Translational Research and Prevention Science Network (2007). Fetal and neonatal experience on child and adolescent mental health. *Journal of Child Psychology and Psychiatry*, 48, 245–61.

Tarullo, A. R., & Gunnar, M. R. (2006). Child maltreatment and the developing HPA axis. *Hormones and Behavior*, 50, 632–39.

Teicher, M. H., Anderson, S. L., Polcari, A., Anderson, C. M., Navalta, C. P., & Kim, D. M. (2003). *Neuroscience and Biobehavioral Reviews*, 27, 33–44.

Tout, K., de Haan, M., Kipp-Campbell, E., & Gunnar, M. R. (1998). Social behavior correlates of adrenocortical activity in daycare: Gender differences and time-of-day effects. *Child Development*, 69(5), 1247–62.

van Oers, H. J. J., de Kloet, E. R., & Levine, S. (1997). Persistent, but paradoxical, effects on HPA regulation of infants maternally deprived at different ages. *Stress: The International Journal on the Biology of Stress*, 1(4), 249–63.

Veenema, A. H., Blume, A., Niederle, D., Buwalda, B., & Neuman, I. D. (2006). Effects of early life stress on male aggression and hypothalamic vasopressin and serotonin. *European Journal of Neuroscience*, 24, 1711–20.

Vythilingam, M., Anderson, G. M., Owens, M. J., Halaszynski, T. M., Bremner, J. D., Carpenter, L. L. et al. (2000). Cerebrospinal fluid corticotropin-releasing hormone in healthy humans: Effects of yohimbine and naloxone. *Journal of Clinical Endocrinology and Metabolism, 85,* 4138–45.

Wadhwa, P. D., Porto, M., Garite, T. J., Chica-DeMet, A., & Sandman, C. A. (1998). Maternal corticotropin-releasing hormone levels in the third trimester predict length of gestation in human pregnancy. *American Journal of Obstetrics and Gynecology, 179,* 1079–85.

Wadhwa, P. D., Sandman, C. A., & Garite, T. J. (2001). The neurobiology of stress in human pregnancy: Implications for prematurity and development of the fetal central nervous system. *Progress in Brain Research, 30,* 131–42.

Watamura, S. E., Donzella, B., Alwin, J., & Gunnar, M. (2003). Morning to afternoon increases in cortisol concentrations for infants and toddlers at childcare: Age differences and behavioral correlates. *Child Development, 74,* 1006–20.

Watamura, S. E., Sebanc, A. M., & Gunnar, M., 2002. Rising cortisol at childcare: Relations with nap, rest, and temperament. *Developmental Psychobiology, 40,* 33–42.

Weinstock, M. (1997). Does prenatal stress impair coping and regulation of the hypothalamic-pituitary-adrenal axis? *Biobehavioral Review, 21,* 1–10.

5

Biology and Context: Symphonic Causation and the Distribution of Childhood Morbidities

W. THOMAS BOYCE

INTRODUCTION

The Nature-Nurture Culture Wars

By and large, investigators and scholars contributing to the Millennium Dialogue on Early Child Development (described in more detail in the Acknowledgments at the beginning of this volume), including the author of this chapter, were academically reared within a scientific generation marked by a confluence of two irreconcilable views on the origins of human disorders. Within a single generation, physicians, clinical and developmental psychologists, social workers, and laboratory investigators were steeped in the twin, sequential agendas of environmental and biological determinism. In the former of these views, prominent in the scientific world of the 1960s and 1970s, disease and disorder were held to be products of contextual exposures and adversities. Human afflictions, it was believed, were due almost exclusively to the acute and chronic, cumulative influences of environmental agents of disease. Such agents included psychological stressors, impoverished living conditions, physical toxins, infectious pathogens, and insufficient or malevolent parenting. Prevention and treatment were taken to mandate alterations in these causative environmental exposures. Thus, schizophrenia was viewed as the product of psychological "double-binds" within dysfunctional family units, autism was regarded as the legacy of cold, distant mothers, and maternal overprotectiveness figured prominently in the presumed etiology of childhood anxiety disorders.

Within a second canonical view, emerging at the pinnacle of the first by virtue of a revolution in molecular biology, human disorder was thought principally the result of biological frailties built into the structure of the heritable genome. Individual genetic differences, occurring on average in one of every thousand nucleotide base pairs, became seen as likely biological substrates for many human disorders, and quantitative trait loci (QTLs) were thought to constitute the multiple gene systems that, in isolation, code for variation in complex behavioral traits (Plomin & Crabbe, 2000). Most if not all disease, it was posited by some, would one day become explicable within the frameworks of evolutionary biology and human genetics, and eagerly anticipated "gene therapies" would transform the treatment and prevention of disordered biology. Thus, Alzheimer's disease would one day be accounted for by mutations in the gene coding for apolipoprotein E4; a polymorphism in the promoter region of the serotonin transporter gene would eventually elucidate anxiety disorders and phobias; and the long repeat allele of a dopamine receptor gene (DRD4) would explain the etiology of attention-deficit/hyperactivity disorder (ADHD) and other externalizing behavior problems. Within three decades' time, two opposing and mutually exclusive causal orthodoxies captured and held the high ground of scientific discourse on the genesis of human disease.

New Evidence for Biology-Context Interactions

Vulnerabilities in both positions first became evident as three sets of research findings emerged in the final years of the twentieth century. First, the nascent field of behavior genetics increasingly documented shared genetic and environmental accounts for variance in the incidence and severity of a broad array of behavioral and psychopathological disorders (Plomin, DeFries, & Loehlin, 1977; Plomin, DeFries, McClearn, & McGuffin, 2001). Using heritability statistics derived from genetically informative research designs, behavior geneticists became able to decompose phenotypic variance into genetic and environmental components. The genetic component was further subdivided into additive and nonadditive effects, and the environmental component was parsed into shared and nonshared influences (Goldsmith, Gottesman, & Lemery, 1997). Although the neuroscience of schizophrenia provided compelling evidence for a heritable disorder of the brain, epidemiological studies revealed that, even among monozygotic twins, the concordance rate for a diagnosis of schizophrenia never exceeded 50 percent. Schizophrenia was thus demonstrably a biological disorder with heritable components, but as much as half of the variance in its rate of occurrence was attributable to environmental exposures. The science of behavior and development thus faced a Kierkegaardian dilemma (Kierkegaard, [1843] 1986), in which an uncompromising "either/or" became a more illuminating and elegant "both/and." Schizophrenia, like virtually all forms of human behavior (Rutter, 2007), became understood as a product of *both* biological *and* contextual etiologies.

A second set of findings revealed the previously unsuspected bidirectional influences of biology and context, each on the other (Rutter, Moffitt, & Caspi, 2006). Whereas it had become apparent that genes affected both normative and disordered human behavior, new evidence indicated a) the heritability of environmental experience, and b) the potentially profound regulatory effects of social environmental exposures on the transcription of DNA. Discoveries of such bidirectional effects led to a recognition of genotype-environment correlations and to the categorization of such correlations into passive, reactive, and active forms (Rutter, 2006).

Studies by Plomin and colleagues demonstrated heritable influences on parental behavior and the home environment (Braungart, Fulker, & Plomin, 1992; O'Connor, Hetherington, Reiss, & Plomin, 1995), and other work showed that even so-called random misfortune, such as stressful events (Kendler, Neale, Kessler, Heath, & Eaves, 1993) and the loss of friends (Bergeman, Plomin, Pederson, McClearn, & Nesselroade, 1990), was partly attributable to genetic variation.

In a ground-breaking demonstration of environmental regulatory effects on gene expression, animal research by Meaney, Szyf, and colleagues (Meaney, Szyf, & Seckl, 2007; Szyf, Weaver, Champagne, Diorio, & Meaney, 2005; Weaver, Cervoni, Champagne, D'Alessio, Sharma, Seckl, Dymov, Szyf, & Meaney, 2004) showed that brief maternal separations of rat pups alter their mothers' licking and grooming behavior, leading to a downregulation of the pups' corticotropin-releasing hormone (CRH) system and diminished adrenocortical reactivity extending into adult life. The same research group further established that these alterations of hypothalamic-pituitary-adrenocortical (HPA) reactivity were attributable to epigenetic modifications of the glucocorticoid receptor (GR) gene, which plays a role in the moderation of HPA activation in the central nervous system. The offspring of mothers showing greater caretaking behavior had higher GR expression in the hippocampus due to demethylation of a promoter region of the GR axon. Thus, early experience, in the form of maternal caretaking behavior, in effect changed the pups' adrenocortical responsivity to stressors through epigenetic modification of gene expression.

Other research documenting social contextual effects on biological processes includes a study showing that disruption of social hierarchies in mice led to viral infection-related mortality, which was attributable to an overexpression of genes for cytokine proteins (Sheridan, Stark, Avitsur, & Padgett, 2000), and another revealing that caloric restriction prevented aging-related alterations in the expression of genes governing protein metabolism (Lee, Klopp, Weindruch, & Prolla, 1999). Although both environmental and biological determinism assumed impenetrable divisions between contexts and genes, new evidence revealed a capacity for the genome to influence environmental experience and for social and physical environments to switch on and off the decoding of genetic material (Robinson, Fernald, & Clayton, 2008). Social experience became a heritable predisposition, and the transcription of genes became a process governable by the character of social experience.

The third source of vulnerability in the twin dogmas of biological and environmental determinism was the more recent and accelerating emergence of biology-context interactions in studies of disease etiology. Much of the work examining biological and contextual contributions to pathogenesis was initially preoccupied with partitioning variance into the two distinctive categories of causal factors. Other studies began to illuminate how an *interplay* between biology and context serves a central etiologic role (Rutter, 2007; Rutter, this volume). Mednick et al. (Mednick, Gabrielli, & Hutchings, 1987), for example, found that a child's rearing by an adoptive parent with a criminal conviction did not increase the likelihood of criminality in the child, unless one of the child's biological parents had also been convicted of a crime. Similarly, a polymorphism in the promoter region of the gene encoding CRH was shown to moderate the association between maternal stress and preterm birth (Wang, Zuckerman, Pearson, Kaufman, Chen, Wang, Wise, Bauchner, & Xu, 2001). Self-perceived maternal stress was associated overall with pregnancies that were an average of 1.3 weeks shorter, but when a mother possessed a variant allele of the CRH gene, perceived stress was associated with a 2.9-week shortening. A third example is the finding by van Herwerden et al. (van Herwerden, Harrap, Wong, Abramson, Kutin, Forbes, Raven, Lanigan, & Walters, 1995) that a polymorphism in the $Fc_\varepsilon R1$-ß gene on chromosome 11q13 alters bronchiolar reactivity to house dust mites and environmental tobacco smoke. Although dust mites and smoke are known causes of airway inflammation and allergic symptoms (Palmer & Cookson, 2000), polymorphic markers within this gene are associated with the serum IgE (immunoglobin class E) production that is thought to mediate airway reactivity. Explorations of gene-environment interactions have also flourished in the field of environmental toxicology, where as many as 100 candidate genes have been identified with potential for metabolically disposing of carcinogens with known links to lung, brain, and bladder cancers (Albers, 1997). Even in so-called monogenic diseases such as phenylketonuria, phenotypic expression of the disease is highly variable, suggesting that modifier genes and environmental factors may influence disease severity and course (Scriver & Waters, 1999).

Most compelling of all, however, were the studies of Caspi, Moffitt, and colleagues on gene-environment interactions in the Dunedin Longitudinal Study. Noting the previously reported association of allelic variation in the promoter region of the serotonin transporter gene with risk for major depression (Lesch, Bengel, Heils, Sabol, Greenberg, Petri, Benjamin, Muller, Hamer, & Murphy, 1996), Caspi et al. (Caspi, Sugden, Moffitt, Taylor, Craig, Harrington, McClay, Mill, Martin, Braithwaite, & Poulton, 2003) examined the incidence of depressive symptoms, major depression, and suicidal ideation among individuals with varying numbers of stressful life events in early adulthood and with different polymorphisms in the serotonin transporter gene. They found a significant and substantial interaction, with stressful events predicting depression more strongly among individuals with one or two copies of the short promoter allele. Similarly and using the same longitudinal data, the genetic specificity of the depression finding was confirmed by the discovery of an additional interaction between childhood maltreatment and polymorphic variation in the monoamine oxidase A (MAOA) gene in the prediction of antisocial behavior in adulthood (Kim-Cohen, Caspi, Taylor, Williams, Newcombe, Craig, & Moffitt, 2006). Again, the significant risk of antisocial behavior was borne by the subsample with a specific functional variant in the MAOA promoter and severe childhood maltreatment. Common to each of these examples is a biogenetic vulnerability that amplifies the pathological effects of a social or environmental exposure. Given recent evidence that not all such studies yield similar effects, future research will need to focus on the identification of the specific pathways hypothesized (Risch, Herrell, Lehner, Liang et al., 2009).

Taken together, these three lines of evidence constituted a convincing empirical threat to a mono-deterministic view of pathogenesis. Whereas human morbidities may still be identified that have their origins in solely genetic or environmental causes, it appears highly plausible at present to anticipate that variation in the onset and course of most human disorders will be ultimately explained by interactions among biological and environmental forces.[1] It is now time to move past questions of how much of the

---

[1] Some would claim, for example, that morbidities such as phenylketonuria (PKU) or a bicycle accident are purely genetic or purely environmental, respectively. Just as PKU requires both a defective gene and dietary phenylalanine, however, "accidental" morbidities may be sometimes co-determined by both an environmental agent (e.g., a car) and a genetically derived predisposition to risk taking.

variation in outcomes is attributable to genes versus environment and into the more complex and potentially rewarding questions of genetic change and continuity, relations among dimensional symptoms and diagnosable psychopathology, and the interplay between biology and context (Plomin & Crabbe, 2000). As noted by Goldsmith et al. (Goldsmith, Gottesman, & Lemery, 1997) and Rutter et al. (Rutter, Moffitt, & Caspi, 2006; Rutter, this volume), there is pressing need for broader metaphors of gene-environment interaction that extend beyond "black box analyses" to hypotheses regarding specific interactive processes.

Impediments to Discovery

Given the difficulties of discovering biology-context interactions (Wachs & Plomin, 1991), it is all the more remarkable that such interactions are being increasingly identified and characterized in studies of human disease. Gene-environment interactions are assessed as the differential risk effects of an exposure among individuals with different genotypes, or as the differing effects of a genotype among those who are heterogeneous with respect to exposure (Kraemer, Stice, Kazdin, Offord, & Kupfer, 2001). As noted by McClelland and Judd (1993), field studies tend to underestimate the magnitude of interaction effects due to the nonoptimal distributions of the component variables. Interactions accounting for as little as 1 percent of outcome variance in field research may conceal a true proportion of variance (R-square [R^2]) many times that size that would be found using an optimal, experimental design. When demonstrated, biology-context interactions in human populations may disproportionately exist at the extremes of genetic and environmental variation (McGue & Bouchard, 1998). Animal studies are far more likely to reveal etiologic interactions because of their capacity both for true experimentation and for creating genotypes and exposures that are dimensionally more extreme. Indeed, animal work in three different laboratories demonstrated that environments may modify the expression of genotypically derived behavioral differences, even when aspects of context are rigorously controlled (Crabbe, Wahlsten, & Dudek, 1999). Mathematically, analyses of variance and related statistical tests often fail to identify nonadditive effects because they have much less power in tests of interactions than in tests of main effects (Wahlsten, 1990). Newer genetic research designs (Andrieu & Goldstein, 1998), along with statistical approaches that are more sensitive to interaction effects (Kraemer, Stice, Kazdin, Offord, & Kupfer, 2001), are now abetting the discovery of complex interactions between and among biologically and environmentally based factors.

Symphonic Causation and the Interaction of Biology and Context

What such methods are poised to uncover, it is argued here most centrally, is a kind of *symphonic causation*, a set of lawful regularities in the character and effects of biology-context interplay that might together comprise a foundational principle for understanding the origins of human disease. Our early forays into the territories of etiologic complexity seem to suggest that both necessity and insufficiency are the twin hallmarks of the biological and experiential factors that contribute to disease causation. It may plausibly be the case that few if any human disorders are the products of unilateral pathogenesis and that one of the core features of disease is its essential dependence upon a confluence of biologically grounded vulnerabilities and contextually based pathogenic influences. If so, then the causation of human morbidities must be viewed as symphonic, in the sense that their etiologies necessitate a coming together or a conjoining of diverse, categorically distinctive etiologic elements. Symphonic means "sounding together," a chordal synthesis by which two or more elemental sounds create, in their confluence, a new presence not derivable from the attributes of its parts. Disease may thus imply a lawful but unanticipated convergence of the interior self, in the form of heritable or acquired biological susceptibilities, with the exterior environment, in the form of pathogenic agents of disease.

It is essential to note that this argument is focused neither upon the statistical evaluation of cross-product interaction terms nor upon the benefits of multilevel analysis. As noted earlier, there are important mathematical impediments to the detection and characterization of interaction effects, and there is ground to be gained both in statistical approaches to evaluating potential interactions and in research designs that maximize their detection. Nonetheless, more is needed than simply better, more pliable, and diverse statistical tools. Rather, there is a deeper, more fundamental need for a theory of disease that incorporates observed invariances in the character of biology-context interactions. Similarly, although multilevel analysis (i.e., concurrent measurement at multiple levels of organizational complexity) is a heuristic and often revelatory tool, a different but complementary strategy is being advanced here. A search for invariances in biology-context interplay may indeed involve scientific analysis at very different levels of organization (e.g., the levels of gene and family environment), but the promise of such a search lies not in its implicit penetration of multiple organizational strata, but rather in its *binocularity*, that is, the conjoining of individual biology and shared context into a unitary image of disease and health.

There are at least two corollaries of such a vision of health and disease. The first is that the preservation of health and the genesis of disease are seen, within such a framework, as two facets of the same phenomenon rather than as distinctive and oppositional biological processes. Historically, for example, medicine and public health have been mutually self-regarded as separable and discontinuous fields, medicine addressing pathogenesis and the curative treatment of diseased individuals and public health expounding prevention and the safeguarding of population health. Adoption of a symphonic causation model, on the other hand, implies a reconciliation of medical and public health perspectives and invokes an explicit recapitulation of the ancient Greek conception of harmonious and disharmonious relations. Disease is the morbid conjunction of individual susceptibility and external threat, whereas health is a state of harmony that exists between individual and context. Thus, the prevention of disease occurrence becomes viewed as a stewardship of harmony, a preservation of the *symphonic* congruity between the needs and vulnerabilities of the individual and the provisions and perils of the environment.

A second corollary is the potentially nonstochastic character of such confluences of biology and context. If biology is capable of invoking and changing context, and if context is capable of altering the expression of individual biology, then the possibility arises for the production of autocatalytic, self-promoting feedback loops involving biology-context-biology chains of influence. Examples are the findings that adolescents with conduct disorder affiliate actively with other conduct-disordered peers (Dishion, Patterson, Stoolmiller, & Skinner, 1991) and that young people with internalizing symptoms show a predisposition to form relationships with others who are similarly affected (Hogue & Steinberg, 1995). An array of multiple alleles predisposing to social anxiety, as another example, may lead to a child's marginalization within early social groups, which may turn on or off the transcription of genes responsible for the regulation of affect, leading to the development of major depression. Such chains of influence could arguably account for the highly nonrandom distribution of disease and for the persistence and acceleration of morbidities within a subset of human populations. The most well-replicated finding in all of child and adult health services research is that a small minority of any given population, usually about 15 to 20 percent, sustains more than half the morbidity within the population and is responsible for the majority of health services utilization (Boyce, 1992; Starfield, Katz, Gabriel, Livingston, Benson, Hankin, Horn, & Steinwachs, 1984). This disproportionate disease burden, borne by a small subpopulation, has been found repeatedly, among the wealthy and the impoverished, for both industrialized and developing societies, in Western as well as Eastern cultures. It appears to be a universal reality that ill health accumulates and multiplies over time in a relatively small group of disproportionately affected individuals. One plausible account for such misfortune is the existence of positive feedback systems in the interplay of biology and context, in which biological susceptibilities invoke new environmental risks, which in turn induce new biological proclivities to disease, and so on. Such failures of homeostatic control might credibly account for a majority of morbidity, disability, and premature mortality within the human populations of the world. If so, a search for invariances in symphonic causation might reasonably lead to new insights into the sources and solutions for the afflictions of twenty-first-century societies.

SENSITIVITY TO CONTEXT: AN EMERGING EXEMPLAR OF BIOLOGY-CONTEXT INTERACTIONS

In the developmental psychobiology laboratory that I led with my colleague Abbey Alkon at the University of California, Berkeley, a particular form of invariance in the symphonic interaction between biology and context emerged from our work over the past ten years. In an effort to illuminate and understand the unevenness in childhood morbidities, we have studied the joint influences of naturally occurring adversities and psychobiological reactivity on biomedical and psychiatric endpoints in 3- to 8-year-old children. Stressors occurring in family, school, and community settings have been used to operationalize contextual adversities, and adrenocortical, autonomic, and immunologic responses to highly standardized laboratory challenges have been assessed to index psychobiological reactivity. A remarkably consistent set of findings, from prospective, quasi-experimental, and experimental research designs, has been revealed, in which highly reactive children evince either unusually high or distinctly low rates of morbidity, depending upon the character of the surrounding social environment. This particular lawful regularity in the patterns of biology-context interactions suggests that biological factors may regulate individual, organismic sensitivity to environmental influence.

Stress and Child Health

The social epidemiological research upon which our program of investigation was built is a broad expanse of findings, reported over thirty to forty years, indicating reliable associations between stressors and health in both children and adults. With stress alternately conceptualized as stressful life events (e.g., Miller, 1998), chronic adversities (e.g., Kelly, 2000), or the occurrence of a major natural or manmade disaster (e.g., Pynoos, Goenjian, & Steinberg, 1998), cohort studies with strong prospective designs have repeatedly documented unconfounded associations between stressors and health endpoints, including acute infections (e.g., Cohen, 1999), injuries (e.g., Boyce, 1996), chronic medical conditions (e.g., Boekaerts & Röder, 1999), psychopathological morbidities (e.g., Jensen, Richters, Ussery, Bloedau, & Davis, 1991), and health care utilization (e.g., Ward & Pratt, 1996). Such evidence has been extended to experimental designs, in which exposures to pathogenic agents of disease have been randomly allocated and prior psychological stressors shown predictive of disease onset and severity (Cohen, Hamrick, Rodriguez, Feldman, Rabin, & Manuck, 2002).

Despite the increasing rigor of such studies and the elegance of the more recent research designs, epidemiological findings have consistently indicated that stress-illness associations, although reliable and significant, are universally modest in magnitude, never accounting for more than about 10 percent of the variance in health outcomes. Such significant but modest associations suggest that contextual stressors may play a predisposing but not sufficient or directly etiologic role in pathogenesis. The findings further suggest that individual susceptibility to stressors may operate as an additional, perhaps interactive, factor in the pathways leading to disease or disorder.

Psychobiological Reactivity to Environmental Adversity

Originating in studies of risk factors for the development of coronary heart disease in adults, systematic investigation of individual differences in stress reactivity began with protocols examining heart rate and blood pressure changes during the completion of stressful laboratory-based tasks (Matthews, Weiss, Detre, Dembroski, Falkner, Manuck, & Williams, 1986). Broad variability in such reactivity was documented in early studies of adult subjects, and predictive relations were described with incident hypertension and coronary disease (Cacioppo, Berntson, Malarkey, Kiecolt-Glaser, Sheridan, Poehlmann, Burleson, Ernst, Hawkley, & Glaser, 1998). The work of Matthews and colleagues extended such research into studies of adolescents and young adults (Matthews, Woodall, & Allen, 1993; Matthews, Woodall, & Stoney, 1990), and Cacioppo, Berntson, and colleagues developed new methods for ascertaining not just integrative, "downstream" physiologic measures such as heart rate and blood pressure, but direct assessments of sympathetic and parasympathetic activation (Berntson, Cacioppo, Quigley, & Fabro, 1994; Cacioppo, Berntson, Sheridan, & McClintock, 2000). Adrenocortical responses to laboratory challenges were pursued by Gunnar and her colleagues using measures of salivary cortisol (Gunnar, 1987; Gunnar & Loman, this volume), and given the known regulatory effects of catecholamines and cortisol on immune targets, other researchers began examining cellular (Manuck, Cohen, Rabin, Muldoon, & Bachen, 1991) and humeral (Cohen, Miller, & Rabin, 2001) immunologic responses in standardized reactivity paradigms.

In our Berkeley laboratory, a series of studies was conducted employing adrenocortical, autonomic, and immunologic reactivity to a set of standardized social, cognitive, physical, and emotional challenges completed over a 20- to 30-minute protocol. As reported in detail elsewhere (Alkon, Goldstein, Smider, Essex, Kupfer, & Boyce, 2003; Boyce, 2007; Boyce & Ellis, 2005; Boyce, Quas, Alkon, Smider, Essex, & Kupfer, 2001; Ellis, Essex, & Boyce, 2005), individual differences commensurate with those found in adults have been identified in children's reactivity to stressors. In addition, developmental trends in the magnitude of such reactivity have been noted, and direct associations have been found with behavior problems indicating risk for internalizing and externalizing forms of developmental psychopathology. Children showing high magnitude reactivity to laboratory stressors appear to be characterized by predispositions to a shy, inhibited behavioral phenotype (Boyce, Barr, & Zeltzer, 1992; Kagan, 1994); precocious capacities for delay of gratification (O'Hara & Boyce, 2001); poorer recall of acutely stressful events (Stein & Boyce, 1997); subordinate positions in early social groups (Goldstein, Trancik, Bensadoun, Boyce, & Adler, 1999); and high risk for affective and anxiety-related behavioral symptoms and low risk for externalizing behaviors (Kraemer, Stice, Kazdin, Offord, & Kupfer, 2001).

Stress-by-Reactivity Interactions

Most pertinent to the interplay of biology and context, however, our studies have almost invariably found significant interaction effects between indicators of psychobiological reactivity and measures of social environmental adversity (Boyce, 1996; Boyce, Chesney, Alkon-Leonard, Tschann, Adams, Chesterman, Cohen, Kaiser, Folkman, & Wara, 1995). As revealed

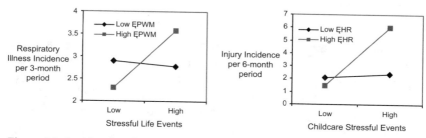

Figure 5.1. Incidences of respiratory illnesses and injuries by level of contextual stress and immunologic (left panel: pokeweed mitogen response) or cardiovascular (right panel: heart rate) reactivity. From Boyce, W. (1996). Biobehavioral reactivity and injuries in children and adolescents. In M. Bornstein & J. Genevro (Eds.), *Child Development and Behavioral Pediatrics: Toward Understanding Children and Health* (pp. 35–58). Mahwah, NJ: Erlbaum Associates. Reproduced with permission.

in Figure 5.1, such interactions display a crossover configuration in which rates of morbidity among low reactive children are essentially indifferent to the surrounding social context, whereas rates among high reactive children are highly dependent on contextual effects. Children with low laboratory reactivity have approximately the same incidences of infectious illnesses and injuries in both low- and high-stress settings. By contrast, children with high laboratory reactivity sustain inordinately high rates of illness and injury in stressful settings, but inordinately *low* rates in minimally stressful, more nurturant settings.

Although there are several credible accounts for the character of such reactivity-by-context interactions, the interpretation we have deemed most plausible is that high reactivity is an index of psychobiological sensitivity to social environmental influence. If autonomic and adrenocortical reactivity reflect heightened susceptibility to the nature of the social environment, then highly reactive children could be expected to sustain more of both the pathogenic effects of stressful, unsupportive social circumstances *and* the protective, salutary effects of highly supportive and predictable social settings. Highly reactive children would thus be expected to reveal, as they have in each of our studies, the best or the worst of observed health outcomes, depending upon the character of the ambient social environment. Such an interpretation is commensurate, moreover, with the work of Belsky and colleagues proposing differential susceptibility to both positive and negative social contexts (Belsky, 2005).

Replication and Extension

Importantly, neither of the prospective studies whose results are summarized here was experimental in design. Thus, reactive children in low stress settings were different from the reactive children in high stress settings; in neither study were the same high and low reactivity children exposed to both low and high stress contexts. Because of this inherent deficiency in research design, two subsequent studies employed quasi-experimental and experimental designs, in which low and high reactive individuals were observed under both stressful and supportive conditions. The first study examined injury rates before, during, and after a crowding stressor in a troop of semi-free-ranging rhesus macaques; the second was a randomized controlled trial of memory for an acute stressful event under supportive and unsupportive conditions.

The first replication was conducted in a troop of thirty-six macaques living in a 5-acre natural habitat on the grounds of the National Institutes of Health Primate Center in rural Maryland (Boyce, O'Neill-Wagner, Price, Haines, & Suomi, 1998). One year prior to the study, construction on the grounds of the habitat had imposed, for reasons of animal safety, a 6-month period of troop confinement in a small, 1,000-square foot building. The crowded conditions of the confinement produced severe stress in the troop, and the incidence of violent attacks and injuries escalated fivefold relative to pre- and postconfinement levels. The animals had been previously assessed, in many hours of individual and group observation, for individual differences in levels of biobehavioral reactivity. As has been observed more generally in rhesus monkeys (Suomi, 1997), approximately 15 to 20 percent of the animals showed a biobehavioral phenotype marked by predispositions to behavioral inhibition and exaggerated psychobiological reactivity to challenge and novelty. Blinded examination of veterinary records revealed the same form of interaction effect noted in the earlier human studies. Low reactive monkeys showed little if any escalation in injury rates during confinement, whereas highly reactive, inhibited monkeys sustained dramatic increases in injury rates, but strikingly lower rates before and after the stressor. Three of the monkeys died in the period during or immediately following the confinement; all three were high in biobehavioral reactivity. Thus, commensurate with observational results from human studies, the high reactivity individuals sustained unusually low rates of injuries in normative, low-stress conditions, but inordinately high rates under conditions of stress.

A remarkably similar result was obtained in a more recent, formally experimental study conducted in our laboratory by postdoctoral fellow

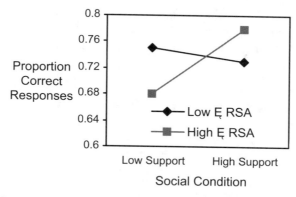

Figure 5.2. Proportion of correct responses on children's memory for stressful event by level of support and autonomic (respiratory sinus arrhythmia) reactivity.

Jodi Quas (Quas, Bauer, & Boyce, 2004). In this experiment, a sample of fifty-six 4- to 6-year-old children completed a standardized reactivity protocol in a laboratory setting, following which a fire alarm was putatively activated by steam during the preparation of hot chocolate with the child (see Stein & Boyce, 1997, for details of the fire alarm paradigm). The fire alarm episode was highly scripted, ensuring that each child experienced the same verbalizations, events, and sequence. During an interim 2-week period following the first laboratory session, the child's psychobiological data were scored and allocated into categories of high and low reactivity; children were then randomized, for the second session, into high and low support conditions, with stratification by reactivity status. High support consisted of a warm, engaging examiner who requested from the child both spontaneous and probed recall of the events surrounding the reactivity testing and the fire alarm episode; in the low support condition, the same examiner was cold and distant in manner. As shown in Figure 5.2, a significant interaction between reactivity and social condition was again detected. Low reactivity children showed approximately the same level of memory performance under both conditions. High reactivity subjects, on the other hand, revealed the highest memory performance of the sample under supportive conditions, but the lowest recall accuracy in unsupportive conditions. Results of the quasi-experimental study of injuries in rhesus monkeys and the randomized experiment examining memory effects in young children both confirm significant interactions between biology and context and suggest that reactivity to stressors indexes an openness or "permeability" to social environmental effects. Such findings further suggest a form of constancy in the character of biology-context interactions; that is, one recurring pattern in the interplay between individual biology and social context is that *biology controls sensitivity to environmental influence.*

SYMPHONIC CAUSATION: A PROVISIONAL NOSOLOGY
OF BIOCONTEXTUAL INTERPLAY

The Need for a Nosological Framework

The existence and importance of biology-context interactions in the genesis of human morbidities have become an article of faith. A 2000 report from the National Academy of Science said it especially well:

> ... it is time for a new appreciation of the coactivity of nature and nur-
> ture in development. Beginning at the moment of conception, hered-
> itary potential unfolds in concert with the environment. The dynamic
> interplay between gene action and environmental processes continues
> throughout life. (Shonkoff & Phillips, 2000)

Beyond this new canon of interactionism, however, there is a need for more systematic research and reflection on *regularities* within the character of biology-context interactions. It is the observation of such regularities that will begin to reveal deeper truths about *how* genes and environments co-determine human outcomes and *how* individual susceptibility and resilience are conjoined with contextual threat and protection to traverse the severe discontinuity between health and disease. A symphonic view of disease causation would not only acknowledge the bipartite contributions of biology and context, but would begin, as well, to codify the systematic congruities in their interactions.

The concept of a symphonic causation of disease, within a nascent "philosophy" of medicine, bears much in common with Percy's view of the triadic character of human language (Percy, 1989). In the same manner in which a conjunction of self and environment produces the experienced *Welt* or "world," the confluence of individual biology and environmental peril produces affliction and disease. Pathogenesis and prevention are thus intrinsically triadic; it is out of a symphonic fusion of individual biology and contextual influence that the possibilities for health or disease arise. Among the implications of such a view is a re-alliance with ancient Platonic and Hippocratic conceptions of health as a harmonious balance between self and society. Thus, the Hippocratic Corpus summarized a view of health as " ... the expression of a harmonious balance between the various components of man's nature (the four humors that control all of human activities) and the environment and ways of life" (cited in Dubos, 1965). Even the pre-Socratic philosopher Alcmaeon of Croton noted the interactive nature of "internal" and "external" causes:

> Disease occurs sometimes from an internal cause such as excess of heat or cold, sometimes from an external cause such as excess or deficiency of food. It may occur in a certain part, such as blood, marrow, or brain; but these parts also are sometimes affected by external causes, such as certain waters, or a particular site, or fatigue, or constraint, or similar reasons. Health is the harmonious mixture of the qualities. (translated by Miller, 1962)

Within such a rubric, health and disease came to be seen as the integrative product of the fit or concordance between individual biological proclivities and dimensions of environmental threat and support. What is now needed is a nosological framework that assembles and classifies observable invariances in the operation of symphonic causation.

A Provisional Nosology

The provisional five-part nosology presented here makes no claim to comprehensiveness in its description of such interactions, and indeed, emerging research blending assessments of both genetic and environmental risk will almost certainly reveal new varieties of interaction. There is also no implication, in the ordering of categories, that one form predominates over another. Rather, this provisional nosological framework is offered as a first effort to aggregate and order emerging observations of biology-context interactions. As such, the nosology is an attempt to advance the same project begun by Lewontin in a 1974 paper on the analysis of causes (Lewontin, 1974), by Meehl in a 1977 exposition on specific etiology and strong influence (Meehl, 1977), and by Kendler and Eaves in their 1986 paper on models for joint effects of genotype and environment (Kendler & Eaves, 1986). Particular but not exclusive attention has been paid here to interactions predicting developmental psychopathology.

Five forms of biology-context interaction are proposed. First, *potentiation* describes individual biology's amplification of the pathogenicity of contextual forces. Because each category describes a form of interaction, such potentiation implies statistical nonadditivity of effects and a situation in which individual biology affects the direction or strength of the relation between environmental factors and child outcomes (Baron & Kenny, 1986;

Kraemer, Stice, Kazdin, Offord, & Kupfer, 2001). A second form of interaction is *protection*, that is, that in which biology amplifies the protectiveness or salutogenic aspects of context. Again, nonadditivity is implied, but the effect of biology is to augment the essentially preventive character of the environmental factor. Third, *sensitization* is the capacity of biological factors to control susceptibility or "permeability" to environmental influence. This is the form most closely defining the character of psychobiological reactivity-by-social-environmental-adversity interactions, which were summarized in the prior section. Fourth, *specification* suggests that biology is capable of selecting the particular symptomatic expression of contextual pathogenesis. This is the interpretation of interaction promulgated by Steinberg and Avenevoli, in their paper noting the tendency for environmental risk to exert generic effects and biological susceptibility to narrow the range of possible disordered outcomes (Steinberg & Avenevoli, 2000). Finally, a fifth form of interaction, *facilitation*, is that in which biology controls exposures to contextual pathogenicity. Here, the role of individual biology is not one of altering the amplitude of environmental effects, but rather one of determining the presence or degree of exposure to pathogenic or protective contextual elements.

Ottman (1996) has also proposed a five-model nosology of gene-environment interactions, which bears little commonality with the categories presented here. Rather, her models comprise four possible combinations of genotype and exposure in terms of their individual effects on disease risk, along with a noninteractive model in which the genotype increases the production of a nongenetic risk factor (e.g., phenylketonuria). The impetus for formulation of Ottman's system, however, was entirely commensurate with that motivating the present discussion; that is, the evident need for an interim taxonomy under which categories of interaction could be usefully and perhaps heuristically organized.

Ottman's nosology also lucidly emphasizes a point important to both categorical systems, that is, the reciprocity between biological and contextual perspectives in describing a given interaction. In defining the character of interaction categories, each has been described in terms of biology's influence on the magnitude, direction, or expression of contextual effects. Although this convention serves to standardize the format of the nosological categories, it implies no precedence of biology over context, or the inverse, and each class of interaction can be interchangeably characterized in terms of contextual modification of biological effects. Thus, it is in the nature of biology-context interactions that "biology amplifies the pathogenicity of context" can be reciprocally interpreted as "context amplifies the pathogenicity of biology." In Wallen's description of sex differences in rhesus monkey behavior, for example, the interaction of prenatal hormones and social contexts could alternately, and with equal accuracy, be described either in terms of hormonal influences on sex-specific social-rearing effects or as social environmental shaping of prenatal hormonal effects (Wallen, 1996). In the section that follows, specific exemplars of biology-context interplay are explored, and examples within each interaction category are presented.

Examples of Biology-Context Interplay

A 1996 lecture by Robert Hinde offered an elegant model of the complexity inherent in interactions among environment, behavior, and biological state within a vertebrate animal species (Hinde, 1998). In order for the timing of nest completion to coincide precisely with the laying of eggs, the female canary engages in an elaborately unfolding "dialogue" among multiple interactive factors, including the physical environment of the nest, reproductive endocrine activity, and maternal behavior. Contextual factors acting on the HPA system result in gonad development and estrogen release, leading to a heightened responsiveness to the male's courtship behaviors. Changes in hormonal state, in interaction with day length and exposures to the male's song, also induce the construction of a nest and the development of a "brood patch," a tactilely sensitive area on the female's ventral surface that loses its feathers and develops increasing vascularity. Stimulation of the brood patch by the physical components of the nest successively initiate alterations in nest-building material from grasses to feathers, a diminution in nest-building behaviors, and, finally, further reproductive development of the mother bird. In an exquisite, sequentially rendered interaction among biology, context, and behavior, the nest's construction is completed just as the eggs are ready to be laid.

Biology-context interactions in human and nonhuman primate species are, at present, in a far more primitive state of elucidation than that portrayed by Hinde, but not surprisingly, early glimpses into the character of such interactions are revealing a picture of equally elegant and exacting complexity. For each of the five categories of interaction – potentiation, protection, sensitization, specification, and facilitation – examples are rapidly accruing, in both experimental studies of laboratory animals and observational studies of human subjects.

The first nosological category, potentiation, is a more general version of the diathesis-stress model of developmental psychopathology (Rosenthal,

1970), which posited that many disorders are the product of a vulnerability or predisposition combined with a set of external pathogenic conditions. Within this interactive model, biologically grounded susceptibilities accentuate or amplify the inherent pathogenicity of a given contextual exposure. In addition to the previously noted tendency for the criminality of biological parents to augment the risk engendered by the criminality of adoptive parents, studies by Cloninger (Cloninger, Bohman, & Sigvardsson, 1981) and Cadoret (Cadoret, Yates, Troughton, Woodworth, & Stewart, 1995) have documented the interactive roles of genetics and developmental environments in accelerating children's risk for alcoholism and drug dependency. Similarly, a positive family history of coronary heart disease has been shown to augment cardiovascular risk associated with smoking (Williams, Hopkins, Wu, & Hunt, 2000), and a particular allele of the dopamine receptor gene is known to potentiate cognitive decline in children experiencing severe family stress and adversity (Berman & Noble, 1997). Other examples include the effect of female sex on the link between psychological stressors and autoimmune thyroiditis (Yoshiuchi, Kumano, Nomura, Yoshimura, Ito, Kanaji, Ohashi, Kuboki, & Suematsu, 1998) and the amplification of dominance/subordination behaviors in specific social settings by the administration of amphetamines to rhesus macaques (Haber & Barchas, 1983).

The second form of interaction, protection, involves a biological amplification of a protective environmental factor. For example, the gene encoding cholesteryl ester transfer protein (CTEP) plays a key role in lipid (fat) metabolism. CTEP decreases the concentration of protective HDL-cholesterol (high-density lipoprotein), and patients with a deleterious mutation in the CETP gene have abnormally high HDL cholesterol. HDL cholesterol has also been shown to increase with alcohol consumption because of reduced activity of CTEP, but this is only the case among individuals with certain CTEP polymorphisms (Corbex, Poirier, Fumeron, Betoulle, Evans, Ruidavets, Arveiler, Luc, Tiret, & Cambien, 2000). Another example is the differential influences of gender on both the protective effects of social support (House, Landis, & Umberson, 1988) and the capacity for provision of such support. In one recent study by Glynn et al. (Glynn, Christenfeld, & Gerin, 1999), blood pressure reactivity to a stress-inducing speech-making task was significantly downregulated, in both men and women, by the presence of a supportive female audience, but no such effects were found for supportive male audiences. The authors conclude that the sex of supportive companions plays a key role in determining the utility and efficacy of the support provided.

Sensitization, the third category of biology-context interaction, represents an amalgam of potentiation and protection, except that the same biological factor is capable, through an increased sensitivity or permeability to the environment, of alternately raising or lowering risk. Such interactions are evident in the studies of Boyce and colleagues, reviewed earlier, on autonomic reactivity as an indicator of susceptibility to positive and negative environmental conditions of early childhood. Strikingly similar findings have been published by Wahlberg et al. (Wahlberg, Wynne, Oja, Keskitalo, Pykäläinen, Lahti, Moring, Naarala, Sorri, Seitamaa, Läksy, Kolassa, & Tienari, 1997), who demonstrated inordinately high or low rates of schizophreniform thought among the adopted children of schizophrenic mothers, depending on the level of disordered family communication, and Cadoret et al. (1996), who found exceptionally high or exceptionally low rates of depression in adoptees of alcoholic biological parents, depending on the psychiatric health of the adoptive parents. Finally, Stephen Suomi presented cross-fostering studies among rhesus monkey infants showing that highly reactive infant macaques, when placed in the care of maximally nurturant adult monkeys, evince a dramatic reversal of developmental course, rising to the top of the dominance hierarchy in the adult troop, becoming developmentally precocious, and, if male, being among the first to emigrate out of the natal troop (Suomi, 1987). Evidence for sensitization interactions suggest that biological, perhaps genetic, factors can calibrate the organism's susceptibility not only to health-diminishing adversities but also to aspects of health-preserving social supports.

Specification is the fourth form of biology-context interaction. In this form, environmental exposures raise generic, global risk for disorder, and biological or genetic factors focus that risk on to specific forms of morbidity. Context thus produces undifferentiated liability, whereas biology narrows the liability to particular somatic, emotional, or behavioral categories. An example is the study of Winokur and colleagues (Winokur, Cadoret, Dorzab, & Baker, 1971) showing that the relatives of individuals with early-onset major depression were more likely to develop psychopathology and that the form of disorder was more likely to be depression among females and antisocial behavior among males. Thus, common family environments produced an elevated risk for mental disorder, but the form of disorder was determined by the sex of the individual. In a second example, Butzlaff et al. (Butzlaff & Hooley, 1998) showed in a meta-analysis that expressed emotion in the family environment is associated with relapse in affective disorders, eating disorders, and schizophrenia, but that each form of disorder had its particular biological substrate.

The fifth and final pattern of interaction in this provisional nosology is facilitation, in which biology implements or controls exposures to contextual exposures. Perhaps the most persistent and tragic example of this interactive regularity is the manner in which genes regulating the melanization of skin determine the likelihood of an individual's chronic exposure to social adversities in North American society (Williams, 1997). Other examples include the possible role of serotonin transporter gene polymorphisms in the presumed selective emigration of rhesus macaques across the Himalayans into China (Champoux, Higley, & Suomi, 1997; Heinz, Higley, Gorey, Saunders, Jones, Hommer, Zajicek, Suomi, Lesch, Weinberger, & Linnoila, 1998; personal communication, Suomi, 2001) and the role of the iron absorption gene HFE in the development of hemochromatosis (Cogswell, McDonnell, Khoury, Franks, Burke, & Brittenham, 1998). Genes and biology may thus interact with environmental conditions and pathogens by fostering or facilitating an individual's exposure to such conditions.

Taken together, these categories of interactive, symphonic causation trace a perimeter of means by which biology and context act together in pathogenesis and prevention. Although there are both experimental and statistical obstacles to the discovery of such interactions, researchers are progressively assembling a repository of examples, some of which are offered here, that collectively present an early but informative picture of how genes and environments work together in causing disease and preserving health. That picture, coming gradually into view, tells a story with vast and immediate implications for the future of health care, biomedical research, and public health policy.

THE WAY FORWARD: OUT OF THE WOODS AND INTO THE WILDERNESS

Implications: Care, Research, and Policy

The implications of biology-context interactions for the future of human health are legion. Advances in our capacity for sampling, analyzing, and employing genetic material in research and practice will arguably revolutionize not only the science of human biology but also the science of human behavior. The capacity for assessing multiple genes using microarray patterns of DNA or RNA sequences will alter fundamentally the ease and practicality of ascertaining human genotypes (Lee & Lee, 2000). DNA sequence variations comprising single-nucleotide polymorphisms (SNPs), which occur once every 1,000 base pairs between two individuals picked at random, will offer the genetic means of "fingerprinting" genotypic disparities that form the most basic biological substrate of disease (Evans, Muir, Blackwood, & Porteous, 2001). These advances thus offer the means of fulfilling what Bronfenbrenner and Ceci have proposed as the need to study proximal processes of organism-environment interaction in order to understand the development of phenotypes (Bronfenbrenner & Ceci, 1994). As noted by Reiss (Reiss, Plomin, & Hetherington, 1991), this new genetic knowledge will advance understanding not only of biological pathogenesis, but of human contexts and social environments, equally and as well. As researchers become increasingly capable of taking genetic variation into account, the roles of social and physical contextual factors in disease will achieve unprecedented clarity. Although much has been achieved by studying social processes in their own right, eliminating the "noise" of biological variation will produce a level of resolution and focus in studying context never previously attained or imagined.

Technological breakthroughs in the ascertainment of genetic variation will also facilitate new understandings of biology-context interactions. Much of the difficulty in discovering and analyzing such interactions has been due to the imprecision inherent in using multiply determined, "downstream" biological measures that are only distantly reflective of the genes that are their sources. The ability to assess DNA polymorphisms directly will allow not only greater precision in the characterization of genetic variation but will also expedite the choice of more extreme genetic representation in research designs such as case-control studies. The implications of more rigorous study of gene-environment interactions are vast. Wilmoth and colleagues have demonstrated, for example, that contrary to prevailing assumptions, the human life span may not be fixed and, in fact, has been slowly advancing for more than a century (Wilmoth, Deegan, Lundström, & Horiuchi, 2000). Wildner noted a similar phenomenon in drosophila and suggested that the slowing of mortality among the older age groups of both species may be related to a favorable gene-environment interaction (Wildner, 2000). Maximizing the "fit" between individual genotypes and the environments that most favor health and well-being could thus have immense consequences for the public health of the world.

Among the implications of greater knowledge of biology-context interactions is the opportunity to construct surveillance processes and interventions that are specifically targeted to individuals at greatest biological risk for a given category of disease (Khoury, Burke, & Thomson, 2000b). Within the field of developmental psychopathology, for example, knowledge of the gene polymorphisms associated with risk for anxiety and affective disorders could potentially lead to school-based interventions capable of preventing entry onto trajectories toward such disorders. In another example, Omenn (2000) pointed out that persons with a genetic trait that decreases the catabolism of nicotine were less likely to be tobacco dependent, suggesting the possibility of preventive interventions targeting the metabolically at-risk individuals. It is important to note that, as argued by Turkheimer (1998), the establishment of genetic contributions to disease etiology does not necessitate genetic, or even biological, interventions. In Suomi's studies, cited earlier, the provision of expert, highly nurturant foster care for vulnerable young monkeys created dramatically beneficial effects, even though the means of identifying such monkeys involved sophisticated biological assays of stress response. The potential for psychosocial and societal intercessions into human adversities could paradoxically be served in unparalleled ways by the study of interactions between biology and context.

Far from a "yellow brick road" into ascendant knowledge, the path we are now following is rife with new perils and ethical conundrums. Khoury et al. (Khoury, Burke, & Thomson, 2000a) have reviewed comprehensively the scope of such issues that science and human societies must now face, highlighting the dilemmas surrounding privacy and confidentiality, fair use of genetic information, and the prospects for individual and group stigmatization. An outstanding precedent for attending to and addressing such dilemmas is the Ethical, Legal, and Social Implications (ELSI) Program that was established as an integral component of the Human Genome Project (National Human Genome Research Institute, 2005). As new research on gene-environment interactions proceeds, it will be critical for universities and agencies at both national and local levels to follow the Human Genome Project's responsible lead by addressing forthrightly the human and ethical problems that necessarily accompany such research.

CONCLUSION

Despite the difficulties, both philosophical and technical, that will surely attend the continuing study of biology-context interaction, it is increasingly clear that such interactions will be among the keys that unlock one of the most ancient and compelling of human mysteries, the etiology of disease and disorder. Having begun the process of documenting and cataloguing the range of biology-context interactions, it is now time to explore symphonic causation: the patterned, intrinsic regularities in *how* biology and context work together, and what genes and environments *do* in their conjoint elicitation of disease. In seeking to understand the complexity of developmental psychopathology, no problem is more salient and no solution more suffused with potential and hidden rewards.

REFERENCES

Albers, J. W. (1997). Understanding gene-environment interactions [news]. *Environ Health Perspect*, 105(6), 578–80.

Alkon, A., Goldstein, L. H., Smider, N., Essex, M., Kupfer, D., & Boyce, W. T. (2003). Developmental and contextual influences on autonomic reactivity in young children. *Dev Psychobiol*, 42(1), 64–78.

Andrieu, N., & Goldstein, A. M. (1998). Epidemiologic and genetic approaches in the study of gene-environment interaction: an overview of available methods. *Epidemiol Rev*, 20(2), 137–47.

Baron, R., & Kenny, D. (1986). The moderator-mediator variable distinction in social psychological research: Conceptual, strategic, and statistical considerations. *J Person Soc Psych*, 51(6), 1173–82.

Belsky, J. (2005). Differential susceptibility to rearing influence: An evolutionary hypothesis and some evidence. In B. J. Ellis & D. F. Bjorklund (Eds.), *Origins of the Social Mind: Evolutionary Psychology and Child Development* (pp. 139–63). New York: Guilford.

Bergeman, C. S., Plomin, R., Pederson, N. L., McClearn, G. E., & Nesselroade, J. R. (1990). Genetic and environmental influences on social support: The Swedish Adoption/Twin Study of Aging (SATSA). *J Gerontol*, 45, 101–106.

Berman, S. M., & Noble, E. P. (1997). The D2 dopamine receptor (DRD2) gene and family stress: Interactive effects on cognitive functions in children. *Behav Genet*, 27, 33–43.

Berntson, G. G., Cacioppo, J. T., Quigley, K. S., & Fabro, V. T. (1994). Autonomic space and psychophysiological response. *Psychophysiology*, 31(1), 44–61.

Boekaerts, M., & Röder, I. (1999). Stress, coping, and adjustment in children with a chronic disease: A review of the literature. *Disabil Rehabil*, 21(7), 311–37.

Boyce, W. T. (1992). The vulnerable child: New evidence, new approaches. *Adv Pediatrics*, 39, 1–33.

Boyce, W. T. (1996). Biobehavioral reactivity and injuries in children and adolescents. In M. H. Bornstein & J. Genevro (Eds.), *Child Development and Behavioral Pediatrics: Toward Understanding Children and Health* (pp. 35–58). Mahwah, NJ: Erlbaum Associates.

Boyce, W. T. (2007). A biology of misfortune: Stress reactivity, social context, and the ontogeny of psychopathology in early life. In A. Masten (Ed.), *Multilevel Dynamics in Developmental Psychopathology: Pathways to the Future* (34th ed., pp. 45–82). Minneapolis, MN: University of Minnesota.

Boyce, W. T., Barr, R. G., & Zeltzer, L. K. (1992). Temperament and the psychobiology of childhood stress. *Pediatrics*, 90(3), 483–90.

Boyce, W. T., Chesney, M., Alkon-Leonard, A., Tschann, J., Adams, S., Chesterman, B., Cohen, F., Kaiser, P., Folkman, S., & Wara, D. (1995). Psychobiologic reactivity to stress and childhood respiratory illnesses: Results of two prospective studies. *Psychosom Med*, 57, 411–22.

Boyce, W. T., & Ellis, B. J. (2005). Biological sensitivity to context: I. An evolutionary-developmental theory of the origins and functions of stressreactivity. *Dev Psychopathol*, 17(2), 271–301.

Boyce, W. T., O'Neill-Wagner, P., Price, C. S., Haines, M., & Suomi, S. J. (1998). Crowding stress and violent injuries among behaviorally inhibited rhesus macaques. *Health Psychol*, 17(3), 285–9.

Boyce, W. T., Quas, J., Alkon, A., Smider, N., Essex, M., & Kupfer, D. J. (2001). Autonomic reactivity and psychopathology in middle childhood. *Br J Psychiatry*, 179, 144–50.

Braungart, J. M., Fulker, D. W., & Plomin, R. (1992). Genetic mediation of the home environment during infancy: A sibling adoption study of the HOME. *Dev Psychol*, 28, 1048–55.

Bronfenbrenner, U., & Ceci, S. J. (1994). Nature-nurture reconceptualization in developmental perspective: A bioecological model. *Psychol Rev*, 101, 568–86.

Butzlaff, R. L., & Hooley, J. M. (1998). Expressed emotion and psychiatric relapse: A meta-analysis. *Arch Gen Psychiatry*, 55, 547–52.

Cacioppo, J. T., Berntson, G. G., Malarkey, W. B., Kiecolt-Glaser, J. K., Sheridan, J. F., Poehlmann, K. M., Burleson, M. H., Ernst, J. M., Hawkley, L. C., & Glaser, R. (1998). Autonomic, neuroendocrine, and immune responses to psychological stress: The reactivity hypothesis. *Ann NY Acad Sci*, 840, 664–73.

Cacioppo, J. T., Berntson, G. G., Sheridan, J. F., & McClintock, M. K. (2000). Multilevel integrative analyses of human behavior: Social neuroscience and the complementing nature of social and biological approaches. *Psychol Bull*, 126(6), 829–43.

Cadoret, R. J., Winokur, G., Langbehn, D., Troughton, E., Yates, W. R., & Stewart, M. A. (1996). Depression spectrum disease, I: The role of gene-environment interaction. *Am J Psychiatry*, 153(7), 892–9.

Cadoret, R. J., Yates, W. R., Troughton, E., Woodworth, G., & Stewart, M. A. (1995). Adoption study demonstrating two genetic pathways to drug abuse. *Arch Gen Psychiatry*, 52(1), 42–52.

Caspi, A., Sugden, K., Moffitt, T. E., Taylor, A., Craig, I. W., Harrington, H., McClay, J., Mill, J., Martin, J., Braithwaite, A., & Poulton, R. (2003). Influence of life stress on depression: moderation by a polymorphism in the 5-HTT gene. *Science*, 301(5631), 386–9.

Champoux, M., Higley, J. D., & Suomi, S. J. (1997). Behavioral and physiological characteristics of Indian and Chinese-Indian hybrid rhesus macaque infants. *Dev Psychobiol*, 31(1), 49–63.

Cloninger, C. R., Bohman, M., & Sigvardsson, S. (1981). Inheritance of alcohol abuse. Cross-fostering analysis of adopted men. *Arch Gen Psychiatry*, 38(8), 861–8.

Cogswell, M. E., McDonnell, S. M., Khoury, M. J., Franks, A. L., Burke, W., & Brittenham, G. (1998). Iron overload, public health, and genetics: evaluating the evidence for hemochromatosis screening. *Ann Intern Med*, 129(11), 971–9.

Cohen, S. (1999). Social status and susceptibility to respiratory infections. *Ann NY Acad Sci*, 896, 246–53.

Cohen, S., Hamrick, N., Rodriguez, M. S., Feldman, P. J., Rabin, B. S., & Manuck, S. B. (2002). Reactivity and vulnerability to stress-associated risk for upper respiratory illness. *Psychosom Med*, 64(2), 302–10.

Cohen, S., Miller, G. E., & Rabin, B. S. (2001). Psychological stress and antibody response to immunization: a critical review of the human literature. *Psychosom Med*, 63(1), 7–18.

Corbex, M., Poirier, O., Fumeron, F., Betoulle, D., Evans, A., Ruidavets, J. B., Arveiler, D., Luc, G., Tiret, L., & Cambien, F. (2000). Extensive association analysis between the CETP gene and coronary heart disease phenotypes reveals several putative functional polymorphisms and gene-environment interaction. *Genet Epidemiol*, 19(1), 64–80.

Crabbe, J. C., Wahlsten, D., & Dudek, B. C. (1999). Genetics of mouse behavior: interactions with laboratory environment. *Science*, 284(5420), 1670–2.

Dishion, T., Patterson, G., Stoolmiller, M., & Skinner, M. (1991). Family, school, and behavioral antecedents to early adolescent involvement with antisocial peers. *Dev Psychol*, 27, 172–80.

Dubos, R. J. (1965). *Man Adapting*. New Haven: Yale University Press.

Ellis, B. J., Essex, M. J., & Boyce, W. T. (2005). Biological sensitivity to context: II. Empirical explorations of an evolutionary-developmental hypothesis. *Dev Psychopathol*, 17(2), 303–28.

Evans, K. L., Muir, W. J., Blackwood, D. H., & Porteous, D. J. (2001). Nuts and bolts of psychiatric genetics: Building on the Human Genome Project. *Trends Genet*, 17(1), 35–40.

Glynn, L. M., Christenfeld, N., & Gerin, W. (1999). Gender, social support, and cardiovascular responses to stress. *Psychosom Med*, 61(2), 234–42.

Goldsmith, H. H., Gottesman, I. I., & Lemery, K. S. (1997). Epigenetic approaches to developmental psychopathology. *Dev Psychopathol*, 9(2), 365–87.

Goldstein, L. H., Trancik, A., Bensadoun, J., Boyce, W. T., & Adler, N. E. (1999). Social dominance and cardiovascular reactivity in preschoolers: Associations with SES and Health. *Ann NY Acad Sci*, 896, 363–6.

Gunnar, M. R. (1987). Psychobiological studies of stress and coping: An introduction. *Child Dev*, 58, 1403–7.

Haber, S. N., & Barchas, P. R. (1983). The regulatory effect of social rank on behavior after amphetamine administration. In P. R. Barchas (Ed.), *Social Hierarchies: Essays toward a Sociophysiological Perspective* (pp. 119–32). Westport, CT: Greenwood Press.

Heinz, A., Higley, J. D., Gorey, J. G., Saunders, R. C., Jones, D. W., Hommer, D., Zajicek, K., Suomi, S. J., Lesch, K.-P., Weinberger, D. R., & Linnoila, M. (1998). In vivo association between alcohol intoxication, aggression, and serotonin transporter availability in nonhuman primates. *Am J Psychiatry*, 155(8), 1023–8.

Hinde, R. A. (1998). Integrating across levels of complexity. In D. M. Hann, L. C. Huffman, I. I. Lederhendler, & D. Meinecke (Eds.), *Advancing Research on Developmental Plasticity: Integrating the Behavioral Science and Neuroscience of Mental Health* (pp. 165–73). Washington, DC: National Institute of Mental Health.

Hogue, A., & Steinberg, L. (1995). Homophily of internalized distress in adolescent peer groups. *Dev Psychol,* 31, 897–906.

House, J. S., Landis, K. R., & Umberson, D. (1988). Social relationships and health. *Science,* 241, 540–5.

Jensen, P. S., Richters, J., Ussery, T., Bloedau, L., & Davis, H. (1991). Child psychopathology and environmental influences: discrete life events versus ongoing adversity. *J Am Acad Child Adolesc Psychiatry,* 30, 303–9.

Kagan, J. (1994). *Galen's Prophecy.* New York: Basic Books.

Kelly, J. B. (2000). Children's adjustment in conflicted marriage and divorce: a decade review of research. *J Am Acad Child Adolesc Psychiatry,* 39(8), 963–73.

Kendler, K. S., & Eaves, L. J. (1986). Models for the joint effect of genotype and environment on liability to psychiatric illness. *Am J Psychiatry,* 143(3), 279–89.

Kendler, K. S., Neale, M., Kessler, R., Heath, A., & Eaves, L. (1993). A twin study of recent life events and difficulties. *Arch Gen Psychiatry,* 50, 789–96.

Khoury, M. J., Burke, W., & Thomson, E. J. (2000a). Genetics and public health: A framework for the integration of human genetics into public health practice. In M. J. Khoury, W. Burke, & E. J. Thomson (Eds.), *Genetics and Public Health in the 21st Century: Using Genetic Information to Improve Health and Prevent Disease* (Vol. 40, pp. 3–23). Oxford: Oxford University Press.

Khoury, M. J., Burke, W., & Thomson, E. J. (Eds.). (2000b). *Genetics and Public Health in the 21st Century: Using Genetic Information to Improve Health and Prevent Disease* (Vol. 40). Oxford: Oxford University Press.

Kierkegaard, S. ([1843] 1986). *Either/Or* (S. L. Ross & G. L. Stengren, Trans.). New York: Harper & Row.

Kim-Cohen, J., Caspi, A., Taylor, A., Williams, B., Newcombe, R., Craig, I. W., & Moffitt, T. E. (2006). MAOA, maltreatment, and gene-environment interaction predicting children's mental health: New evidence and a meta-analysis. *Mol Psychiatry,* 11(10), 903–13.

Kraemer, H. C., Stice, E., Kazdin, A., Offord, D., & Kupfer, D. (2001). How do risk factors work together? Mediators, moderators, independent, overlapping, and proxy-risk factors. *Am J Psychiatry,* 158, 848–56.

Lee, C.-K., Klopp, R. G., Weindruch, R., & Prolla, T. A. (1999). Gene expression profile of aging and its retardation by caloric restriction. *Science,* 285, 1390–3.

Lee, P. S., & Lee, K. H. (2000). Genomic analysis. *Curr Opin Biotechnol,* 11(2), 171–5.

Lesch, K. P., Bengel, D., Heils, A., Sabol, S. Z., Greenberg, B. D., Petri, S., Benjamin, J., Muller, C. R., Hamer, D. H., & Murphy, D. L. (1996). Association of anxiety-related traits with a polymorphism in the serotonin transporter gene regulatory region. *Science,* 274(5292), 1527–31.

Lewontin, R. C. (1974). The analysis of variance and the analysis of causes. *Am J Hum Genet,* 26, 400–11.

Manuck, S. B., Cohen, S., Rabin, B. S., Muldoon, M. F., & Bachen, E. A. (1991). Individual differences in cellular immune response to stress. *Psychol Sci,* 2(2), 111–15.

Matthews, K. A., Weiss, S. M., Detre, T., Dembroski, T. M., Falkner, B., Manuck, S. B., & Williams, R. B. (1986). *Handbook of Stress, Reactivity, and Cardiovascular Disease*. New York: John Wiley & Sons.

Matthews, K. A., Woodall, K. L., & Allen, M. T. (1993). Cardiovascular reactivity to stress predicts future blood pressure status. *Hypertension*, 22, 479–85.

Matthews, K. A., Woodall, K. L., & Stoney, C. M. (1990). Changes in and stability of cardiovascular responses to behavioral stress: Results from a four-year longitudinal study of children. *Child Dev*, 61, 1134–44.

McClelland, G. H., & Judd, C. M. (1993). Statistical difficulties in detecting interactions and moderator effects. *Psychol Bull*, 114(2), 376–90.

McGue, M., & Bouchard, T. J., Jr. (1998). Genetic and environmental influences on human behavioral differences. *Ann Rev Neurosci*, 21(5), 1–24.

Meaney, M. J., Szyf, M., & Seckl, J. R. (2007). Epigenetic mechanisms of perinatal programming of hypothalamic-pituitary-adrenalfunction and health. *Trends Mol Med*, 13(7), 269–77.

Mednick, S. A., Gabrielli, W. F., & Hutchings, B. (1987). Genetic factors in the etiology of criminal behavior. In S. A. Mednick, W. F. Gabrielli, & B. Hutchings (Eds.), *The Causes of Crime: New Biological Approaches* (pp. 74–91). Cambridge, England: Cambridge University Press.

Meehl, P. E. (1977). Specific etiology and other forms of strong influence: Some quantitative meanings. *J Med Philosophy*, 2(1), 33–53.

Miller, G. (1962). Airs, waters, and places. *J Hist Med*, 17, 129–40.

Miller, T. W. (Ed.). (1998). *Children of Trauma: Stressful Life Events and Their Effects on Children and Adolescents*. Madison, CT: International Universities Press, Inc.

National Human Genome Research Institute. (2005). Ethical, legal, and social implications of human genetics research. Located at: http://www.genome.gov/10001618.

O'Connor, T. G., Hetherington, E. M., Reiss, D., & Plomin, R. (1995). A twin-sibling study of observed parent-adolescent interactions. *Child Dev*, 66, 812–29.

O'Hara, K., & Boyce, W. T. (2001). *Behavioral and psychobiological predictors of preschoolers' delayed approach during resistance to temptation*. Paper presented at the Biannual Meeting of the Society for Research in Child Development, Minneapolis, MN.

Omenn, G. S. (2000). Genetics and public health: Historical perspectives and current challenges and opportunities. In M. J. Khoury, W. Burke, & E. J. Thomson (Eds.), *Genetics and Public Health in the 21st Century: Using Genetic Information to Improve Health and Prevent Disease* (Vol. 40, pp. 25–44). Oxford: Oxford University Press.

Ottman, R. (1996). Gene-environment interaction: Definitions and study designs. *Prev Med*, 25, 764–70.

Palmer, L. J., & Cookson, W. O. C. M. (2000). Genomic approaches to understanding asthma. *Genome Res*, 10, 1280–7.

Percy, W. (1989, Summer). The divided creature. *Wilson Quarterly*, 13, 77.

Plomin, R., & Crabbe, J. (2000). DNA. *Psychol Bull*, 126(6), 806–28.

Plomin, R., DeFries, J. C., & Loehlin, J. C. (1977). Genotype-environment interaction and correlation in the analysis of human behavior. *Psychol Bull*, 84, 309–22.

Plomin, R., DeFries, J. C., McClearn, G. E., & McGuffin, P. (2001). *Behavioral Genetics* (Fourth ed.). New York: Worth Publishers.

Pynoos, R. S., Goenjian, A. K., & Steinberg, A. M. (1998). A public mental health approach to the postdisaster treatment of children and adolescents. *Child Adolesc Psychiatr Clin N Am*, 7(1), 195–210.

Quas, J. A., Bauer, A., & Boyce, W. T. (2004). Physiological reactivity, social support, and memory in early childhood. *Child Dev*, 75(3), 797–814.

Reiss, D., Plomin, R., & Hetherington, E. M. (1991). Genetics and psychiatry: An unheralded window on the environment. *Am J Psychiatry*, 148, 283–91.

Risch, N., Herrell, R., Lehner, T., Liang, K., Eaves, L., Hoh, J., Griem, A., Kovacs, M., Ott, J., Merikangas, K. (2009). Interaction between the serotonin transporter gene (5-HTTLPR), stressful life events, and risk of depression: A meta-analysis. *JAMA*, 301(23), 2462–71

Robinson, G. E., Fernald, R. D., & Clayton, D. F. (2008). Genes and social behavior. *Science*, 322(5903), 896–900.

Rosenthal, D. (1970). *Genetic Theory and Abnormal Behavior*. New York: McGraw-Hill.

Rutter, M. (2006). *Genes and Behaviour: Nature/Nurture Interplay Explained*. Oxford, UK: Blackwell Publishing.

Rutter, M. (2007). Gene-environment interdependence. *Dev Sci*, 10(1), 12–18.

Rutter, M., Moffitt, T. E., & Caspi, A. (2006). Gene-environment interplay and psychopathology: Multiple varieties but real effects. *J Child Psychol Psychiatry*, 47(3–4), 226–61.

Scriver, C. R., & Waters, P. J. (1999). Monogenic traits are not simple: lessons from phenylketonuria. *Trends Genet*, 15(7), 267–72.

Sheridan, J. F., Stark, J. L., Avitsur, R., & Padgett, D. A. (2000). Social disruption, immunity, and susceptibility to viral infection. Role of glucocorticoid insensitivity and NGF. *Ann NY Acad Sci*, 917, 894–905.

Shonkoff, J. P., & Phillips, D. A. (Eds.). (2000). *From Neurons to Neighborhoods: The Science of Early Child Development*. Washington, DC: National Academy Press.

Starfield, B., Katz, H., Gabriel, A., Livingston, G., Benson, P., Hankin, J., Horn, S., & Steinwachs, D. (1984). Morbidity in childhood: A longitudinal view. *N Engl J Med*, 310, 824–9.

Stein, N. L., & Boyce, W. T. (1997). *The role of individual differences in reactivity and attention in accounting for memory of a fire-alarm experience*. Paper presented at the Society for Research in Child Development Biannual Meeting, Washington, DC.

Steinberg, L., & Avenevoli, S. (2000). The role of context in the development of psychopathology: A conceptual framework and some speculative propositions. *Child Dev*, 71, 66–74.

Suomi, S. J. (1987). Genetic and maternal contributions to individual differences in rhesus monkey biobehavioral development. In N. Krasnagor (Ed.),

Psychobiological Aspects of Behavioral Development (pp. 397–419). New York: Academic Press.

Suomi, S. J. (1997). Early determinants of behaviour: Evidence from primate studies. *Br Med Bull*, 53(1), 170–84.

Szyf, M., Weaver, I. C., Champagne, F. A., Diorio, J., & Meaney, M. J. (2005). Maternal programming of steroid receptor expression and phenotype through DNA methylation in the rat. *Front Neuroendocrinol*, 26(3–4), 139–62.

Turkheimer, E. (1998). Heritability and biological explanation. *Psychol Rev*, 105(4), 782–91.

van Herwerden, L., Harrap, S., Wong, Z., Abramson, M., Kutin, J., Forbes, A., Raven, J., Lanigan, A., & Walters, E. (1995). Linkage of high-affinity IgE receptor gene with bronchial hyperreactivity, even in the absence of atopy. *Lancet*, 346, 1262–5.

Wachs, T. D., & Plomin, R. (Eds.). (1991). *Conceptualization and Measurement of Organism-Environment Interaction.* Washington, DC: American Psychological Association.

Wahlberg, K. E., Wynne, L. C., Oja, H., Keskitalo, P., Pykäläinen, L., Lahti, I., Moring, J., Naarala, M., Sorri, A., Seitamaa, M., Läksy, K., Kolassa, J., & Tienari, P. (1997). Gene-environment interaction in vulnerability to schizophrenia: Findings from the Finnish Adoptive Family Study of Schizophrenia. *Am J Psychiatry*, 154(3), 355–62.

Wahlsten, D. (1990). Insensitivity of the analysis of variance to heredity-environment interaction. *Behav Brain Sci*, 13(1), 109–61.

Wallen, K. (1996). Nature needs nurture: The interaction of hormonal and social influences on the development of behavioral sex differences in rhesus monkeys. *Horm Behav*, 30, 364–78.

Wang, X., Zuckerman, B., Pearson, C., Kaufman, G., Chen, C., Wang, G., Wise, P. H., Bauchner, H., & Xu, X. (2001). *Maternal stress, CRH genotype, and preterm birth.* Paper presented at the Pediatric Academic Societies Annual Meeting, Baltimore, MD.

Ward, A., & Pratt, C. (1996). Psychosocial influences on the use of health care by children. *Aust NZ J Pub Health*, 20(3), 309–16.

Weaver, I. C., Cervoni, N., Champagne, F. A., D'Alessio, A. C., Sharma, S., Seckl, J. R., Dymov, S., Szyf, M., & Meaney, M. J. (2004). Epigenetic programming by maternal behavior. *Nat Neurosci*, 7(8), 847–54.

Wildner, M. (2000). Gene-environment interaction and human lifespan. *Lancet*, 356(9247), 2103.

Williams, D. R. (1997). Race and health: Basic questions, emerging directions. *Ann Epidemiol*, 7, 322–33.

Williams, R. R., Hopkins, P. N., Wu, L. L., & Hunt, S. C. (2000). Applying genetic strategies to prevent atherosclerosis. In M. J. Khoury, W. Burke, & E. J. Thomson (Eds.), *Genetics and Public Health in the 21st Century: Using Genetic Information to Improve Health and Prevent Disease* (pp. 463–85). Oxford: Oxford University Press.

Wilmoth, J. R., Deegan, L. J., Lundström, H., & Horiuchi, S. (2000). Increase of maximum life span in Sweden, 1861–1999. *Science*, 289(5488), 2366–8.

Winokur, G., Cadoret, R., Dorzab, J., & Baker, M. (1971). Depressive disease: A genetic study. *Arch Gen Psychiatry*, 24(2), 135–44.
Yoshiuchi, K., Kumano, H., Nomura, S., Yoshimura, H., Ito, K., Kanaji, Y., Ohashi, Y., Kuboki, T., & Suematsu, H. (1998). Stressful life events and smoking were associated with Graves' disease in women, but not in men. *Psychosom Med*, 60(2), 182–5.

6

Understanding Within-Family Variability in Children's Responses to Environmental Stress

JENNIFER JENKINS AND ROSSANA BISCEGLIA

INTRODUCTION

Perhaps surprisingly, children can be exposed to very similar life experiences and yet they will be affected by these experiences in very different ways. In other words, similar environmental experiences do not result in children developing more similarly to one another. Several types of evidence suggest this. One type of evidence comes from twin studies in which it is possible to partition variance into genetic and environmental influence. Such studies show that once genetic effects have been controlled, siblings tend to be more dissimilar than similar on emotions and behavior (Plomin & Daniels, 1987). This is the case even though siblings are raised in the same home and exposed to, we assume, many of the same environmental influences. This suggests enormous variability in the ways in which individuals respond to environmental influences. Surprisingly, this is also the case at high levels of psychosocial adversity. We might think that being raised in a highly stressful environment would have an adverse effect on all children. It is clear, however, that this is not the case (Luthar, Cicchetti, & Becker, 2000). Even under highly adverse conditions such as living through wars in which loved ones are killed (Howard & Hodes, 2000) or being raised by parents with serious mental health problems (Jaffee et al., 2003; Niemi et al., 2004) there is still variability in children's responses to such stressors. For some, the exposure is associated with compromised development. For others, no evidence for behavioral or emotional compromise is evident. Characteristics of some children or the contexts in which they live allow them to cope more effectively with adverse events, such that development is not compromised.

These findings raise important questions about the ways in which environmental stressors affect children. Bronfenbrenner (1977) stressed the importance of conceptualizing development within a multilevel framework. He argued that children are influenced both directly and indirectly by environmental influences at different levels of their social context: the family, school, broader society, and so on. The family provides one example of a multilevel structure: as siblings are nested within the family some of their experiences are shared between siblings, whereas some are unique to individual siblings. Given that some family experiences are shared, sibling studies provide a particularly powerful design for investigating environmental influence. Why does one child in a family show an adverse reaction to a family event such as the parents divorcing, whereas another child seems unaffected?

Through a comparison of siblings' responses to influences shared by all siblings, we have a degree of control that we do not have in single-child-per-family designs. Furthermore, we can examine the relative importance of family-wide and child-specific influences in the development of children. Family-wide risks refer to risks to which all children in the family are exposed such as parental depression, poverty, and marital conflict. Child-specific risks refer to unique experiences of siblings including the quality of the parent-child relationship and relationships with peers. The relative importance of these factors at different levels of the child's environment and the ways in which they operate independently or in interaction with one another provide important insights into the ways in which adverse environments influence child behavior.

The goal of this chapter is to describe processes in families that contribute to sibling differences in development. We do this by presenting results from sibling studies that have been carried out in our laboratory. We examine four processes that speak to this issue. These processes include (1) differentiating between family-wide and child-specific influences, (2) examining the extent to which siblings are exposed equally to family-wide risks, (3) the role of environmental stress in exacerbating individual differences between siblings, and (4) contingent relationships that explain within-family vulnerability to stress.

THE VALUE OF THE MULTILEVEL SIBLING DESIGN IN UNDERSTANDING WITHIN-FAMILY VARIABILITY

The observation that siblings exposed to similar risks showed very different outcomes (Plomin & Daniels, 1987) has led to an important distinction between the objective and the effective environment (Turkheimer & Waldron, 2000). The objective environment refers to those characteristics of

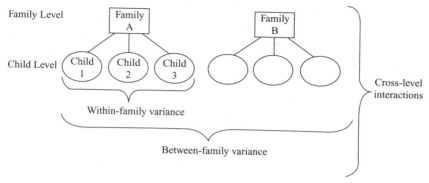

Figure 6.1. Data structure for multilevel, sibling studies.

people's lives that are directly observable and measurable and encompasses both family-wide and child-specific processes. Examples of this include measures of poverty, parental depression, and parental hostility directed toward children (Jenkins, 2008). The effective environment, on the other hand, is not directly observable. This refers to the *presumed* effect of the environment on children based on the extent of similarity between children sharing that environment (i.e., siblings). Statistical models that allow for the partitioning of variance into *within-* and *between-*family variance allow for the measurement of the "effective" environment as they result in an estimate of the degree of similarity shown by siblings (the intraclass correlation). Analytic techniques that allow for means and variances to be modeled as a function of measured predictor variables allow for the assessment of the objective environment. Multilevel or hierarchical linear models (Bryk & Raudenbush, 1993; Goldstein, 1995) are very useful in the analysis of family data because they allow for the simultaneous modeling of means and variance. This means that findings that pertain both to objective and effective environments can be derived from the same analysis.

In Figure 6.1, we illustrate the data structure of multilevel sibling studies. Imagine that we have scores on child behavior for multiple children per family in multiple families. The variance at the child level indicates the extent to which siblings in the same family vary from one another on their behavioral outcomes, and the variance at the family level indicates the extent to which families differ from one another on the family average of the outcomes in question. Higher variance estimates indicate greater *dissimilarity* within and between families. It is these variance estimates that provide data on the "effective" environment. Predictor variables are measured at the child and family levels, and these measurements provide data on the objective environment. We can use these measured predictor variables to explain mean differences between children and also differences in variance. For instance, a characteristic of a family such as family income (objective environment) may be associated with within-family variance (effective environment). This would be the case if siblings were more varied on their behavioral outcomes in poorer families.

DIFFERENTIATING BETWEEN FAMILY-WIDE AND CHILD-SPECIFIC INFLUENCES

A significant advantage of multilevel modeling for the study of risk and resilience is the ability to examine the impact of family-wide and child-specific processes simultaneously in families. We can determine the extent to which outcomes of childhood are attributable to processes that involve the whole family and how much they are attributable to processes that are specific to individual children. This is theoretically important for several reasons. First, as siblings have been found to develop very differently from one another, this may indicate that the child-specific environment plays a much larger role in child well-being than the shared environment. As other interpretations are also possible (e.g., children react to family-wide stressors in different ways), it is important to examine the relative importance of family-wide and child-specific predictors to differentiate between these explanations. Second, the holism idea in family systems theory states that the family environment is more than the sum of its parts (Minuchin, 1981; Nichols & Schwartz, 2001). Family systems theorists have argued that the whole family environment (as opposed to individual, dyadic, or triadic processes in the family that we refer to as component processes) is critical to child well-being. Translated into family-wide and child-specific, the holism concept refers to the idea that family-wide experiences (and their measurement) capture something that is not captured by taking account of component processes at the dyad- or child-specific levels. This has been a very difficult proposition to test because it can only be done if component processes lower down in the hierarchical structure of the family (dyads and individuals) are simultaneously accounted for. This necessitates a multi-level sibling design (otherwise family and child-specific measurement are confounded). When two or more children are included in the design, the family average captures the holism tenet, whereas the child-specific versions of the variables ensure that the lower level component processes are accounted for.

To illustrate the value of this, we draw on a study that investigated the predictors of negativity between siblings (Jenkins, Dunn, O'Connor, Rasbash, & Behnke, 2005) using data from the Avon Brothers and Sisters Study (ABSS). This study included families with two or more children. The sample was drawn from a much larger population study: the Avon Longitudinal Study of Parents and Children (ALSPAC). The target child was 4 years old at Time 1, and their siblings ranged between 6 and 17 years old. The sample consisted of 171 families; 71 families had three children, whereas 100 families had two children. Thus, 71 families had three sibling dyads, whereas 100 families had just one sibling dyad. Mothers were interviewed on two occasions, with a 2-year interval between visits, and reported on all measures. Sibling negativity at Time 2 was the outcome variable. The Time 1 scores of sibling negativity were entered into the model prior to any other predictor variables. Thus all subsequent predictors predict *change* in sibling negativity from Time 1 to Time 2. The use of the autoregressive design makes for a stronger argument with respect to causal process.

We use this study to explore within-family differences, but a word of clarification is in order. In this particular study the unit of analysis is not the child, but instead the dyadic sibling relationship. The nested structure of the data is the same as that presented in Figure 6.1. The difference is that the unit of analysis is the sibling relationship as opposed to an individual's behavior. All the arguments about the prediction of within-family variation remain the same, except that in this case within-family refers to between-dyad rather than between-individual variation.

The first question involved the breakdown of variance in families. Fifty-two percent of the variance was found to be at the family level. This demonstrated that sibling dyads within the family showed a considerable degree of similarity to one another on their negativity. This suggests several things. First, this high degree of clustering within families makes it likely that family-wide factors play an important role in the development of the sibling relationship. Second, dyad-specific influences must also be important, given that within-family variance accounted for 48 percent of the total variance. Thus, we went on to consider the dyad-specific and family-wide processes that might account for the quality of the sibling relationship.

Of particular interest was the construct of maternal negativity. In previous studies, it has been shown that when mothers direct negativity to their children, siblings in turn show more negativity to one another (Brody et al., 1992; Furman & Lanthier, 2002; Stocker, 1999). Previous studies, investigating only one sibling-dyad per family, have not been able to distinguish between the effects of ambient or family-wide maternal negativity, and negativity directed specifically at the dyad. Given the inclusion in the present study of multiple dyads per family, we were able to test the holism hypothesis. We calculated the family average for maternal negativity as well as each dyad's deviation from the average to take account of component processes.

The influence of other family-level (i.e., lone-parent status; Deater-Deckard et al., 2002) and dyad-specific characteristics, previously shown to explain sibling negativity, was also assessed. Dyad-specific characteristics included genetic relatedness between siblings in the dyad (Deater-Deckard et al., 2002), age of oldest child in the dyad (Buhrmester, 1992), age gap, the gender composition of the dyads (Dunn, 2002), and maternal differential treatment. Maternal differential treatment is the extent to which two children in a family are treated differentially. It has been found to be an important predictor of sibling negativity (see Brody, 1998, for review).

Results were as follows. First, the family average score of maternal negativity explained more variance in the change of sibling negativity than the dyad-specific score. This suggested that ambient maternal negativity was a more important predictor of sibling negativity than the dyad-specific exposure to maternal negativity. Given that dyad-specific maternal negativity had been controlled within the same model, this provided evidence in support of the holism hypothesis. Second, both family-wide and child-specific factors were important in explaining the variance in sibling negativity. Family-wide factors included lone-parent status and the family average maternal negativity. Dyad-specific characteristics included differential maternal negativity, genetic relatedness, the gender composition of the dyad, and the age of the oldest child in the dyad. These dyad-specific factors explained 12.5 percent of the within-family variance in sibling negativity.

These findings illustrate the importance of a multilevel sibling design for understanding the ways in which environments operate to influence children's development. Some influences are shared by siblings, whereas others are unique to individuals or dyads. One of the reasons why siblings differ so much in their behavior is because many aspects of their environment are unique. Here we demonstrated the range of dyadic experiences that have an impact on sibling negativity, including the degree of genetic relatedness of the dyad, differential treatment in the parent-child relationship, and the gender composition of the sibling dyad. Furthermore, we have illustrated the value of the multilevel sibling design for testing the holism hypothesis. After controlling for maternal negativity at the dyadic level, the family-wide measurement of maternal negativity was found to be a stronger predictor of sibling negativity than the dyad-specific measure.

ARE SIBLINGS EXPOSED EQUALLY TO FAMILY-WIDE RISKS?

The second issue to consider with respect to the enormous variability of children within families is the extent to which family-wide risks are really family-wide. Are all children exposed to the risk in question to the same degree? We explored the extent to which marital conflict was a family-wide or child-specific risk, using the ABSS data set described earlier (Jenkins, Simpson, Dunn, Rasbash, & O'Connor, 2005). We had three measures of marital conflict. One was measured at the family level. Mothers reported on the frequency of argument between herself and her husband on finance, recreation time, sex, friends, in-laws, conventional behavior, affection, and philosophy of life. All children received the same score for this variable because the construct measured a characteristic of the partner relationship held in common by all children. Argument about children was, in contrast, a child-specific measure. Mothers rated the frequency with which they argued with their partner about each child in the family. A third measure of marital conflict, also child-specific, assessed the frequency with which each child in the family was present in the room when parents argued. Because of the nature of the questions, it is possible to imagine that the latter two measures could show enormous variability across children. A more difficult child might be the source of a lot more argument between parents and might be likely to be in the room more when parents argued, *because* he or she played more of a role in generating the argument. The sample consisted of 137 biological and stepparent families who were cohabiting or married at Times 1 and 2. Teachers reported on children' externalizing behavior at both time points, using the Teacher's Report Form (Achenbach, 1991).

First, we examined the extent to which the two child-specific measures really represented child-specific phenomena. If these scores were to cluster in families to a high degree, irrespective of their measurement at the child-specific level (i.e., measures had been taken separately for each child), they would more appropriately be viewed as family-wide phenomena. Argument about children was the dependent measure in a multilevel model with "child" and "family" as levels. The data structure is the same as that described for Figure 6.1. The between-family variance represents the extent to which families differ from one another on the family average for arguing about children. The within-family variance represents the extent to which different children in the family are differentially argued about by their parents. The intraclass correlation coefficient demonstrates the extent of the sibling clustering (calculated as the between-family variance/total variance: between+within). Zero represents no clustering within families and 1.00 represents perfect clustering. The intraclass correlation for argument about children was found to be .49 (= .21/.21 + .22). This demonstrates substantial clustering but also substantial unique experience for individual children. Results for exposure to conflict were similar except the degree of clustering was greater, ICC = .57, than that seen for arguments about children.

The three measures of marital conflict were used to predict change over time in teachers' ratings of children's behavior problems. Argument about children was the marital conflict measure that most strongly predicted change in children's externalizing behavior. Neither the overall frequency of conflict nor children's exposure to conflict significantly predicted change in child behavior. Others have also found that the frequency of arguments that are focused on children is a better predictor of children's behavioral problems than a more general measure of marital conflict (Grych & Fincham, 1993). Note that argument about children in this analysis predicted the change in externalizing behavior from Time 1 to Time 2, so that the initial level of the child's externalizing behavior is controlled.

Taking these findings together, we can reflect on the extent to which the family-wide risk of marital conflict was really shared by all children in the family. Given the degree of clustering found for the child-specific measures of marital conflict, we can say that there is evidence for a family style of (1) arguing about children and (2) exposure of children to conflict. What do we mean by a family style? Given the extent of clustering, it is not just that particularly difficult children are the source of arguments between children. Rather, some parents show a tendency to argue about their children, which is a tendency expressed across multiple children in the family. However, it is also clear that some children elicit much more argument than other children given the large amount of within-family variation. This suggests a "child effect," an issue that we take up further later in the chapter. Given that child-focused arguments are more deleterious than arguments on other topics, and that there is variability on how much parents fight about different children in the family, we conclude that risks that at first glance appear to be shared by all children in the family, are not in fact shared. Children are being differentially exposed to family risks, in part accounting for sibling differences.

The Role of Child Effects in Family Risk

"Child effects" refer to the genetic and constitutional aspects of child behavior and personality that influence the caregiving experience that the child receives. Children evoke responses from those in their environment, and these responses, in turn, influence their development. In the study of marital conflict, as well as considering the effects of marital conflict on children, we also examined the effects of children on marital conflict (Jenkins et al., 2005). We found that children's externalizing behavior at Time 1 predicted a change in parental argument about children from Time 1 to Time 2. The parents of children with higher levels of externalizing behavior at Time 1 argued more about these more difficult children over time. This research shows us that child effects influence the expression of family risks, such that over and above the exposure to the family-wide risk, there is an added element of child-specific risk, in part attributable to a child effect. It is likely that individual children bear the brunt of certain risks because of elements of their behavior and personalities that augment their exposure to risk.

Let us imagine how such a process might play out for other family-wide risks. A child-specific element of poverty may be the way in which children make consumer demands on their parents. In parents who respond to child demands, if one sibling demands vociferously and another demands nothing, the less demanding child may be the recipient of many fewer economic resources than the demanding child. A similar scenario can be imagined for parental depression. If a depressed parent is responsive to child demands, then for a child who responds to the parent's depression by increasing their own demands, they may end up receiving more care than a child who withdraws in response to this stressor.

In conclusion, risk exposure constitutes a complex mixture of experiences that are both exogenous (e.g., the parents' tendency to argue, a parent's poverty, parental responsiveness to child demand) and endogenous to the child (demanding child behavior). Thus it is not, as some have argued, that family-wide experiences are unimportant to children's adjustment (Harris, 1998). It is more that children make their own contribution to the expression of these risks, such that some will be more directly and intensely exposed than others.

THE ROLE OF ENVIRONMENTAL STRESS IN EXACERBATING INDIVIDUAL DIFFERENCES BETWEEN SIBLINGS

The third process to consider in explaining within-family variability in adjustment between siblings is that environmental risks to which the whole family is exposed may operate to increase variability within families. As many behavioral genetic studies do not include direct measures of the environment, relatively few have examined whether the degree of sibling similarity changes as a function of environmental adversity. Two sets of findings bring this issue to mind, and at first glance the findings seem to contradict one another. On the one hand, as previously described, the behavioral genetic evidence suggests that siblings develop very differently compared to one another. On the other hand, substantial evidence from single-child-per-family designs shows that environmental risks affecting whole families, such as poverty (Costello et al., 2003) and being raised in poor neighborhoods (Caspi et al., 2000; Xue et al., 2005), have negative consequences for development. As these environmental studies have been done using sophisticated designs that allow for causal inference, including natural experiments, autocorrelation, genetic controls, and so on, it is clear that at least in some children, family-wide adversities do play a causal role in the development of behavioral problems.

Given these two sets of findings, one possibility is that because of preexisting individual differences, family members react differentially to family-wide stress exposure. Thus family members may spread out in their adjustment, becoming increasingly different from one another, when exposed to a family-wide risk. The stress-diathesis model provides an account of the ways in which individual vulnerabilities condition an individual's response to stress (Davidson & Neale, 2001). In the final section of the chapter, we review evidence for these individual vulnerabilities. Here, however, we describe findings that suggest that within-family variance does indeed increase with family-wide environmental adversity.

We start by describing a within-family study of maternal depression and children's externalizing behavior using Canada's National Longitudinal Survey of Children and Youth (NLSCY), Cycle 3 (Bisceglia & Jenkins, 2007).[1] This database is a nationally representative sample of Canadian children and includes multiple children per family. The subsample chosen for this study consisted of 2,439 children, including 382 sibling pairs, 7 to 11 years of age. For reasons described in the next section, only children who had a score on a reading test administered in school were included in the analysis. In the NLSCY, items related to emotional and behavioral problems were factor analyzed by Statistics Canada and six factors were derived (NLSCY, 1995). For this analysis, we used the factor indexing property-related, externalizing behaviors. This is an aggregate score based on six items. Parents were asked to rate, on a scale of 0 to 10, the extent to which their child (1) destroys their own things; (2) steals at home; (3)

---

[1] This analysis is based on Statistics Canada's National Longitudinal Survey of Children and Youth Cycle 3, which contains anonymous data. All computations on these microdata were prepared by Dr. Jenny Jenkins and Rossana Bisceglia. The responsibility for the use and interpretation of these data is entirely that of the author(s).

Table 6.1. *Fixed effects (top half) and within- and between-family variance estimates (bottom half) in the prediction of children's externalizing behavior. Standard errors are in brackets (N = 2,439)*

	Model 1	Model 2	Model 3	Model 4
Intercept	0.76 (0.025)	0.73 (0.023)	0.73 (0.022)	0.73 (0.022)
Family-level predictors				
Socioeconomic status		−0.04 (0.015)	−0.04 (0.015)	−0.04 (0.015)
Low-income-to-needs ratio		−0.04 (0.017)	−0.04 (0.016)	−0.04 (0.016)
Neighborhood safety		−0.03 (0.009)	−0.03 (0.009)	−0.03 (0.009)
Maternal depression		0.30 (0.071)	0.32 (0.093)	0.29 (0.093)
Child-specific predictors				
Parental hostility		0.14 (0.006)	0.13 (0.006)	0.13 (0.006)
Reading achievement		−0.03 (0.005)	−0.03 (0.005)	−0.02 (0.005)
Female		−0.11 (0.042)	−0.10 (0.041)	−0.10 (0.041)
Cross-level interaction				
Reading achievement-by-maternal depression				−0.06 (0.020)
Variances				
Between family	0.54 (0.072)	0.37 (0.056)	0.28 (0.053)	0.28 (0.053)
Within family	0.88 (0.068)	0.70 (0.054)		
Nondepressed-Within family			0.67 (0.052)	0.67 (0.052)
Depressed-Within family			1.16 (0.181)	1.10 (0.176)
−2*log like	7727.747	7062.145	6971.036	6962.343
Change in model fit (χ^2) from prior model to present, degrees of freedom		665.602 7	91.109 1	8.693 1

destroys things belonging to others; (4) tells lies or cheats; (5) vandalizes; and (6) steals outside of the home. Maternal depression was assessed using the Center for Epidemiology Depression Scale. Mothers scoring one standard deviation above the mean were designated as the depressed group (cutoff of nine). Thus maternal depression was dummy coded. The reference group was families in which the mother was not depressed. Data were analyzed using a two-level model with separate variance estimates within and between families.

Results can be found in Table 6.1. (As this dataset is also used to describe other processes in this chapter, results of Model 2 will be further explicated in the next section.) Results for family-level and dyad-specific predictor variables can be found in the top half of Table 6.1, and variance estimates are detailed in the bottom half. For the predictor variables, estimates and standard errors are interpreted as in a regression model. An estimate that is approximately twice the size of its standard error is significant ($p < .05$). As predictors are entered into the model, the between- and within-family variance estimates drop depending on whether the predictor variable explains variance within or between families (this is the value of including Model 1 in Table 6.1). Multilevel models are run sequentially. Each model is compared with the previous model (using change in the log likelihood) to determine whether the addition of the new parameter(s) improved the fit of the model. In the last line of the table, the change in the model fit from the previous to the present model is given. Model 1 is the model without predictor variables. This shows (a) the mean for children's externalizing behavior (see Intercept), and (b) the breakdown of between- and within-family variance from which the intraclass correlation can be derived. Predictor variables are entered in Model 2. In Model 3, we examine whether the within-family variance differs for families with and without a depressed mother. Model 4 includes a cross-level interaction term that is discussed in the section "Contingent relationships that index processes of vulnerability."

The results in Table 6.1 show that there is some similarity in externalizing behavior across siblings (intraclass correlation = .38). Children with depressed mothers showed higher levels of externalizing behavior than children with nondepressed mothers (see coefficient labeled "maternal depression" under "family-level predictors"). More important for the present discussion is the finding that the within-family variance on externalizing behavior is significantly higher for families with a depressed mother than families without a depressed mother (see "within-family variance, depressed mother"). We have plotted this effect in Figure 6.2. This suggests that certain children within families are more vulnerable to the deleterious effects of maternal depression than other children with respect to externalizing behavior. We follow up this finding in the next section as we try to identify the sources of siblings' increased vulnerability.

Two other findings from our research suggest increased within-family variance on children's behavioral problems as a function of family-wide stressors. In the study on the effects of marital conflict on children (Jenkins et al., 2005), we found that within-family variance on change in externalizing behavior (over a 2-year period) was greater when marital conflict was higher at Time 1. In another study, we examined the effects of parental separation on children (O'Connor & Jenkins, 2000). The design involved

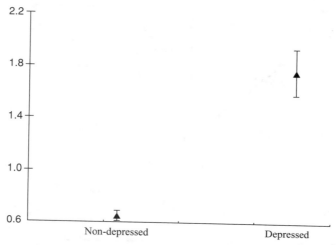

Figure 6.2. Within-family variance for families without and with a depressed mother.

a follow-up of children in the NLSCY seen at Time 1 and 2 (two years later. In this time period, most families remained married but some experienced separations, allowing us to examine changes in children's behavior within families as a function of separation. Separation was associated with an increase in children's emotional problems, as reported by both parents and teachers, as well as higher levels of within-family variation on teacher-reported, externalizing behavior.

We provide one further example of family-wide stressors being associated with increased within-family variability (Jenkins, Rasbash, & O'Connor, 2003). Instead of modeling children's behavioral problems, our dependent variable in this example is parenting. We were interested in the extent to which parents treated siblings differentially as a function of environmental stress. Differential parenting is an important process in families because it has been shown to have negative consequences for the child in a sibling pair who receives less positivity and more negativity (Conger & Conger, 1994; McGuire, Dunn, & Plomin, 1995). It has also been shown to be associated with higher levels of behavioral problems for the sibling group as a whole (Boyle et al., 2004). We hypothesized that low socioeconomic status, marital conflict, and large family size would increase the risk of differential positivity and negativity expressed by parents to their children. The sample consisted of 8,476 children in 3,860 families: 63 families had four children, 630 families had three children, and 3,167 families had

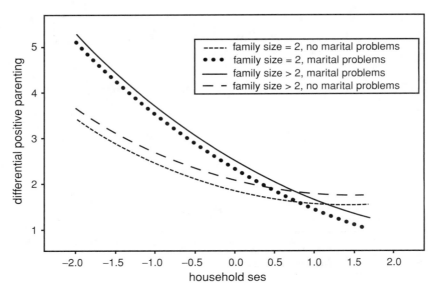

Figure 6.3. Differential parental positivity as a function of socioeconomic status, marital problems, and family size. From Jenkins, J. M., Rasbash, J., & O'Connor, T. (2003). The role of the shared context in differential parenting. *Developmental Psychology*, 39, 99–113. Adapted with permission.

two children between ages 4 and 11. The mother rated the extent to which her interaction with individual children was characterized by affection, hostility, and so on. We present data from results for positivity.

Data were analyzed in a multilevel model with two levels to the data structure: child and family. We allowed the within-family variance on parental positivity to vary as a function of three family-wide stressors (socioeconomic status, marital conflict, and family size) after taking into account a variety of child-specific factors known to be associated with positivity (see Jenkins et al., 2003, for a full description). Results are plotted in Figure 6.3. Higher values on the *y* axis denote higher levels of differential positive parenting. It is possible to see from this figure that the amount of differential parenting is higher in families with low socioeconomic status (SES) than in families with high SES. Both marital dissatisfaction and larger family size are also associated with higher levels of differential positivity. Furthermore, marital dissatisfaction is more strongly associated with differential parental positivity when SES is low rather than when SES is high. These findings show that variability in the parenting that children receive is higher when environmental adversity is high. There was also evidence that these stressors potentiated one another. Marital dissatisfaction was more strongly associated with differential positivity at low levels of SES than it was at high levels of SES.

In summary, we suggest that one of the ways in which family-wide stressors operate is to increase within-family variability. We see this with respect to two different family processes: child behavior and parenting. In both cases, it is likely that preexisting diatheses contribute to the patterns. If a child has a tendency toward externalizing behavior, this may get much worse when the environment is less supportive. If a parent has a greater attachment and interest in one child (Caspi et al., 2004), this may get more strongly expressed as financial and marital stressors increase. In the final section, we try to identify underlying diatheses that help to explain individual vulnerability.

CONTINGENT RELATIONSHIPS THAT EXPLAIN THE INCREASED VULNERABILITY OF CHILDREN WITHIN FAMILIES

If environmental stressors function to increase the variation among members of the family as proposed earlier, why does this happen? The stress-diathesis model suggests that individuals are vulnerable in idiosyncratic ways, both physically and psychologically. Preexisting individual vulnerabilities (diatheses) become exacerbated when the system is put under pressure. Within a family, exposure to the same stress will have a differential effect on different family members depending on the nature of these preexisting vulnerabilities. Thus, people who are shy by temperament may find themselves withdrawing from others when they are faced with environmental adversities. Those who do not have a tendency toward feeling shy may experience no increase in social withdrawal as a function of stress.

Individual differences in children's responses to stress have been widely investigated over the past thirty years (see Masten et al., 1999; Rutter, 1999). The designs of such studies tend to be naturalistic and correlational. Inevitably, overlapping and simultaneous influences occur in children's lives. The main issue in testing a vulnerability hypothesis is to isolate a diathesis for a specific stressor from all the other sources of variability in children's lives. Consider a circumstance such as children living in poverty. When we use a correlational design, our sample of children who are exposed to poverty are also differentially exposed to *co-occurring risks*. Thus, a child whose family is poor may also live in a violent neighborhood with a depressed mother and an alcoholic father. This child is exposed to poverty plus three other risks known to increase the likelihood of disturbance. The pattern of co-occurring risks has been found to be very important in explaining differential adjustment of children exposed to the same risk (Luthar et al., 2000; Masten & Coatsworth, 1998). Indeed, these co-occurring risks usually explain more variance in children's outcomes than the variance explained by the statistical interactions that we discuss later (Rutter & Pickles, 1991). However, although such co-occurring risks tell us a lot about differences between children in their adjustment, they do not tell us about why one child is more vulnerable than another to a specific stressor. In order to investigate this, it is essential to test the statistical interaction between the stressor and the hypothesized vulnerability factor, having accounted for all possible risks that might explain variation in children's outcomes.

The sibling design and a multilevel treatment of the data represent a valuable contribution to the study of vulnerability for three reasons. First, they enable a more accurate control of co-occurring risks, which are both child-specific and family-wide. This allows for increased specificity in isolating the contingent relationship between the specific stress and the vulnerability factor. Second, risks that are shared across siblings provide a particularly provocative test of the vulnerability hypothesis as multiple children in the family are exposed to the same risk. Third, if previous steps in the analysis have accounted for co-varying risks and an examination of the extent to which the risk is really shared (as discussed in the section, "Are siblings exposed equally to family-wide risks?"), then examining the cross-level interaction (see Figure 6.1) explains within-family variability to a family-wide risk.

We present several examples from sibling studies to illustrate the value of this design for understanding within-family variability to stress. In the previous section, we presented initial findings from a within-family study of maternal depression and children's externalizing behavior, showing that within-family variability on children's externalizing behavior increased at higher levels of maternal depression. In this section, we seek to explain this increase in within-family variance by examining vulnerability factors. In single-child-per-family studies, cognitive factors such as lower intelligence (Masten et al., 1999) and developmental delay (Werner & Smith, 1982) have been identified as vulnerability factors. Consequently, we hypothesized that lower reading achievement might make particular siblings more vulnerable to the effects of maternal depression than other siblings, such that they developed higher levels of externalizing behavior. Reading ability was assessed in the school, using questions from the Canadian Achievement Tests, Second Edition (CAT/2). Item Response Theory was used by

Statistics Canada to develop the reading score (Statistics Canada, 1996). For the present analysis, the reading score was divided by 10 prior to mean-centering to reduce the number of decimal points needed to accurately represent the data. Hostility in the parent-child relationship was derived from six items that measured the extent to which parents disciplined with anger, expressed disapproval of their children, offered praise to their children (reversed), and so on. Neighborhood safety was measured using an aggregate score from various questions that assessed the level of cohesiveness and safety in the neighborhood. The SES of the family was derived from five sources including parental education (spouse and person most knowledgeable [PMK] about the child), prestige of employment (spouse and PMK), and household income. All variables were grand mean-centered.

The first step in the analysis was to control for co-occurring risks at the child and family levels. Results of this can be seen in Model 2, Table 6.1. At the family level, we controlled for SES, family income (using the low-income-to-needs ratio), and the safety of the neighborhood in which the child lived. These variables explained 31 percent of the variance at the family level. At the child level, we controlled for gender, age, children's reading, as well as negativity in the mother-child relationship. All of these variables were found to raise the risk of showing externalizing behavior except children's age, which was nonsignificant and dropped from the model. It is possible to see that the within-family variance drops substantially (20 percent) when these child-specific attributes are taken into account. These variables explain differences between children within the same family, but they do not say anything (yet) about why one child is more vulnerable to the effects of maternal depression than is another child. In Model 3, we look specifically at the role of reading achievement by entering the interaction term of interest: maternal depression by reading achievement. This was significant, and the interaction is plotted in Figure 6.4. As one can see in that figure, lower reading achievement is associated with a greater vulnerability to externalizing behavior at high levels of maternal depression than at low levels of maternal depression. This interaction term explains 5 percent of the within-family variance in families characterized by maternal depression.

In another study done in our laboratory using the ABSS (Frampton, Jenkins, & Dunn, 2010) we examined the role of children's unique cognitive perspectives in explaining within-family differences in children's response to maternal depression. Cognitive perspectives have been identified as a vulnerability factor in single-child-per-family studies (e.g., Hammen, 1988). Our interest here was to extend these studies by examining whether unique cognitive perspectives would explain siblings' differential reaction to the

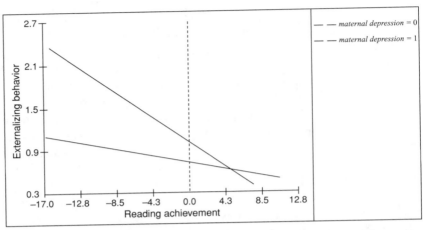

Figure 6.4. Children's externalizing behavior as a function of maternal depression and children's reading achievement.

family-wide stress of maternal depression. Thirty-two sibling pairs took part in the study. Each sibling had a score on unique cognitive perspectives based on a comparison of children's and mothers' reports on family relationships at Time 1. This was used to predict within-family differences in the development of internalizing problems over the next 2 years when children were exposed to maternal depression. Siblings who were more positive in their perspective on the family showed a reduction in their internalizing problems over time in spite of exposure to maternal depression. Children who were more negative in their perspective on the family showed an increase in internalizing disturbance, when exposed to maternal depression. Children who were not exposed to maternal depression showed very little change in their internalizing behavior from Time 1 to Time 2.

These two studies on the effects of parental depression on siblings are helpful in showing that exposure to the same environmental stress influences the behavior of siblings differentially based on the presence of pre-existing vulnerabilities. It is important to remember, however, that in the examples described here single outcomes were assessed. It may be that environmental stressors increase levels of disturbance in most children in a family, but they do so based on the preexisting behavioral organization of the child. Thus, one child may be vulnerable to an increase in aggression, another to an increase in internalizing behavior, and a third to a decrease in achievement at school. Capturing such variability in within-family response would require a hierarchical multivariate response model (Rasbash, Steele,

Browne, & Prosser, 2004; Thum, 1997), which was beyond the scope of the studies described in this chapter. Such models, however, will be very important in further understanding the ways in which family-wide stressors influence child behavior.

In the two studies described in this section, results for cognitive vulnerability factors were described. From results based on single-child-per-family designs, biological processes are also likely to be very important in explaining differential sibling vulnerability to stress. Biological vulnerabilities may be endogenous (e.g., genetic effects), or they may arise from uterine environmental adversities such as smoking (Fergusson, 1999) or nutrition (Liu, Raine, Venables, & Mednick, 2004) that affect the developing brain (Repetti, Taylor, & Seeman, 2002). Polymorphisms of genes that affect neurotransmitters have been found to increase vulnerability to stress for certain types of disturbance. Caspi et al. (2002) found that males with a particular form of the MAOA gene were at high risk for conduct disorder when they had also been exposed to maltreatment during childhood. The presence of this particular form of the gene and the exposure to maltreatment did not, on their own, raise the risk of disorder. In a second study on the same sample Caspi et al. (2003), found that individuals exposed to life events were more likely to develop depression if they had a short rather than a long form of a gene involved in the production of serotonin, although this finding has not always been replicated in other studies (Risch, Herrell, Lehner, Liang et al., 2009). This may suggest moderating influences that have not yet been identified. Vagal tone is another biological factor that has been shown to moderate the effects of stressful environments. For example, El-Sheikh (2005) found that for children with high vagal tone, exposure to parental alcohol abuse did not predict an increase in externalizing or internalizing problems. For children with low vagal tone, however, exposure to parental alcohol abuse was a predictor of an increase in both types of symptomatology over time.

In conclusion, cognitive and biological factor vulnerabilities of individuals within families leave them more susceptible to the adverse effects of family-wide stressors. Such processes are likely to explain the increased within-family variability associated with family-wide stressors.

CONCLUSIONS

The goal of this chapter was to illustrate the value of multilevel family studies for examining processes within families that explain sibling differences. We have identified four processes that contribute to understanding why siblings develop differently in spite of the fact that they are raised in similar environments.

There are methodological challenges to try to solve in the next generation of multilevel family studies. The most pressing are measurement issues. The questionnaire tends to be the measurement method of choice in multilevel family studies. This is because relatively large samples are needed to investigate the kinds of processes outlined in this chapter. Problems with single-informant data are well recognized. To ensure that correlations between predictor and outcome variables cannot be explained by the same person reporting on both, in our field we require that different informants report on each. The measurement of family and child-specific processes brings a new challenge, however. Most of the studies described in this chapter rely on maternal report for the assessment of family and child-specific processes. Thus, mothers report on the extent to which they argue with their partner about different children in the family, or the amount of affection that they give to different children. Child-specific and family-wide measurements derive from these scores. The effect of distortions that come from single-informant reports on family and child-specific predictors is not yet understood. It may be that some parents are insensitive to differences between siblings, whereas others exaggerate such differences, with both processes resulting in obfuscation of the true measurement of family-wide and child-specific processes. The use of multiple methods (e.g., observation and questionnaire) and multiple informants will be necessary to estimate the impact of such measurement issues. The value of short-term longitudinal data will also be considerable for the next generation of multilevel family studies. As we have identified, as well as family risks affecting children, children affect the development and expression of these risks. This makes it essential that we examine bidirectional processes between children and families. If data are naturalistic, these bidirectional processes are best identified by tracking changes in child effects and family risks over time. We must also recognize that these solutions represent an added measurement burden for families and increased expenditure for funding agencies.

As we move forward in understanding environmental influences on children, we think that the combination of biological and family process data is likely to hold particular promise. The last few years have shown the value of combining molecular genetic data with measures of environmental risk to examine gene-environment interaction (Caspi et al., 2002, 2003). Investigating the biological diatheses that come from early environmental experience (both pre- and postnatal) is also likely to be important for understanding within-family vulnerability to family-wide stressors (Rutter & O'Connor, 2004).

REFERENCES

Achenbach, T. M. (1991). *Manual for the Teacher's Report Form*. Burlington, VT: Department of Psychiatry, University of Vermont.

Bisceglia, R., & Jenkins, J. M. (2007, April). *Risk and Resilience within a Multilevel Framework*. Paper presented at the Society for Research in Child Development, Boston.

Boyle, M. H., Jenkins, J. M., Georgiades, K., Cairney, J., Duku, E., & Racine, Y. (2004). Differential-maternal parenting behavior: Estimating within- and between-family effects on children. *Child Development*, 75(5), 1457–76.

Brody, G. H. (1998). Sibling relationship quality: Its causes and consequences. *Annual Review of Psychology*, 49, 1–24.

Brody, G. H., Stoneman, Z., & McCoy, J.K. (1992). Associations of maternal and parental direct and differential behavior with sibling relationships: Contemporaneous and longitudinal analyses. *Child Development*, 63, 82–92.

Bronfenbrenner, U. (1977). Toward an experimental ecology of human development. *American Psychologist*, 32(7), 513–31.

Bryk, A. S., & Raudenbush, S. W. (1993). *Hierarchical Linear Models*. Newbury Park, CA: Sage.

Buhrmester, D. (1992). The developmental courses of sibling and peer relationships. In F. Boer & J. Dunn (Eds.), *Children's Sibling Relationships: Developmental and Clinical Issues*. Hillsdale, NJ: Lawrence Erlbaum.

Caspi, A., McClay, M., Moffitt, T. E., Mill, J., Martin, J., Craig, I. W., Taylor, A., & Poulton, R. (2002). Role of the genotype in the cycle of violence in maltreated children. *Science*, 297, 851–4.

Caspi, A., Moffitt, T. E., Morgan, J., Rutter, M., Taylor, A., Arseneault, L., Tully, L., Jacobs, C., Kim-Cohen, J., & Polo-Tomas, M. (2004). Maternal expressed emotion predicts children's antisocial behavior problems: Using MZ twin differences to identify environmental effects on behavioral development. *Developmental Psychology*, 40, 149–61.

Caspi, A., Taylor, A., Mofitt, T. E., & Plomin, R. (2000). Neighborhood deprivation affects children's mental health: Environmental risks identified in a genetic design. *Psychological Science*, 11(4), 338–42.

Caspi, A., Sugden, K., Moffitt, T. E., Taylor, A., Craig, I. W., Harrington, H., McClay, J., Mill, J., Martin, J., Braithwaite, A., & Poulton, R. (2003). Influence of life stress on depression: Moderation by a polymorphism in the 5-HTT gene. *Science*, 301, 386–9.

Conger, K., & Conger, R. (1994). Differential parenting and change in sibling differences in delinquency. *Journal of Family Psychology*, 8, 287–302.

Costello, E. J., Compton, S. N., Keeler, G., & Angold, A. (2003). Relationships between poverty and psychopathology. *Journal of the American Medical Association*, 290(15), 2023–9.

Davidson, G. C., & Neale, J. M. (2001). *Abnormal Psychology* (8th ed.). New York: Wiley.

Deater-Deckard, K., Dunn, J., & Lussier, G. (2002). Sibling relationships and social-emotional adjustment in different family contexts. *Social Development*, 11(4), 571–90.

Dunn, J. (2002). Sibling relationships. In P. K. Smith & C. H. Hart (Eds.), *Blackwell Handbook of Childhood Social Development. Blackwell Handbooks of Developmental Psychology* (pp. 223–37). Malden, MA: Blackwell Publishing.

El-Sheikh, M. (2005). Does vagal tone exacerbate child maladjustment in the context of parental problem drinking? A longitudinal examination. *Journal of Abnormal Psychology*, 114, 735–41.

Fergusson, D. M. (1999). Prenatal smoking and antisocial behavior. *Archives of General Psychiatry*, 56, 223–4.

Frampton, K., Jenkins, J. M., & Dunn, J. (2010). Within-family differences in internalizing behaviors: The role of children's perspectives of the mother-child relationship. *Journal of Abnormal Child Psychology*, 38, 557–68.

Furman, W., & Lanthier, R. (2002). Parenting siblings. In M. H. Bornstein (Ed.), *Handbook of Parenting* (2nd ed., Vol. 1, pp. 165–88). Mahwah, NJ: Lawrence Erlbaum.

Goldstein, H. M. (1995). *Multilevel Statistical Models*. London: Edward Arnold.

Grych, J. H., & Fincham, F. D. (1993). Children's appraisals of marital conflict: Initial investigations of the cognitive-contextual framework. *Child Development*, 64, 215–30.

Hammen, C. (1988). Self-cognitions, stressful events, and the prediction of depression in children of depressed mothers. *Journal of Abnormal Child Psychology*, 16(3), 347–60.

Harris, J. R. (1998). *The Nurture Assumption: Why Children Turn out the Way They Do*. New York: Free Press.

Howard, M., & Hodes, M. (2000). Psychopathology, adversity, and service utilization of young refugees. *Journal of the American Academy of Child & Adolescent Psychiatry*, 39, 368–77.

Jaffee, S. R., Moffitt, T. E., Caspi, A., & Taylor, A. (2003). Life with (or without) father: The benefits of living with two biological parents depend on the father's antisocial behavior. *Child Development*, 74, 109–26.

Jenkins, J. M. (2008). Psychosocial adversity and resilience. In M. Rutter, D. Bishop, D. Pine, S. Scott, J. Stevenson, E. A. Taylor, & A. Thapar (Eds.), *Rutters' Handbook of Child and Adolescent Psychiatry* (5th ed., pp. 377–91). Oxford: Blackwell.

Jenkins, J., Dunn, J., O'Connor, T. G., Rasbash, J., & Behnke, P. (2005) Change in maternal perception of sibling negativity: Within and between family influences. *Journal of Family Psychology. Special Issue on Sibling Relationships*, 19, 533–41.

Jenkins, J. M., Rasbash, J., & O'Connor, T. G. (2003). The role of the shared family context in differential parenting. *Developmental Psychology*, 39, 99–113.

Jenkins, J. M., Simpson, A., Dunn, J., Rasbash, J., & O'Connor, T. G. (2005). The mutual influence of marital conflict and children's behavior problems: Shared and nonshared family risks. *Child Development*, 76, 24–39.

Liu, J., Raine, A., Venables, P. H., & Mednick, S. A. (2004). Malnutrition at age 3 years and externalizing behavior problems at ages 8, 11, and 17 years old. *American Journal of Psychiatry*, 161, 2005–13.

Luthar, S. S., Cicchetti, D., & Becker, B. (2000). The construct of resilience: A critical evaluation and guidelines for future work. *Child Development*, 71, 543–62.

Masten, A., & Coatsworth, J. D. (1998). The development of competence in favorable and unfavorable environments. Lessons from research on successful children. *American Psychologist*, 53, 205–20.

Masten, A., Hubbard, J., Gest, S., Tellegen, A, Garmezy, N., & Ramirez, M. (1999). Competence in the context of adversity: Pathways to resilience and maladaptation from childhood to late adolescence. *Developmental Psychopathology*, 11, 143–69.

McGuire, S., Dunn, J., & Plomin, R. (1995). Maternal differential treatment of siblings and children's behavioral problems: A longitudinal study. *Development and Psychopathology*, 7, 515–28.

Minuchin, S. (1981). *Family Therapy Techniques*. Cambridge, MA: Harvard University Press.

Nichols, M. P., & Schwartz, R. C. (2001). *Family Therapy: Concepts and Methods*. Boston: Allyn & Bacon.

Niemi, L. T., Survisaari, H., & Huakka, J. K. (2004). Cumulative incidence of mental disorder among offspring of mothers with psychotic disorder: Results from the Helsinki High-Risk Study. *British Journal of Psychiatry*, 185, 11–17.

NLSCY. (1995). *Overview of Survey Instruments for 1994–1995*. Ottawa, Canada: Statistics Canada & Human Resources Canada.

O'Connor, T., & Jenkins, J. M. (2000). *Coping with Parents' Marital Transitions: Understanding Why Children in the Same Family Show Different Patterns of Adjustment*. Ottawa: Human Resources Canada.

Plomin, R., & Daniels, D. (1987). Why are children in the same family so different from each other? *Behavioural and Brain Sciences*, 10, 1–16.

Rasbash, J., Steele, F., Browne, W. J., & Prosser, B. (2004). *A User's Guide to MLwiN, Version 2.0*. (Chapter 14). London: Institute of Education.

Repetti, R. L., Taylor, S. E., & Seeman, T. E. (2002). Risky families: Family social environments and the mental and physical health of offspring. *Psychological Bulletin*, 128, 330–66.

Risch, N., Herrell, R., Lehner, T., Liang, K., Eaves, L., Hoh, J., Griem, A., Kovacs, M., Ott, J., & Merikangas, K. (2009). Interaction between the serotonin transporter gene (5-HTTLPR), stressful life events, and risk of depression: A meta-analysis. *JAMA: Journal of the American Medical Association*, 301(23), 2462–71.

Rutter, M. (1999). Resilience concepts and findings: Implications for family therapy. *Journal of Family Therapy*, 21, 119–44.

Rutter, M., & O'Connor, T. G. (2004). Are there biological programming effects for psychological development? Findings from a study of Romanian adoptees. *Developmental Psychology*, 40, 81–94.

Rutter, M., & Pickles, A. (1991). Person-environment interactions: Concepts, mechanisms, and implications for data analysis. In T. D. Wachs & R. Plomin (Eds.), *Conceptualization and Measurement of Organism-Environment Interaction* (pp. 105–41). Washington, DC: American Psychological Association.

Statistics Canada. (1996). *National Longitudinal Survey of Children and Youth, Cycle 2, User Guide*. Ottawa, Canada: Statistics Canada.

Stocker, C. M. (1999). Marital conflict and parental hostility: Links with children's sibling and peer relationships. *Journal of Family Psychology*, 13(4), 598–609.

Thum, Y. M. (1997). Hierarchical linear models for multivariate outcomes. *Journal of Educational and Behavioral Statistics*, 22, 77–108.

Turkheimer, E. & Waldron, M. (2000). Statistical analysis, experimental method, and causal inference in developmental behavioral genetics. *Human Development*, 43(1), 51–2.

Werner, E. E., & Smith, R. S. (1982). *Vulnerable but Invincible: A Longitudinal Study of Resilient Children and Youth*. New York: McGraw-Hill.

Xue, Y., Leventhal, T., Brooks-Gunn, J., & Earls, F. J. (2005). Neighborhood residence and mental health problems of 5- to 11-year-olds. *Archives of General Psychiatry*, 62(5), 554–63.

7

Origins, Development, and Prevention
of Aggressive Behavior

RICHARD E. TREMBLAY

What is aggressive behavior? When does it start? How does it develop? Can we prevent the development of chronic physical aggression? These are the four main questions that are addressed in this chapter. Research on the development and prevention of violent behavior is a growing industry. Paradoxically, the more humans become "civilized," the more they appear to be preoccupied by violence. The development of aggressive behavior is central to most theories of human behavior and is addressed by disciplines as varied as zoology, psychiatry, economy, psychology, public health, and political sciences.

WHAT DO WE MEAN BY "AGGRESSIVE BEHAVIOR"?

It may be a surprise that the major problem with this area of research is one of definition. Investigators have been putting their finger on this problem over and over again, but no simple solution has been found (Berkowitz, 1962; Burt, 1925; Buss, 1961; Cairns, 1979; Coie & Dodge, 1998; Hartup & de Wit, 1974; Parke & Slaby, 1983; Pitkanen, 1969). The best way to observe the difficulties faced by investigators is to examine the content of the "aggression" scales that have been used over the past decades. The content of the scales defines what is measured. The popular scales used to assess children's and adolescents' aggressive behavior contain a mix of behaviors that range from physical aggression to attention seeking and disobedience. One of the most frequently used scales in which parents rate "aggression" (Achenbach & Edelbrock, 1983) includes the following items: "argues, brags, demands attention, disobeys, poor peer relations, jealous, lies, shows off, stubborn, moody, sulks, loud." The common denominator of these items is that they describe annoying behavior. Should we classify annoying or irritating children in the same "aggressive" category as those who physically attack?

169

The most often-used[9] "aggression" scale for children and adolescents has twenty-three items with only two that clearly refer to physical aggression, and two others that could be interpreted as pertaining to physical aggression. Peer rating "aggression" scales have the same problems (Huesmann et al., 1984; Lefkowitz et al., 1977). Such scales are regularly used to identify either "aggressive," "externalizing," "conduct problem," or "antisocial" individuals in clinical practice. They have also been used to identify genetic influences on phenotypes that are alternatively referred to as "aggressive," "externalizing," "conduct disordered," or "antisocial" (Eley et al., 1999; O'Connor et al., 1998; Slutske et al., 1997; Arsenault et al., 2003).

One of the important problems with defining aggressive behavior is identifying its different forms of expression and differentiating them from other phenomena that are associated but different. For example, hyperactivity and opposition are highly associated with physical aggression (Nagin & Tremblay, 1999; Pulkkinen & Tremblay, 1992), but their aggregation cannot lead to a better understanding of the development of each of these types of behaviors or to a better understanding of their association.

Aggressive behavior and antisocial behavior are often aggregated because research on the development of aggressive behavior has concentrated mainly on behaviors that are socially undesirable. Is aggressive behavior always antisocial? Is the aim of research on aggressive behavior to prevent the development of aggressive behavior in the human species? These questions highlight the importance of clearly defining what we mean by aggressive behavior and aggression. Most parents would be proud to hear their child described as an aggressive tennis player. Most sales managers want aggressive salespeople. Most political parties want leaders who can be aggressive when needed. What do we consider an aggression?

The most frequently used definition for research on the development of human aggressive behavior over the last half of the twentieth century is a "moral" judgment approach where an observer decides that the behavior he or she observed was or was not intended to be "harmful to another person." However, although a major part of the research on children's aggressive behavior over the past two decades was inspired by a social learning approach, the majority of the studies did not rely on judges' attribution of intent to harm or injure. Most studies used ratings by adults or peers on items that described behaviors the investigator had *a priori* labeled "aggressive": hits, gives dirty looks, or fights (Nagin & Tremblay, 1999; Cairns et al., 1989; Crick & Grotpeter, 1995; Dodge et al., 1997; Huesmann et al., 1994). Direct observation studies were influenced by animal behavior studies and generally used behavior coding procedures that minimize raters' attribution of intent (Archer & Browne, 1989; McGrew, 1972; Patterson, 1982; Rubin et al., 1998; Strayer & Strayer, 1976). The "intent" criterion is especially problematic for understanding early aggressive behavior development. For example, Kagan (1974) argued "that a young child cannot be aggressive until he has some psychic intention of injuring another" (109). This "psychological" approach to aggression led him to the conclusion that "onset" of aggressive behavior was "well into the second year" after birth, when a child can "put himself into the psychic state of another." What would become of research on aggressive behavior in nonhuman animals if such a criterion were used? Does a dog intend to hurt when it bites another dog that tries to take its food away? Does a 12-month-old boy intend to hurt when he hits the peer who grabbed the toy in his hand? Anger and fear lead to reactions that are clearly not under the control of one's will even in mentally healthy adults. It could be argued that many, if not most, of the aggressive behaviors following intense frustration are impulsive behaviors, which were not "intended." Gray (1971, 1982) proposed that a "fight-flight system" controlled the behavioral reactions to unconditioned punishment and nonreward. Anger has clearly been shown to be expressed soon after birth (Lewis et al., 1990; Tremblay, 2008). These same reactions are expressed more clearly by the limbs of children a few months later when motor maturation enables the child to hit and kick.

The careful description of the development of aggressive behaviors during early childhood should help us understand the social and moral value of these behaviors. If we *a priori* decide that aggressive behaviors cannot exist before a given age, we of course prevent the falsification of the hypothesis. Similarly, if we start by taking a moral stance to define aggressive behavior, we bias our observations.

WHEN DO HUMANS START TO USE AGGRESSION?

The word *onset* can be used to describe the start of a phenomenon that has a relatively short duration (e.g., onset of a cold, onset of a panic attack). However, from a developmental perspective, the word "onset" generally refers to the age at which an individual first starts to engage in a type of behavior that will endure for a relatively long period of time. For example, we can study the age of onset of standing on two feet without support, the age of onset of using words or phrases, the age of onset of walking and running, as well as the age of onset of self-awareness. The American Psychiatric Association (DSM-IV-TR, 1994) introduced the concept of early and late onset conduct disorder for children in its fourth edition of the *Diagnostic*

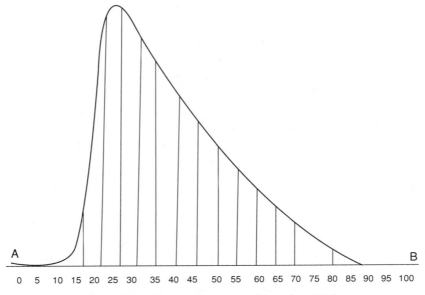

0   5   10   15   20   25   30   35   40   45   50   55   60   65   70   75   80   85   90   95   100

Figure 7.1. Male age-crime curve (Quetelet, 1833).

and Statistical Manual of Mental Disorders. They decided that there was an early and late onset form (i.e., before or after age 10).

In 1998, a review of misconceptions concerning the development of aggression (Loeber & Stouthamer-Loeber, 1998) concluded that "the age of onset of aggression in male populations is not concentrated only in the preschool years" (246). Data from a longitudinal study of boys were presented showing that less than 5 percent of the boys had "onset" of minor aggression before age 5, whereas close to 40 percent had onset of minor aggression by age 13. These results confirm the general impression that as children grow older they are more likely to start using physical aggression. The 2001 USA Surgeon General report on Youth Violence (www. surgeongeneral.gov/library/youthviolence/toc.html) and the World Health Organization Report on Violence and Health (2002) both concluded that onset of violent behavior occurred mainly during adolescence.

The apparent rapid increase in deviant behavior during adolescence, which is followed by an equivalently rapid decrease, has been labeled the age-crime curve (Farrington, 1986). It was described by the Belgian mathematician-astronomer-biosocial scientist Adolphe Quetelet in his 1833 book, *Research on the Propensity for Crime at Different Ages* (Figure 7.1). Quetelet concluded that "age is without contradiction the cause which acts with the most energy to develop or moderate the propensity for crime. This fatal propensity seems to develop in proportion to the intensity of physical strength and passions in man" (65). Late-twentieth-century scientists have suggested that the rise in testosterone levels during adolescence explained both the increase in strength and in physically aggressive passions (Ellis & Coontz, 1990; Eysenck & Gudjonsson, 1989). The alternative explanation has of course been environmental. As children become older they are more and more influenced by their environment, and are thus more likely to learn to aggress from such bad environmental influences as deviant families, deviant peers, and the media (Patterson, 1982; Dishion et al., 1991; Eron & Huesmann, 1986; Farrington, 1998; Huesmann & Miller, 1994; McCord, 1991; Vitaro et al., 1997).

Systematic and unsystematic observations of single cases had long led philosophers (Hobbes, [1651] 1958; Locke, [1693] 1996) and child specialists (Sully, 1895) to the conclusion that young children react with anger, physically aggress, and need to be socialized. Part of the confusion concerning the concept of "onset" appears to come from the concept of "seriousness." Physical aggression is usually considered the most serious form of aggression. Courts will generally sanction a physical aggression more severely than a verbal one. Teachers and parents are more likely to do the same when they have to sanction a child who hit another and one who gave "dirty looks." Both observational data from small samples of children (Strayer & Strayer, 1976; Hartup, 1974; Bridges, 1931; Dawe, 1934; Goodenough, 1931; Murphy, 1937; Restoin et al., 1985; Strayer & Trudel, 1984) and recent parent reports, and self-reports of large samples (Tremblay et al., 1996, 1999, 2004; Choquet & Ledoux, 1994; Alink et al., 2006; Côté et al., 2006; NICHD, 2004) indicate that children start by physically aggressing during infancy and go on to verbal aggression once they have learned to talk.

The cross-sectional data collected by Goodenough (1931) at the end of the 1920s clearly indicated that the form of aggressive behavior that is generally considered more "serious" or "socially unacceptable" (physical aggression) is ontogenetically antecedent to less "serious" forms of aggressive behavior, such as verbal aggression. However, the "onset" of physical aggression during infancy is often not accepted as a "valid age of onset" because it does not meet the criterion of "intentionality" or the one of "harm to the victim" of the aggression. These two objections appear to be the result of two biased views of child development. The first bias was discussed earlier: infants cannot intend to aggress; thus physical aggressions at that age do not count as onset as they would for puppy dogs.

The second bias is related to the harm done to the victim by the aggression. The literature on delinquency, antisocial behavior, and aggression generally uses the term *serious* to qualify the harm done to the victim. Infants who hit adults do not hurt adults; thus it is not an aggression. When infants seriously hurt, the behavior is attributed to an accidental act. For example, in the case of a serious bite during a fight for a toy the aggressive behavior will be attributed to a rage provoked by the pain of growing teeth. Infants kick, hit, bite, throw objects at others, and will use objects to hit others. Most parents and daycare personnel know this, and thus give plastic toys, hide objects that could be used to hurt seriously, and supervise groups of infants.

THE DEVELOPMENT OF AGGRESSION DURING EARLY CHILDHOOD

The first developmental trajectories studies of physical aggression started to follow children at age 6 (Nagin & Tremblay, 1999; Broidy et al., 2003). These studies led to three important observations. First, the majority of children used physical aggression less and less frequently as they grow older; second, there were no statistically significant groups of children who started to show stable high levels of physical aggression after age 6; and finally, each of the four developmental trajectories of physical aggression indicates that children were at their peak in frequency of physical aggression when they were in kindergarten.

These observations clearly showed that we must study the preschool years to understand the onset and development of physical aggression. However, surprisingly few longitudinal studies had tried to chart the development of physical aggression during the preschool years. This lack of attention to physical aggression during the early years appears to be the result of a long-held belief that physically violent behavior "onsets" during late childhood and early adolescence because of bad peer influences, television violence, and increased levels of male hormones (Tremblay, 2006).

Recent studies in Canada, the Netherlands, and the United States show that most children substantially increase the frequency of physical aggressions from 9 to 48 months, and then decrease the frequency of use until adolescence (Côté et al., 2006; NICHD, 2004). Figure 7.2 shows the different developmental trajectories of physical aggression from ages 2 to 11 for a random sample of Canadian children ($N = 10,658$) first assessed in 1994. We clearly see that the frequency of physical aggressions among children decreases substantially from the preschool years to preadolescence except for a small group who use physical aggression most often throughout that period.

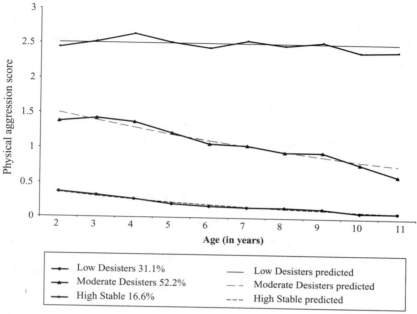

Figure 7.2. Physical aggression from ages 2 to 11. With kind permission from Springer Science+Business Media: Côté S., Vaillancourt T., LeBlanc J. C., Nagin D. S., Tremblay R. E. The development of physical aggression from toddlerhood to preadolescence: A nationwide longitudinal study of Canadian children. *Journal of Abnormal Child Psychology* 2006; 34, 71–85.

PREDICTORS OF CHRONIC PHYSICAL AGGRESSION DURING EARLY CHILDHOOD

The studies of antisocial behavior during adolescence and preadolescence clearly show that family characteristics and parental behavior are good predictors of antisocial behavior (McCord et al., 2001; Wasserman & Seracini, 2001). Is this also true for preschool children on trajectories of chronic physical aggression? From the results of the preschool longitudinal studies discussed earlier, the answer is a clear yes. For example, results from the Quebec longitudinal study of a birth cohort (Tremblay et al., 2004 showed that parent separation before birth and low income, two of the classic family risk factors, also predict high physical aggression during early childhood. It will be no surprise that mother characteristics before birth were among the best predictors: frequent antisocial behavior during adolescence, giving birth before age 21, not having finished high school, and smoking during pregnancy. Smoking apparently affects the development of the brain because of nicotine reaching the fetus's brain (Wakschlag et al., 2002). Of course, males were more at risk than females of being on the high physical aggression trajectory, even when the assessment started at seventeen months of age. After controlling for prenatal assessments, the five months after birth assessments revealed two significant predictors: family dysfunction and coercive-hostile parenting by the mothers. Thus, the traditional predictors of adolescent antisocial behavior are predicting chronic physical aggression during the preschool years. Interestingly, a twin study also showed that at age 17 months, more than half of the variation in frequency of physical aggression is explained by genetic factors (Dionne et al., 2003).

Why Are the Early Years so Important?

How can the increase and decrease in physically aggressive behavior during early childhood be explained? It is very likely that physical, cognitive, and emotional development play an important role. Within the first twenty-four months after birth, babies grow in height by more than 70 percent, and almost triple their weight. At birth, babies can hardly lift their heads; nine months later they can move on all fours, by twelve months they can walk, and by twenty-four months they can run and climb stairs. Compared to other periods of human development, the early years are on "fast forward."

The ability to grasp objects is an important development for social interactions. At birth, babies do not control their arms; at 6 months they can reach, grasp, and voluntarily let go objects. If they see an interesting toy in the hands of another 6-month-old baby, they will reach and grasp the toy. A struggle for the toy will occur if the other child does not let go. Note that at 6 months, the child does not have the language ability to ask the other child for the toy – this ability will come much later – but the frequency and complexity of interactions between babies and other persons in their environments increase at least as rapidly as their physical growth. Infants' waking time is spent exploring their physical environments. Before 12 months of age, they spend most of their playtime exploring one object at a time. Between 12 and 18 months, they imitate real-life activities alone. By the end of the second year, they are "pretend playing" with others as if they were professional comedians (Rubin & Clark, 1983).

Thus, over the first fifteen months after birth, with increased motor coordination and cognitive competence, children also become more and more able to discover their environment by themselves. The frequency of their interactions with peers increases with age, and playing with others increases dramatically from the end of the first year to the end of the second year (Ross & Goldman, 1977). This is the period when the rate of physical aggression increases most dramatically. At this age, children are exploring social interactions with their newly acquired walking, talking, running, grasping, pushing, kicking, and throwing skills. Most of their interactions are positive, but conflicts become more frequent (Restoin et al., 1985). Most of these conflicts are over possession of objects. During these conflicts, children learn that they can hurt and be hurt. Most children will quickly learn that a physical attack on a peer will be responded to by a physical attack, and that adults will not tolerate these behaviors. Most children will learn to wait for the toy to be free. They are also learning that asking for toys rather than taking them away from someone will more likely prevent negative interactions.

Learning to wait for something you want (Mischel et al., 1989) and learning to use language to convince others to satisfy your needs may be the most important protective factors against chronic physical aggression. Stattin and Klackenberg-Larsson (1993) showed that language skills between 18 and 24 months were a good predictor of adult criminality in a Swedish sample of males followed from birth to adulthood. In fact, numerous studies have shown an inverse correlation of verbal skills with impulsivity and criminal behavior (Moffitt, 1993). We need to understand the mechanisms underlying these associations. They are clearly operating in the first two years of life.

EPIGENETIC PROGRAMMING OF CHRONIC PHYSICAL AGGRESSION?

By 12 months of age, children have the physical, cognitive, and emotional means of being physically aggressive toward others (Tremblay, 2008). Before their second birthday most children will have "onset" of hitting, biting, or kicking another child or even an adult. However, their environment will play an important role in the developmental trajectory of these newly acquired skills. If children are surrounded by adults and other children who are physically aggressive, they will probably learn that physical aggression is part of everyday social interactions. On the other hand, if a child lives in an environment that does not tolerate physical aggression, and rewards pro-social behavior, it is likely that the child will acquire the habit of using means other than physical aggression to obtain what he or she wants, or for expressing frustration. This is the case for most children. All but a very small minority are using physical aggression more often in early childhood than later on. Apparently we do not need to learn to aggress. We need to learn not to aggress. Humans and many other animal species learn to use alternatives to physical aggression (see Tremblay, Hartup, & Archer, 2005; Tremblay & Nagin, 2005).

Thus, physical aggression is not a behavior children learn like reading or writing nor an illness children "catch" like poliomyelitis or smallpox. It is rather a behavior like crying, eating, grasping, throwing, and running, which young humans do when the physiological structure is in place. The young human learns to regulate these "natural" behaviors with age, experience, and brain maturation. The learning-to-control process implies regulating your needs to adjust to those of others, and this process is generally labeled "socialization."

It is not hard to imagine why the evolutionary process would have given humans a genetic program coding for all the basic mechanisms in order to react to hunger and threat. Young children's muscles are activated to run, push, kick, grab, hit, throw, and yell with extreme force when hungry, angry, or strongly attracted by something. However, stating that humans are genetically programmed to be able to physically aggress when needed is different from stating that the frequency of the physical aggressions they use is genetically programmed. Because all 18 month olds who have developed normally can and possibly do physically aggress, but not all do so at the same frequency and with the same vigor, to what extent are these individual differences due to the genetic program they have inherited or to the environment in which they have been growing? The trajectories shown in Figure 7.2 clearly indicate that these individual differences exist at any given point, starting in early childhood; but the most interesting phenotype is the development over time. There is obviously intraindividual change over time. Most children learn to reduce the frequency of the use of a behavior that they apparently did not need to learn. However, relatively stable differences among individuals remain.

What are the gene-environment mechanisms that explain the change and stability? They are possibly very similar to the mechanisms that explain the developmental trajectories of growth in height. Genes code for the growth mechanisms, but there are individual differences in these codes as well as environmental differences (e.g., access to food) that lead to stable individual differences. Thus the individual differences in the frequency of physical aggression at one point in time, and over time, can be due to a large number of "causes," for example, to individual differences in the genetic coding for serotonin (Pihl & Benkelfat, 2005), testosterone (van Goozen, 2005), language development (Dionne, 2005), or cognitive development (Séguin & Zelazo, 2005), or to environmental differences such as mother's tobacco use during pregnancy (Wakschlag et al., 2002), birth complications (Arsenault et al., 2002), parental care (Gatti & Tremblay, 2005; Raine et al., 1997; Zoccolillo et al., 2005), and peer characteristics (Boivin et al., 2005). However, the individual differences that we observe are likely because of interactions among many of these mechanisms, hence to epigenetic mechanisms (Francis et al., 2002, 2003; Weaver et al., 2004).

Knowledge on gene-environment interactions that could explain the development of chronic physical aggression is perilously close to zero. The first reason is that gene-environment interaction studies are recent. The second reason is that there are very few longitudinal studies that have included repeated assessments of physical aggression from early childhood onward (Rhee & Waldman, 2002) and most do not have genetically informative data. However, the most important problem is that molecular genetic studies and twin studies have concentrated on global antisocial behavior phenotypes, generally assessed at one point in time. This is an old problem in the antisocial behavior literature (Blumstein et al., 1988; Gottfredson & Hirschi, 1990; Tremblay, 2000; Tremblay, 2003; Tremblay et al., 1991). Genetic studies have simply followed the main trend, which tends to rely on measurement scales constructed by lumping items that are shown to correlate at one point in time.

We initiated a study that started at the earliest point during development to assess gene-environment effects on physical aggression with a large sample of 18- to 19-month-old twins. Results showed that variance in mother reports of physical aggression was explained somewhat more by genetic factors (58 percent) than by environmental factors (42 percent), suggesting that there are strong genetic effects on physical aggression during early childhood (Dionne et al., 2003).

Many molecular genetic studies have attempted to identify polymorphisms related to aggressive behavior, mainly with animal and human adult samples (Pihl & Benkelfat, 2005). Caspi et al. (2002) used a longitudinal study to specifically address gene-environment interactions. They observed that the most maltreated males were at higher risk of being convicted of a violent crime before age 27 if they had the short version of the functional polymorphism in the gene coding for monoamine oxidase A (MAOA) activity. The MAOA enzyme metabolizes neurotransmitters linked in previous studies to behavior problems (e.g., dopamine, norepinephrine, and serotonin), and the short version of the allele leads to low activity. Effects were similar for conduct disorder assessed between ten and eighteen years of age, antisocial personality symptoms, and disposition to violence measured at twenty-six years of age. Individuals with a history of chronic physical aggression may be the driving force in these associations, because they are the most likely to be found in each of the assessed categories.

This study is a good illustration of gene-environment issues related to prevention that need to be addressed. First, although the study was a longitudinal study from birth to age 26, the analyses did not provide information on the developmental impact of the gene-environment interaction. Was the effect of the gene-environment interaction on physical aggression present in early childhood? Did it appear later during elementary school, adolescence, and even adulthood? What are the biological mechanisms involved in the so-called gene-environment interactions? These are important questions for preventive interventions, as is the replicability of the findings and the specific developmental factors that come into play (Risch, Herrell, Lehner, Liang et al., 2009). From the developmental data presented in the preceding section, one would expect that the gene-environment effects were present early and may have grown with time.

A second question concerns the intervention strategies. Let's assume that the gene-environment interaction effects appear in early childhood and will increase with time if there are no early interventions. Which type of intervention should we use? For example, we could screen pregnant women soon after conception to identify those at risk of maltreating their child and offer a support program to help prevent the family from abusing the child (Olds et al., 1998). An alternative strategy would be to give the child a chemical treatment that would correct or compensate for the low MAOA activity (Caspi et al., 2002; Weaver et al., 2004). Both strategies could also be used with some cases. To answer these questions, we need to understand the biological mechanisms involved in the so-called gene-environment interactions.

With a longitudinal study of postnatal mothering behavior with rat pups, Weaver et al. (2004) recently showed that the mechanisms involved are likely to be at the level of epigenetic programming of the genome. They showed that frequency of licking has long-term effects because it regulates the expression of genes that influence the development of the hypothalamic-pituitary-adrenal (HPA) axis. Following this work, we are studying the differences in gene expression between boys on a chronic physical aggression trajectory and those on a normal trajectory. Preliminary results indicate that there are important differences in gene expression between the two groups (Szyf, Weaver, Provencal, McGowan, Tremblay, & Meaney, 2009).

Twin studies and molecular genetic studies can address the gene-environment interaction issue. However, concerning causal mechanisms leading to chronic physical aggression, they are still limited to a correlational analysis. To test the causal mechanisms, we need true experiments. Not only are they possible but they will enable us to kill two birds with one stone. Randomized control trials can indeed be used to identify effective preventive interventions and test for causal mechanisms (Schwartz et al., 1981). Cross-fostering experiments like those done with rats, mice, and monkeys (Francis et al., 2003; Suomi, 2005; Weaver et al., 2004) would give better insights into causal mechanisms than studies that are limited to correlational analyses. Obviously, it is hard to conduct such studies with humans, except with adoption studies. However, prevention experiments aimed at helping high-risk families and children can be used to understand the mechanisms that prevent children from chronic physical aggression (Tremblay, 2003).

If the regulation of physical aggression is learned during the preschool years, one would expect that interventions specifically targeting this regulation during this sensitive period would prevent chronic trajectories of physical aggression and increase the likelihood of positive aggression. By targeting different hypothesized causal mechanisms, such as smoking during pregnancy, quality of parenting, language development, emotional regulation, executive functions, and peer influence, we could test these hypotheses more directly than we can with longitudinal studies and simultaneously find the best preventive interventions.

REFERENCES

Achenbach, T. M., Edelbrock, C. (1983). *Manual for the child behavior checklist and revised child behavior profile.* Burlington, VT: University of Vermont, Department of Psychiatry.

Alink, L. R. A., van Zeijl, J., Stolk, M. N. et al. (2006). The early childhood aggression curve: Development of physical aggression in 10- to 50-month-old children. *Child Development, 77,* 954–66.

Archer, J., & Browne, K. (1989). *Human aggression: Naturalistic approaches.* London: Routledge.

Arseneault, L., Moffit, T. E., Caspi, A. et al. (2003). Strong genetic effects on cross-situational antisocial behavior among 5-year-old children according to mothers, teachers, examiner-observers, and twins' self-reports. *Journal of Child Psychology & Psychiatry, 44,* 832–48.

Arseneault, L., Tremblay, R. E., Boulerice, B., & Saucier, J. F. (2002). Obstetrical complications and violent delinquency: Testing two developmental pathways. *Child Development, 73,* 496–508.

Berkowitz, L. (1962). *Aggression: A social psychological analysis.* New York: McGraw-Hill.

Blumstein, A., Cohen, J., & Farrington, D. P. (1988). Criminal career research: Its value for criminology. *Criminology,* 26(11).

Boivin, M., Vitaro, F., & Poulin, F. (2005). Peer relationships and the development of aggressive behavior in early childhood. In R. Tremblay, W. Hartup, & J. Archer, (Eds.) *Developmental origins of aggression* (376–97). New York: Guilford.

Bridges, K. M. B. (1931). *The social and emotional development of the preschool child.* London: Kegan Paul.

Broidy, L. M., Nagin, D. S., Tremblay, R. E. et al. (2003). Developmental trajectories of childhood disruptive behaviors and adolescent delinquency: A six-site, cross-national study. *Developmental Psychology,* 39, 222–45.

Burt, C. L. (1925). *The young delinquent.* London: University of London Press.

Buss, A. H. (1961). *The psychology of aggression.* New York: Wiley.

Cairns, R. B. (1979). *The analysis of social interactions: Method, results, and illustrations.* Hillsdale, NJ: Lawrence Erlbaum.

Cairns, R. B., Cairns, B. D., Neckerman, H. J., Ferguson, L. L., & Gariépy, J. L. (1989). Growth and aggression: 1. Childhood to early adolescence. *Developmental Psychology,* 25, 320–30.

Caspi, A., McClay, J., Moffitt, T. et al. (2002). Role of genotype in the cycle of violence in maltreated children. *Science,* 297, 851–4.

Choquet, M., & Ledoux, S. (1994). *Adolescents: Enquête nationale.* Paris: Les éditions INSERM.

Coie, J. D., & Dodge, K. A. (1998). Aggression and antisocial behavior. In: W. Damon & N. Eisenberg (Eds.), *Handbook of child psychology: Social, emotional, and personality development,* 3rd edition (779–862). Toronto: Wiley.

Côté, S., Vaillancourt, T., LeBlanc, J. C., Nagin, D. S., & Tremblay, R. E. (2006). The development of physical aggression from toddlerhood to preadolescence: A nationwide longitudinal study of Canadian children. *Journal of Abnormal Child Psychology,* 34, 71–85.

Crick, N. R., & Grotpeter, J. K. (1995). Relational aggression, gender, and social-psychological adjustment. *Child Development,* 66, 710–22.

Dawe, H. C. (1934). An analysis of 200 quarrels of preschool children. *Child Development,* 5, 139–57.

Diagnostic and Statistical Manual of Mental Disorders (DSM-IV-TR). (1994). Washington, D.C.: American Psychological Association.

Dionne, G. (2005). Language development and aggressive behavior. In R. Tremblay, W. Hartup, & J. Archer (Eds.), *Developmental origins of aggression.* New York: Guilford.

Dionne, G., Tremblay, R. E., Boivin, M., Laplante, D., & Pérusse, D. (2003). Physical aggression and expressive vocabulary in 19-month-old twins. *Developmental Psychology,* 39, 261–73.

Dishion, T. J., Patterson, G. R., Stoolmiller, M., Skinner, & M. L. (1991). Family, school, and behavioral antecedents to early adolescent involvement with antisocial peers. *Developmental Psychology,* 27, 172–80.

Dodge, K., Lochman, J. E., Harnish, J. D., Bates, J. E., & Pettit, G. S. (1997). Reactive and proactive aggression in school children and psychiatrically impaired chronically assaultive youth. *Journal of Abnormal Psychology,* 106, 37–51.

Eley, T. C., Lichenstein, P., & Stevenson, J. (1999). Sex differences in the etiology of aggressive and nonaggressive antisocial behavior: Results from two twin studies. *Child Development,* 70, 155–68.

Ellis, L., & Coontz, P. D. (1990). Androgens, brain functioning, and criminality: The neurohormonal foundations of antisociality. In L. Ellis & H. Hoffman (Eds.), *Crime in biological, social and moral contexts* (162–93). New York: Praeger.

Eron, L. D., & Huesmann, L. R. (1986). The role of television in the development of pro-social and antisocial behavior. In D. Olweus, J. Block, & M. Ratkie-Iarrow (Eds.), *The development of pro-social and antisocial behavior.* New York: Academic Press.

Eysenck, H. J., & Gudjonsson, G. H. (1989). *The causes and cures of criminality.* New York: Plenum.

Farrington, D. P. (1986). Age and crime. In M. Tonry & N. Morris (Eds.), *Crime and justice: An annual review of research,* Vol. 7. (189–250). Chicago: University of Chicago Press.

Farrington, D. P. (1998). Individual differences and offending. In M. Tonry (Ed.), *The handbook of crime and punishment* (241–68). New York: Oxford University Press.

Francis, D. D., Diorio, J., Plotsky, P. M., & Meaney, M. J. (2002). Environmental enrichment reverses the effects of maternal separation on stress reactivity. *Journal of Neuroscience,* 22, 7840–3.

Francis, D. D., Szegda, K., Campbell, G., Martin, W. D., & Insel, T. (2003). Epigenetic sources of behavioral differences in mice. *Nature Neuroscience,* 6, 445–6.

Gatti, U., & Tremblay, R. E. (2005). Social capital and physical violence. In R. Tremblay, W. Hartup, & J. Archer (Eds.), *Developmental origins of aggression.* New York: Guilford.

Goodenough, F. L. (1931). *Anger in young children.* Westport, CT: Greenwood Press.

Gottfredson, M. R., & Hirschi, T. (1990). *A general theory of crime.* Stanford, CA: Stanford University Press.

Gray, J. A. (1971). *The psychology of fear and stress.* London: Weidenfeld & Nicolson.

Gray, J. A. (1982). *The neuropsychology of anxiety.* New York: Oxford University Press.

Hartup, W. W. (1974). Aggression in childhood. *American Psychologist,* 29, 336–41.

Hartup, W. W., & de Wit, J. (1974). The development of aggression: Problems and perspectives. In J. de Wit & W. W. Hartup (Eds.), *Determinants and origins of aggressive behavior* (595–620). The Hague, Netherlands: Mouton.

Hobbes, T. ([1651] 1958). *The Leviathan.* Indianapolis, IN: Liberal Arts Press.

Huesmann, L. R., Eron, L. D., Guerra, N. G., & Crawshaw, V. B. (1994). Measuring children's aggression with teachers' predictions of peer nominations. *Psychological Assessment,* 6, 329–36.

Huesmann, L. R., Eron, L. D., Lefkowitz, M. M. & Walder, L. O. (1984). Stability of aggression over time and generations. *Developmental Psychology,* 20, 1120–34.

Huesmann, L. R, & Miller, L. S. (1994). Long-term effects of repeated exposure to media violence in childhood. In L. Huesmann (Ed.), *Aggressive behavior: Current perspectives* (153–86). New York: Plenum Press.

Kagan, J. (1974). Development and methodological considerations in the study of aggression. In J. de Wit & W. W. Hartup (Eds.), *Determinants and origins of aggressive behavior* (107–14). The Hague, Netherlands: Mouton.

Lefkowitz, M. M., Eron, L. D., Walder, L. O., & Huesmann, L. R. (1977). *Growing up to be violent. A longitudinal study of the development of aggression.* New York: Pergamon Press.

Lewis, M., Alessandri, S. M., & Sullivan, M. W. (1990). Violation of expectancy, loss of control, and anger expressions in young infants. *Developmental Psychology*, 26, 745–51.

Locke, J. ([1693] 1996). *Some thoughts concerning education.* Indianapolis, IN: Hackett Publishing Co.

Loeber, R., & Stouthamer-Loeber, M. (1998). Development of juvenile aggression and violence. Some common misconceptions and controversies. *American Psychologist*, 53, 242–59.

McCord, J. (1991). Family relationships, juvenile delinquency, and adult criminality. *Criminology*, 29, 397–417.

McCord, J., Widom, C. S., & Crowell, N. E. (2001). *Juvenile crime, juvenile justice.* Washington, DC: National Academy Press.

McGrew, W. C. (1972). *An ethological study of children's behavior.* New York: Academic Press.

Mischel, W., Shoda, Y., & Rodriguez, M. L. (1989). Delay of gratification in children. *Science*, 244, 933–8.

Moffitt, T. E. (1993). Adolescence-limited and life-course persistent antisocial behavior: A developmental taxonomy. *Psychological Review*, 100, 674–701.

Murphy, L. B. (1937). *Social behavior and child personality.* New York: Columbia University Press.

Nagin, D., & Tremblay, R. E. (1999). Trajectories of boys' physical aggression, opposition, and hyperactivity on the path to physically violent and nonviolent juvenile delinquency. *Child Development*, 70, 1181–96.

NICHD, Network ECCR. (2004). Trajectories of physical aggression from toddlerhood to middle school. *Monographs of the Society for Research in Child Development*, Serial 278, 69(4), 1–25.

O'Connor, T. G., McGuire, S., Reiss, D., Hetherington, M. E., & Plomin, R. (1998). Co-occurrence of depressive symptoms and antisocial behavior in adolescence: A common genetic liability. *Journal of Abnormal Psychology*, 107, 27–37.

Olds, D., Henderson, C. R., Cole, R. et al. (1998). Long-term effects of nurse home visitation on children's criminal and antisocial behavior: Fifteen-year follow-up of a randomized controlled trial. *Journal of the American Medical Association*, 280, 1238–44.

Parke, R. D., & Slaby, R. G. (1983). The development of aggression. In P. H. Mussen, (Ed.), *Handbook of child psychology (Vol. 4: Socialization, personality, and social development)* (547–641). New York: John Wiley & Sons.

Patterson, G. R. (1982). *A social learning approach to family intervention: III. Coercive family process.* Eugene, OR: Castalia.

Pihl, R. O., & Benkelfat, C. (2005). Neuromodulators in the development and expression of inhibition and aggression. In R. E. Tremblay, W. W. Hartup, & J. Archer (Eds.), *Developmental origins of aggression* (261–80). New York: Guilford.

Pitkanen, L. (1969). A descriptive model of aggression and nonaggression with applications to children's behaviour. *Jyväskylä Studies in Education, Psychology and Social Research*, No. 19. Jyväskylä: University of Juväskylä.

Pulkkinen, L., & Tremblay, R. E. (1992). Patterns of boys' social adjustment in two cultures and at different ages: A longitudinal perspective. *International Journal of Behavioural Development*, 15, 527–53.

Quetelet, A. ([1833] 1984). *Research on the propensity for crime at different ages.* Sylvester SF, Trans. Cincinnati, OH: Anderson.

Raine, A., Brennan, P., & Mednick, S. A. (1997). Interaction between birth complications and early maternal rejection in predisposing individuals to adult violence: Specificity to serious, early-onset violence. *American Journal of Psychiatry*, 154, 1265–71.

Restoin, A., Montagner, H., Rodriguez, D. et al. (1985). Chronologie des comportements de communication et profils de comportement chez le jeune enfant. In R. E. Tremblay, M. A. Provost, & F. F. Strayer (Eds.), *Ethologie et développement de l'enfant* (93–130). Paris: Editions Stock/Laurence Pernoud.

Rhee, S. H., & Waldman, I. D. (2002). Genetic and environmental influences on antisocial behavior: A meta-analysis of twin and adoption studies. *Psychological Bulletin*, 128, 490–529.

Risch, N., Herrell, R., Lehner, T., Liang, K., Eaves, L., Hoh, J., Griem, A., Kovacs, M., Ott, J., & Merikangas, K. (2009). Interaction between the serotonin transporter gene (5-HTTLPR), stressful life events, and risk of depression: A meta-analysis. *JAMA: Journal of the American Medical Association*, 301(23), 2462–71.

Ross, H. S., & Goldman, B. D. (1977). Infants' sociability toward strangers. *Child Development*, 48, 638–42.

Rubin, K. H., & Clark, M. L. (1983). Preschool teachers' ratings of behavioral problems: Observational, sociometric, and social cognitive correlates. *Journal of Abnormal Child Psychology*, 11, 273–86.

Rubin, K. H., Hastings, P., Chen, X., Stewart, S., & McNichol, K. (1998). Intrapersonal and maternal correlates of aggression, conflict, and externalizing problems in toddlers. *Child Development*, 69, 1614–29.

Schwartz, D., Flamant, R., & Lelouch, J. (1981). *Clinical trials.* London: Academic Press.

Séguin, J. R., & Zelazo, P. (2005). Executive function in early physical aggression. In R. Tremblay, W. Hartup, & J. Archer (Eds.), *Developmental origins of aggression* (307–29). New York: Guilford.

Slutske, W. S., Heath, A. C., Dunne, M. P., et al. (1997). Modeling genetic and environmental influences in the etiology of conduct disorder: A study of 2,682 adult twin pairs. *Journal of Abnormal Psychology*, 106, 266–79.

Stattin, H., & Klackenberg-Larsson, I. (1993). Early language and intelligence development and their relationship to future criminal behavior. *Journal of Abnormal Psychology*, 102, 369–78.

Strayer, F. F., & Strayer, J. (1976). An ethological analysis of social agonism and dominance relations among preschool children. *Child Development*, 47, 980–8.

Strayer, F. F., & Trudel, M. (1984). Developmental changes in the nature and function of social dominance among young children. *Ethology and Sociobiology*, 5, 279–95.

Suomi, S. J. (2005). Genetic and environmental factors influencing the expression of impulsive aggression and serotonergic functioning in rhesus monkeys. In R. Tremblay, W. Hartup, & J. Archer (Eds.), *Developmental origins of aggression* (63–82). New York: Guilford Press.

Sully, J. (1895). *Studies of childhood*. London: Longmans, Green and Co.

Szyf, M., Weaver, I., Provencal, N., McGowan, P., Tremblay, R. E., & Meaney, M. (2009). Epigenetics and behaviour. In R. E. Tremblay, M. A. G. van Aken, & W. Koops (Eds.), *Development and prevention of behaviour problems: From genes to social policy* (25–59). Sussex, United Kingdom: Psychology Press.

Tremblay, R. E. (2000). The development of aggressive behaviour during childhood: What have we learned in the past century? *International Journal of Behavioral development*, 24, 129–41.

Tremblay, R. E. (2003). Why socialization fails?: The case of chronic physical aggression. In B. Lahey, T. Moffitt, & A. Caspi (Eds.) *Causes of conduct disorder and juvenile delinquency* (182–224). New York: Guilford Press.

Tremblay, R. E. (2006). Prevention of youth violence: Why not start at the beginning? *Journal of Abnormal Child Psychology*, 34, 481–7.

Tremblay, R. E. (2008). Anger and aggression. In M. Haith & J. Benson (Eds.), *Encyclopedia of infant and early childhood development*. 2nd ed. New York: Academic Press.

Tremblay, R. E., Boulerice, B., Harden, P. W. et al. (1996). Do children in Canada become more aggressive as they approach adolescence? In Human Resources Development Canada, Statistics Canada (Eds.), *Growing up in Canada: National Longitudinal Survey of Children and Youth* (127–37). Ottawa: Statistics Canada.

Tremblay, R., Hartup, W., & Archer, J. (Eds.) (2005). *Developmental origins of aggression*. New York: Guilford Press.

Tremblay, R. E., Japel, C., Pérusse, D. et al. (1999). The search for the age of onset of physical aggression: Rousseau and Bandura revisited. *Criminal Behavior and Mental Health*, 9, 8–23.

Tremblay, R. E., Loeber, R., Gagnon, C., Charlebois, P., Larivée, S., & LeBlanc, M. (1991). Disruptive boys with stable and unstable high fighting behavior patterns during junior elementary school. *Journal of Abnormal Child Psychology*, 19, 285–300.

Tremblay, R. E., & Nagin, D. S. (2005). The developmental origins of physical aggression in humans. In R. E. Tremblay, W. H. Hartup, & J. Archer (Eds.), *Developmental origins of aggression*. New York: Guilford Press.

Tremblay, R. E., Nagin, D., Séguin, J. R., et al. (2004). Physical aggression during early childhood: Trajectories and predictors. *Pediatrics*, 114, e43–e50.

van Goozen, S. H. M. (2005). Hormones and the developmental origin of aggression. In R. Tremblay, W. Hartup, & J. Archer (Eds.), *Developmental origins of aggression* (281–306). New York: Guilford Press.

Vitaro, F., Tremblay, R. E., Kerr, M., Pagani-Kurtz, L., & Bukowski, W. M. (1997). Disruptiveness, friends' characteristics, and delinquency: A test of two competing models of development. *Child Development*, 68, 676–89.

Wakschlag, L, Pickett, K. E., Cook, E., Benowitz, N. L., & Leventhal, B. (2002). Maternal smoking during pregnancy and severe antisocial behavior in offspring: A review. *American Journal of Public Health*, 92, 966–74.

Wasserman, G. A., & Seracini, A. M. (2001). Family risk factors and interventions. In R. Loeber & D. P. Farrington (Eds.), *Child delinquents: Development, intervention, and service needs* (165–89). Thousand Oaks, CA: Sage Publications.

Weaver, I. C. G., Cervoni, N., Champagne, F. A. et al. (2004). Epigenetic programming by maternal behavior. *Nature Neuroscience*, 7, 847–54.

World Health Organization. (2002). *World report on violence and health.* Geneva: WHO.

Zoccolillo, M., Paquette, D., & Tremblay, R. E. (2005). Maternal conduct disorder and the risk for the next generation. In D. Pepler, K. Masden, C. Webster, & K. Levene (Eds.), *Development and treatment of girlhood aggression* (225–52). Mahwah, NJ: Lawrence Erlbaum Associates.

8

Mental Health Intervention in Infancy
and Early Childhood

ALICIA F. LIEBERMAN AND CHANDRA GHOSH IPPEN

Mental health intervention in the early years is a relatively recent development, spurred by increasing clinical and research evidence that emotional and behavioral disturbances in the first five years of life often involve more than transient developmental upheavals that the child will outgrow. It is increasingly recognized that these disturbances, if severe and enduring enough, may represent the earliest manifestations of clinical disorders that can become entrenched unless effectively addressed (Sameroff & Emde, 1989; Scheeringa, Zeanah, Myers, & Putnam, 2005). This chapter focuses on the factors that need to be addressed in providing effective mental health intervention to infants, toddlers, and preschoolers who are at risk for mental health disorders due to detrimental environmental circumstances. Mental health intervention for children with diagnosable neurodevelopmental or physical disabilities is outside the scope of this chapter.

There is solid empirical evidence to support the effectiveness of intervention programs in infancy and early childhood. As stated in "Neurons to Neighborhoods," a landmark report from the National Research Council and Institute of Medicine Committee on the Science of Early Childhood Development, "The overarching question of whether we can intervene successfully in young children's lives has been answered in the affirmative and should be put to rest" (Shonkoff & Phillips, 2000, 10). The report goes on to caution: "However, interventions that work are rarely simple, inexpensive, or easy to implement" (10). This chapter seeks to elucidate the common themes that characterize successful mental health intervention in infancy and early childhood.

A pivotal moment in the evolution of the field of infant mental health in the United States occurred in 1977, when *Zero to Three: National Center for Clinical Infant Programs* was created by a small group of distinguished clinicians from a variety of disciplines (among them, in alphabetical order: T. Berry Brazelton, Selma Fraiberg, Stanley Greenspan, Ron Lally, Reginald Lourie, Peter Neubauer, Sally Provence, Albert Solnit, and Leon Yarrow) to disseminate knowledge about the early origins of mental health disturbances and about the clinical intervention methods that were being developed. Later renamed *Zero to Three: National Center for Infants, Toddlers, and Families*, the organization has expanded its goals to include the development and dissemination of a national vision about the salience of the gestational period and the first three years of life as the foundations for the child's healthy developmental trajectory. The relevance of this goal is buttressed by the rapid growth in the past thirty years of scientific knowledge about the physical, cognitive, emotional, and social capacities of infants and very young children, including advances in our understanding of early brain development, the central role of early relationships in personality formation, and the early origins of the child's sense of self and moral conscience. Early intervention and treatment programs benefit from this exploding scientific knowledge and, in turn, contribute to it by elucidating the specific manifestations of emotional distress in infants, toddlers, and preschoolers and by exploring strategies that may restore the child's developmental trajectory toward healthier outcomes.

CONTRIBUTING FACTORS TO YOUNG CHILDREN'S EMOTIONAL HEALTH

It is now widely accepted that children's emotional health is shaped by complex transactional processes among a variety of risk and protective factors, with cumulative risk factors increasing the prediction of emotional and behavioral problems (Anda et al., 2006; Dube et al., 2003: Cichetti & Sroufe, 2000; Rutter & Sroufe, 2000; Sameroff, 2000). Risk and protective factors include *parent characteristics* such as their mental health, education level, sense of efficacy, and resourcefulness; *family factors* such as quality of the parent-child relationship, emotional climate, and marital quality; *community connectedness* factors such as parental social support, social resources, and children's peer relationships; and *neighborhood factors* such as availability of resources, adequacy of housing, and violence (Sameroff & Fiese, 2000). The predictive value of these factors across many studies led to the development of transactional-ecological models that attempt to conceptualize the relative contributions of proximal and distal risk and protective factors to children's developmental outcome (Cichetti & Lynch, 1993; Cicchetti, Toth, & Maughan, 2000).

Rapid advances in genetics and the mapping of the human genome have led to hopes that children's developmental outcome and mental health profiles could be predicted from their genetic makeup. This emphasis on genetic factors as the primary engine leading development and mental health is not justified by the scientific data. Genes engage in complex transactions with the variety of internal environments that surround them within the body so that the environment of the cell influences which of the tens of thousands of genes are turned on or off to affect cell characteristics (Boyce, this volume; Greenough, 1991; Greenough & Black, 1992; Rutter, Moffit, & Caspi, 2006; Rutter, this volume). Elegant studies of schizophrenia, child depression, and child abuse indicate that the interaction between the genome and the environment account for more variance in outcome than genetic or environmental factors alone (Tiernari et al., 2004; Caspi et al., 2003; Kaufman et al., 2006). In other words, "It is impossible to think of gene expression apart from the multiple environments in which it occurs. It is impossible to think of the manifestation of hereditary potential independently of the hierarchy of environments that shape its appearance" (Shonkoff & Phillips, 2000, 40).

The concrete ramifications of this intricate interplay of nature and nurture are graphically demonstrated in a longitudinal study of children raised in a variety of environmental conditions (Sameroff, Bartko, Baldwin, Baldwin, & Seifer, 1998). Infants in the first year of life were classified into low and high competence groups based on an integration of thirteen measures that included physical health, cognitive performance scores, and temperamental characteristics. There was no relation between infant competence scores and the child's IQ or mental health scores at age four. Most notably, highly competent infants raised in detrimental environments had worse functioning as preschoolers than infants with low competence who were raised in favorable environments. These findings suggest that attention to environmental risk factors may be, as a rule, a more effective strategy for predicting developmental course than attention to individual factors within the child (Sameroff & Fiese, 2000).

THE CENTRALITY OF PARENTS IN YOUNG CHILDREN'S EMOTIONAL LIFE

A young child's development is determined by the transaction between multiple factors, but in the first years of life parents are the primary emotional funnels for these environmental influences. The baby's emotional well-being is crucially influenced by how the parents respond to their circumstances, and their responses to their circumstances in turn affect the ways they relate to the baby. As the baby becomes a toddler and a preschooler and increasingly learns by observation and imitation, the parents model ways of acting and interacting that become an indelible component of the child's repertoire of feelings, attitudes, and behaviors. Intimate relationships are the building blocks of the child's internal working models of the self and of interpersonal relationships, and constitute the earliest and most persistent organizers of the child's "lived experience" and developmental outcome (Sroufe et al., 2005).

For an infant, the caregiving parent (most frequently the mother) is the first and most encompassing environment. Distal and proximal risk and protective factors are useful abstractions that permit predictions of developmental outcome for groups of children, but these constructs have no meaning for the individual baby. What the baby "knows," at the most visceral level, is whether he or she feels good or distressed and whether experiences of relief, well-being, and pleasure are consistently associated with a specific person or small group of persons who provide reassuringly familiar sights, sounds, tactile sensations, smells, and tastes. The much-cited Winnicott (1965) quotation, "There is no such thing as a baby," has become an almost indispensable condensation of this intuitively understood but difficult to articulate body of knowledge. As the infant becomes more autonomous, the mother and other cherished caregivers become the object of "social referencing," whose behavior is monitored by the baby and becomes a basis for the baby's decisions about what is safe and what is risky and for the appropriate actions that he or she takes accordingly.

Even toddlers and preschoolers know little about distal and proximal risk and protective factors. For them, the parents, siblings, and other family members constitute "the environment." They turn to the parents as a "secure base" from which to explore the world, which is full of exciting possibilities but also of unwelcome surprises that call for a quick return to the safety represented by the parents (Ainsworth, Blehar, Waters, & Wall, 1978). In their roles as the child's primary socializers, the parents also set the standards for what is right and wrong, allowed and forbidden, celebrated or disapproved (Kagan, 1981; Lieberman, 1993).

There are large individual differences in the quality of parenting, and these individual differences make a difference for the child's emotional health. The empirical evidence pertains primarily to mothers, who tend to be the caregivers most readily available for research. Children tend to have insecure attachments to their mothers when their caregiving is characterized by harshness, unpredictability, or emotional withdrawal, whereas sensitive and responsive caregiving is associated with secure attachments (Ainsworth et al., 1978; Sroufe, 1979; Thompson, 1999a, 1999b). These early attachment patterns tend to be stable in relatively stable circumstances, and are associated with different developmental outcomes as late as adolescence. Securely attached infants tend to have parents who are sensitively responsive to their needs, and in the course of their development they are more likely than anxiously attached infants to be cooperative as toddlers, to have more ego-resiliency as preschoolers, and to show lesser incidence of psychopathology as older children and adolescents (Sroufe, Carlson, & Schulman, 1993; Sroufe, Carlson, Levy, & Egeland, 1999). Across multiple studies, caregiver functioning and the quality of the parent-child relationship have been found to be important predictors of young children's functioning, particularly when the child has been exposed to trauma (Bogat, DeJonghe, Levendosky, Davidson, & von Eye, 2006; Scheeringa & Zeanah, 2001).

"ALLOPARENTING": THE IMPORTANCE OF GOOD SUBSTITUTE CARE

Young children are not raised solely by their parents. This was the case long before the changes brought about by rapid industrialization and work by mothers outside the home. Among humans, as among other primates who are "cooperative breeders," infant survival has always depended on the mother being assisted by others, who received the name of alloparents (from the Greek *allo*, or "other than"). Alternative caregivers were essential among humans living in foraging societies to help a mother rear her infant. Ethnographic evidence from foraging peoples such as the Aka and the Efe indicates that alloparents have been a frequently used alternative to continuous mother-infant contact whenever evolutionary pressures favored this collaborative form of child rearing (Hrdy, 1999). Work and motherhood have always entailed compromises and trade-offs between the mother's survival and well-being and those of the child.

In modern societies, maternal work out of the home and away from the child does not endanger infant survival as a rule, but it creates distress in the infant and stress in the mother (see Barr, this volume). These stresses are compounded when the alternative caregivers are paid substitutes who do not share communal bonds with the parents or the child, rather than maternal kin whom the child has known from birth (as was consistently the case for foraging peoples).

The absence of deep emotional bonds between the hired caregiver and the child's parents, which would result in more emotionally committed care of the child, makes it particularly important to select modern-day alloparents for their natural affinity with children and their spontaneous pleasure in taking care of them. These strengths need to be cultivated through extensive training in development and in the art and science of caring for young children in groups. There is abundant empirical evidence that the quality of substitute care, measured by adult-to-child ratios, group size, stability of caregiver, caregiver skill, and quality of caregiver-child relationship, is significantly correlated with positive developmental outcomes in the child (Lamb, 1998; NICHD Early Child Care Research Network, 1996).

THE CHANGING SOCIAL CONTEXT OF MENTAL HEALTH IN INFANCY AND EARLY CHILDHOOD

Paradoxically, advances in knowledge about infants and young children are taking place in a rapidly changing social context that poses major challenges to raising children in ways that support their emotional needs. Just as scientific data accumulate about the developmental importance of emotionally available parenting and of predictable and knowledgeable substitute care, social changes are making these crucial conditions for an emotionally healthy infancy and early childhood increasingly difficult to achieve. Some key characteristics of these social changes are highlighted next.

- In the United States, the acceleration in paid maternal employment is the best known of these social changes. From 1965 to 2000, employment of mothers with children under eighteen increased from 45 percent to 78 percent, and full-time employment increased from 19 percent to 57 percent (Bianchi & Raley, 2005). Between 1975 and 1999, the proportion of children under six years of age with mothers in the labor force increased from 38.8 percent to 61.1 percent, and the proportion of those whose mothers work full time nearly tripled, from 11 percent to 30 percent. The increase in maternal employment is largest for infants in the first year of life when compared to older children (1999 Current Population Survey). The implication of these trends is that a high percentage of infants, toddlers, and preschoolers are no longer raised primarily by their parents, and that child rearing is now frequently shared among a variety of caregivers who may or may not be related to the child.
- Parents have less free time to spend with their children because of work pressures. The average worker now spends 163 hours more at work

than in 1969, or the equivalent of one extra month of work annually (Schor, 1992). Concomitantly, the amount of parental time spent with children declined considerably over a thirty-year span (Wolff, 1996). Insufficient parental availability can have detrimental consequences for children's mental health when it deprives them of appropriate stimulation and supervision and when it creates in children a feeling that they are secondary to work and other priorities in the parents' lives.

• Increased work hours have not resulted in higher standards of living for many families. Poverty among children has increased since 1970 in spite of major gains in Gross Domestic Product (GDP). Census data indicate that in 2005, 17.6 percent of children under age eighteen and 20 percent of those under age six lived in poverty. African American and Latino children, children living with single mothers, and children under age six were more likely to live in poverty (Childtrends Data Bank, 2007).

• Children under age five are the poorest group in the United States, with poverty increasing more for minority than nonminority children (Bennett, Li, Song, & Yang, 1999). Poverty is often associated with other conditions that adversely affect children's development, including parental depression, single parenthood, and poor quality of parenting (Sameroff, 2000). In one study, 60 percent of the poorer children lived in families with more than six adverse circumstances, whereas 25 percent of the wealthier children did (Sameroff, Bartko, Baldwin, Baldwin, & Siefer, 1998).

• Violence in the United States has been characterized as a "public health epidemic" (Centers for Disease Control and Prevention, 1967–1994). A nationally representative study of married or cohabitating couples estimates that in one year, 29.4 percent of children (15.5 million) experienced domestic violence between their parents (McDonald, Jouriles, Ramisetty-Mikler, Caetano, & Green, 2006). Child maltreatment data show that in 2005, 3.6 million children received an investigation by Child Protective Services, and children aged birth to three had the highest rates of victimization (U.S. Department of Health and Human Services, 2007).

• The quality of childcare outside the home is often substandard, with 10 percent to 20 percent of childcare settings falling below criteria for adequate care (Galinsky, Howes, Kontos, & Shinn, 1994; Helburn, 1995). There are no uniform standards for training childcare providers, and the states that have childcare licensing requirements define minimum standards in terms that seek to prevent physical harm rather than promote children's developmental progress. Childcare providers remain among the most underpaid occupations in the nation, leading to rapid turnover of personnel caring for young children at an age where stability and predictability of care are major predictors of sound developmental functioning (Howes, Rodning, Galluzzo, & Myers, 1988).

• Availability of early intervention services for children at risk is well below the documented need. Head Start, long heralded as a model preventive developmental program, is available to less than half of the eligible children, although this program has been in existence since 1965 (Meisels & Shonkoff, 2000). This problem is compounded by increased concerns within Head Start about the mental health of the children and families served by the program (Knitzer, 2000). The concerns involve high levels of aggressive and noncompliant behavior as well as deficits in children's readiness to learn. Studies of Head Start populations also report high levels of poverty-related distress among the children's parents (Johnson & Walker, 1991; Parker, Piotrkowski, Horn, & Greene, 1995).

IMPLICATIONS FOR MENTAL HEALTH INTERVENTION

These social trends highlight the society-wide gap between what we know and what we do. There is a vast disparity between the knowledge we are accumulating and the application of this knowledge to improve developmental outcomes for young children. The ecological/transactional models highlight the detrimental impact of poverty, parental dysfunction, violence, and inadequate substitute care on children's development. In spite of robust empirical data, current social policy efforts to alleviate risk and strengthen protective mechanisms for young children and their families fall short of the scope needed to bring about substantial improvement (Harris, Lieberman, & Marans, 2007).

Intervention programs in the United States must address the fact that families often have a multiplicity of problems that cut across the three main categories of risk factors described earlier. It is not uncommon to find, for example, that a family referred to an intervention program is not only in extreme poverty but also includes a depressed mother, an unemployed and drug-abusing father, inadequate and dangerous housing conditions, domestic violence, an environment lacking in predictable routines for the child, and/or two or more children under five years of age showing clear behavioral problems and developmental delays. The research on the cumulative impact of multiple risk factors indicates that these are precisely the kinds of families where children are most in need of early intervention.

The deteriorating social and economic conditions facing an increasing number of families with young children must influence how mental health intervention is conceptualized and implemented. The traditional "fifty-minute hour" spent in the therapist's office discussing inner experience is patently inadequate as the primary format for addressing the needs of young children and their families who face a multiplicity of external stresses as well as the psychological sequelae of their circumstances. In response to this situation, it is imperative to develop, implement, and evaluate the effectiveness of intervention strategies that attempt to ameliorate risk factors and strengthen protective factors in the family's life.

PARAMETERS OF EFFECTIVE EARLY MENTAL HEALTH INTERVENTION

The common denominator in intervention programs designed to improve young children's mental health is a focus on the quality of parenting, which in practical terms almost invariably means the quality of mothering because mothers are predominantly the primary caregivers. Quality of parenting, in turn, does not exist in a vacuum. It is deeply influenced by the individual emotional experience of the parent, the individual characteristics of the child, and the ways that the parent and the child perceive and relate to one another. These factors, in turn, are profoundly affected by the cultural and socioeconomic context of the family. The stresses impinging on the parents as the result of poverty, social marginalization, and other hardships need to be carefully assessed and incorporated into the treatment plan.

A RELATIONSHIP-BASED CLINICAL MODEL

The emotional quality of the parent-child relationship comprises a natural focus for efforts to promote mental health in infancy and early childhood because it has a well-documented impact on early development. Parenting-oriented interventions strive to bring about positive changes in the parents' commitment to the child and in their ability to understand and respond appropriately to the child's needs. These interventions seek to strengthen the family relationships that will comprise the child's most immediate and influential emotional environment for many years after the intervention has ended. Individual child psychotherapy, by comparison, is more limited in its scope because it places the major responsibility for the young child's psychological well-being on the child himself rather than on the parents. Individual child psychotherapy also emphasizes the curative power of the child-therapist relationship at the expense of the child-parent relationship. It may be said that approaches that focus on joint parent-child intervention emphasize the interconnectedness of the child and the family members, and actively search for ways of increasing feelings of intimacy, safety, and comfortable reciprocity among the child and the parents. Approaches that focus on individual intervention with the child emphasize the emotional primacy of the child's subjective experience of the family situation, even if collateral sessions are also used to encourage changes in the parents' behavior.

Early mental health intervention that includes the parent(s) as integral partners in the process is particularly important for multineed families where the parents are beleaguered by their life circumstances. Parents whose internal resources are depleted by multiple sources of stress have difficulty giving emotional priority to their child or to their parenting role. They may not be motivated to cooperate with an intervention format that excludes them, such as individual child psychotherapy or play therapy. Waiting outside the therapist's office while the child and therapist meet alone can be a lonely and alienating experience for a needy parent. It may rekindle early feelings of being left out, and these feelings may be compounded by the confidentiality of the child-therapist, which precludes the parent's knowing in any detail what transpired during the therapeutic sessions.

Parents from marginalized cultural groups may be especially threatened by the emphasis that individual psychotherapy places on self-disclosure by the child in the absence of the parent. Foster care placement is a very real threat among impoverished minority groups, who are overrepresented in the child protective system. Many parents respond to this situation by warning their children not to talk about family matters in psychotherapy, placing the child in a dilemma about what to tell and what not to tell. (It must be stressed that this problem does not occur only among high-risk families. The seven-year-old daughter of a successful physician announced to the therapist, on the last session of a yearlong treatment: "My daddy told me that I shouldn't tell you about him and my mom"). These problems do not mean that intervention geared individually to the child has no place in early mental health intervention programs. However, it is important to stress that parents need to be integral partners in whatever individual work is done with the child.

The parent's contribution. The parent's psychological state can be a major contributor to conflicts in the parent-child relationship and to the child's emotional problems. For example, maternal depression is associated with increased child risk of serious psychopathology and problems in school and relationships with peers (Cummings & Davies, 1994; Downey & Coyne, 1990; Zeanah, Boris, & Larrieu, 1997). These findings are of particular concern because the prevalence of depression is consistently higher among mothers living in poverty (Lennon, Aber, & Blum, 1998; Olson & Pavetti, 1996), who also have less access to protective resources that may ameliorate the impact of their depression on their children. It follows that children growing up in poverty *and* raised by depressed mothers are at increased risk for mental health difficulties when compared to children whose parents have the financial resources to compensate for the emotional deficits of the home environment. For this reason, parenting interventions need to incorporate sustained attention to the mental health needs of the parents.

The child's contribution. It is often difficult to elucidate whether the clinical symptoms and relationship problems originate in constitutional characteristics of the child, for example through regulatory difficulties that interfere with the acquisition of normal emotional milestones (Greenspan & Wieder, 1993; Zero to Three, 2004). It is safe to assume that, from the beginnings of the child's experience, an intricate interplay of constitutional and parenting factors is set in motion that shape how the parent(s) and the infant perceive themselves, the other, and themselves in relation to the other. The now classical concepts of "goodness of fit" and "poorness of fit" (Thomas, Chess, & Birch, 1968) capture the sense of synchrony and ease versus alienation and discomfort that parent and baby may experience as they strive to form a relationship with each other. When parents can manage to remain sensitively attuned and emotionally available, even temperamentally difficult infants are able to develop secure attachment relationships (Goldberg, 1990; van Ijzendoorn, Juffer, & Duyvesteyn, 1995).

Nevertheless, it is likely that constitutional vulnerabilities such as extreme irritability, difficulty with transitions, and behavioral inhibition may predispose the child to clinical problems when he or she is faced with adverse environmental circumstances such as harsh, insensitive, or neglecting parenting. In this sense, biological susceptibility may be amplified and intensified by environmental adversity. A much-cited intervention study targeted low socioeconomic status (SES) Dutch mothers with irritable neonates on the assumption that mothers with few financial resources who were faced with a difficult baby would welcome intervention in order to alleviate an additional stressor in their lives (van den Boom, 1994, 1995). A brief intervention showing the mothers how to respond sensitively to their babies' crying and how to interact with them resulted in statistically significant and persistent improvements in quality of attachment (van den Boom, 1994, 1995).

This study is an elegant example of effective intervention when biological susceptibility and environmental stress are present simultaneously. At the same time, these relatively prompt and clear-cut effects need to be understood in the context of the day-to-day circumstances of these families. This intervention was conducted with families where the father was present and working, the mothers stayed at home to raise the child, and risk factors such as depression or other mental illness, violence, and substance abuse were not present. The findings must also take into account the reasonably homogeneous cultural environment of the Netherlands, where expectations about child rearing are widely shared, government-sponsored family support services are universally available and free from stigma, and as a result of these state supports, poverty levels are considerably less severe than in the United States. Indeed, when van den Boom's model was used with middle-class adoptive parents and with lower class families whose infants were not selected for irritability, the intervention effects were not replicated (van Ijzendoorn et al., 1995). These findings stress the importance of matching the intervention to the needs of the family.

The parent-child relationship. In complex conditions, when the parents themselves are overwhelmed by their circumstances and would feel marginalized and alienated by the intervenor's unwavering attention to "what is best for the baby," the intervenor needs to move flexibly between the needs of the parent and those of the child, deploying his or her attention to whatever is emotionally salient and meaningful at a particular moment. The target of the intervention is, optimally, the web of meanings that are being constructed between the parent and the child and the behaviors through which these meanings are expressed. At the same time, intervenors needs to be judicious about the timing of parenting interventions in order to remain responsive to the parent's sense of what is needed.

PARENTAL MOTIVATION AND EXPECTATIONS

Families may be referred for early mental health intervention from a variety of sources, and parents vary greatly in their motivation and expectations from the intervention. At one end of the continuum, parents may be referred by a benevolent pediatrician in response to their pleas for help with an intractably difficult young child and are deeply committed to the treatment. At the other extreme, parents may be mandated by the courts to participate in treatment as a last resort before the child's placement in foster care or as a condition for reunification. These parents are likely to perceive the intervention as coercive and potentially destructive to the family's cohesion. Between these two extremes, there are parents who may vacillate in their motivation, at times resenting their child for bringing them into treatment and at other times valuing the intervention for the increased understanding and feeling of competence that it brings to their sense of themselves as parents. Even when the referral is couched as a voluntary service, there is often an implicit message from the referring party that negative consequences might follow from a parent's failure to participate. The parent finding him- or herself in these circumstances is likely to perceive the intervention program, and the intervenor, as punitive authority figures with the power to inflict pain, including removing the child from the home. Among minority populations, particularly African Americans, who are overrepresented in the foster care population, this is a widespread fear that may generate a self-protective reserve in welcoming intervention programs.

An important obstacle to intervention is the parent's resentment of a single-minded focus on improving the child's situation when the parents themselves are suffering. Unless the intervenor is sensitive to the parental perception that "all that matters to the intervenor is the baby," the intervention will be greeted by the parents with the conviction that their needs are being overlooked. This does not bode well for the baby or young child because the parents' competitiveness with the child for the "scarce resources" offered by the intervenor will be manifested either in unwillingness to participate in the intervention or in anger at the child once the session is over.

Intervenors must always keep an eye on the fluctuations of parental motivation and the hidden fears, anxieties, and resentments that the parents may experience toward the therapist. One memorable example of these hidden fears was provided by a mother who, seeing the name of the Infant-Parent Program written on the door after several months of treatment, exclaimed in relief: "Now I see the name! I always thought it was called 'The Unfit Parent Program.'" This touching story illustrates the deep vulnerability of parents when they undertake intervention of behalf of their young children. They feel unfit because the social norms expect that parents should naturally know how to raise their children properly. The intervenor's sympathetic acknowledgment and reassurance for the parents' feelings of incompetence, shame, and guilt are essential to facilitate treatment.

THE ROLE AND USEFULNESS OF DIAGNOSIS

Diagnosis is not always essential in early mental health intervention, but it can be very helpful to inform therapists about the strengths and vulnerabilities of the parents and the child and to keep them alert to possible pitfalls in the unfolding of the intervention.

Diagnosis of the parents presents particular challenges in parenting intervention because parents with such serious diagnoses as schizophrenia, borderline personality disorder, or bipolar disorder may use the intervention well and can become adequate parents if they have adequate environmental supports and good psychopharmacological treatment when appropriate. In contrast, parents with less severe diagnoses may not profit from treatment if they remain rigidly entrenched in a punitive and blaming stance toward their children or cannot overcome their emotional withdrawal. This clinical experience is congruent with research findings that psychiatric diagnosis by itself is not a good predictor of parenting disturbances (Sameroff, Seifer, & Zax, 1982). This is because parental emotional investment in their children and their motivation to make use of treatment in behalf of the child cut across diagnostic categories.

It is also important to remember that diagnoses can be simply incorrect when applied to people from different cultural traditions or in particularly traumatic circumstances. For example, a battered woman may appear to have a thought disorder when emerging from a violent home situation, but her thinking patterns may improve markedly after several months in a safe environment. The diagnostic process must unfold cautiously at best with parents enduring multiple sources of stress and with parents with different cultural traditions. Optimally, diagnosis must be conceptualized whenever possible as an aid to individually tailored intervention rather than as an argument against mental health treatment.

Diagnosing infants, toddlers, and preschoolers is even more challenging. The setting and the relational context in which a young child is assessed are inextricably linked to the behaviors observed, making it difficult to ascertain what is an inherent characteristic of the child and what is a response to the assessment situation. In addition, the rapid rate of development presents a serious obstacle to the categorization of behavior into presumably stable categories of psychopathology. At the same time, diagnostic classifications in infancy and early childhood may be scientifically useful as a way of categorizing observations, monitoring the stability of these categories and their co-occurrence with other conditions, and facilitating communication among practitioners. The pragmatic usefulness of diagnostic instruments in making possible financial reimbursement for mental health intervention with young children needs to be acknowledged.

The diagnostic manual DC: 0–3R (*Zero to Three: Diagnostic Classification of Mental Health and Developmental Disorders of Infancy and Early Childhood*, Revised) (Zero to Three, 2004) represents an effort to present a view of the child's individual functioning from a developmental perspective and in the context of the child's primary relationships. It has five axes, which describe the child's individual diagnosis, the diagnosis of the child's primary relationships, the child's developmental level of functioning, the intensity of disorder in the child-parent relationship, and the environmental stressors encountered by the child. The instrument recommends that the diagnostic assessment process take place in the course of several sessions, preferably in a variety of settings, including those that are most familiar for the child, such as the home and the childcare setting, and in the presence of key caregivers in the child's life.

PORTS OF ENTRY FOR INTERVENTION

In his illuminating discussion of how different clinical approaches conceptualize what needs to be changed in what he calls "the parent-child clinical system," Stern (1995) speaks of two constructs: the "theoretical aim or target," which is the basic element of the parent-child system that the intervention seeks to change, and the "port of entry," which is the basic element of the system that is the immediate object of clinical attention – the pathway chosen by the therapist to enter the clinical system.

As stated earlier, a clinically flexible approach such as infant-parent psychotherapy or child-parent psychotherapy targets the web of mutually constructed meanings in the infant-parent relationship, moving back and forth between the experiences of the child, the parent, and what happens between them to enhance empathic mutuality and developmentally appropriate responsiveness (Lieberman et al., 2000; Pawl & St. John, 1998). This is what is meant when the answer to the question, "What is the target of early mental health intervention: the parent, the child, or the relationship?" is "All three."

To choose a port of entry into the web of meanings created by the parent and the child in the course of forming and sustaining their relationship, the therapist monitors their ongoing exchanges, using the presence, appropriateness, and modulation of affect as guideposts for selecting when and how to intervene. The specific port of entry chosen may vary from family to family or, within a family, from session to session. Commonly used but not exhaustive ports of entry are listed here.

1. Child behavior (e.g., a baby's inconsolable crying or a toddler's reckless darting off in an unfamiliar setting);
2. Parent behavior (e.g., bitter complaint about not sleeping all night because of the baby's crying or about fatigue from having to keep track of the toddler's whereabouts);
3. Parent-child interaction (e.g., parental unresponsiveness to the baby's frantic screaming or obliviousness to the toddler's self-endangerment);
4. Child self-representation (e.g., a preschooler saying: "I don't like myself");
5. Child mental representation of the parent (e.g., a toddler making the mother doll spank the baby doll; a preschooler saying to the parent "You don't love me");
6. Parent self-representation (e.g., a parent saying: "I was never good at anything");
7. Parental mental representation of the child (e.g., a parent saying: "He is a bad seed, just like his father");
8. Child-mother-father interaction (e.g., a baby crying and straining toward the mother while held by the father; a toddler saying to the mother: "not you, daddy do it");
9. Parental conflicts regarding the child (e.g., the father saying: "you spoil him" and the mother replying: "well, you never pay any attention to him, so I have to give him extra love");
10. Parent-therapist relationship (e.g., a parent not showing up for an appointment, a parent responding to the therapist's offer of treatment by saying: "programs are problems")
11. Child-therapist relationship (e.g., a preschooler telling the therapist: "Go away. I hate you"; a toddler crying and holding on to the therapist at the end of a session, saying "no bye, no bye!");
12. Parent-child-therapist relationship (e.g., the parent telling the therapist: "my son likes you better than me").

This list of possible ports of entry illustrates the versatility of clinical approaches that target the meaning of relationships for the parent and for the child. The infant-parent or child-parent psychotherapist can make use of whatever promising port of entry emerges in the session to move seamlessly between behavior and internal representation; between individual, dyadic, and family issues; and between the past and the present to construct a rich joint narrative between parent and child.

Clinical talent is an imponderable skill that is difficult to teach but easy to recognize. Indispensably generic ingredients in a good psychotherapist's makeup are empathy for the plight of others, insight into the possible sources of this pain, and creativity in searching for avenues to alleviate the pain that are compatible with the person's values, worldview, and character structure. Other skills are easier to teach once these abilities are present. Intervenors working in a relationship-based clinical model need to develop the ability to deploy attention flexibly among the different partners in the interaction, always remembering that every partner is simultaneously an actor, a recipient, and an observer of what is transpiring during the session. Relationship-based psychotherapists should understand development as a continuous process, from the initial days of life to the life-transforming milestones represented by marriage and parenthood. They must learn to observe nonverbal behavior and gauge its meaning, whether in the baby and young child or in the parents. The most meaningful avenues for emotional expression between parents and children often take place at the body level, through unconscious, automatic behaviors such as gaze avoidance or a subtle stiffening of the muscles upon physical contact. For a relationship-oriented therapist, the therapeutic relationship becomes the catalyst for change, and it becomes the vehicle for utilizing a combination of intervention modalities that may include insight-oriented interventions, developmental guidance, emotional support, and concrete assistance with problems of living as well as crisis intervention when needed. The underlying assumption is that the therapeutic relationship becomes a "corrective attachment relationship," generated through the therapist's supportive and empathic stance, which coalesces with the new knowledge, self-understanding, and adaptive behaviors fostered and practiced through the different therapeutic modalities. These processes, created jointly by the parent, child, and therapist, can lead to enduring and positive changes in the parent's and child's experiences of each other, the meanings they construct together, and their sense of themselves both individually and in relation to each other.

To address the complex challenges faced by young children and their families requires flexible interventions that incorporate the key principles of infant mental health delineated earlier. In this section, three such interventions are presented: (1) High/Scope Perry Preschool Program, (2) Nurse-Family Partnership, and (3) Child-Parent Psychotherapy. These interventions were selected because they represent distinct approaches for improving the adaptation of at-risk populations of young children, and because they have multiple empirical studies supporting their efficacy.

High/Scope Perry Preschool Program (www.highscope.org)

This project, which began in 1962 and continues to the present day, demonstrates how a secondary prevention approach primarily targeting intellectual development and early childhood educational risk factors can be effective in producing long-term positive outcomes in factors associated with mental health (Nores, Belfield, Barnett, & Schweinhart, 2005; Schweinhart, 2002, 2007). Participants included 123 African American children ages three to four living in Ypsilanti, Michigan. Children were classified as at-risk and selected for program participation based on low parental education, SES, and Stanford-Binet IQ scores, with no biologically based mental deficiencies. Participants were randomly assigned to the program or to a comparison group that did not receive the program. The intervention consisted of 2.5 hours of preschool five days a week. The staff-to-child ratio was one adult for five to six children, and teachers were trained in special education and early childhood. Teachers also visited each child's family for 1.5 hours a week, and parents participated in monthly small group meetings with other parents. Follow-up data were collected annually when children were between the ages of 3 and 11 and at ages 4, 15, 19, 27, and 40.

The data show significant intervention effects. At age forty, significantly more program males were employed and fewer were incarcerated. There was significantly lower lifetime criminal activity, fewer health problems, greater wealth accumulation, and greater educational attainment for the program group. Cost-benefit analyses show an economic benefit to society ranging from $6.14–$6.87 for every dollar spent. These findings and data from numerous other early childhood education programs, such as the Abecedarian Project (Campbell, Ramey, Pungello, Sparling, & Miller-Johnson, 2002) and Chicago Child-Parent Preschool Center Program (Ou & Reynolds, 2006) underscore the importance of interventions that target the child's ecological milieu as a way of enhancing early childhood mental health. As demonstrated by Reynolds, Ou, and Topitzes' (2004) path analysis of data from the Chicago longitudinal study, changes in school quality and support, cognitive functioning, and parental support result-ing from program participation lead to change in long-term functioning, including adult health (see review, Keating & Simonton, 2008).

Nurse-Family Partnership (NFP)

This intervention addresses mental health by targeting a variety of biological and environmental risk factors thought to place the child at increased risk for psychopathology (Olds, 2007). Goals of intervention include improv-ing pregnancy outcomes by focusing on prenatal health, improving child outcomes by supporting competent and sensitive parenting, and helping parents plan for their own and their family's future (e.g., family planning, educational and work goals). The intervention begins during the mother's first pregnancy, continues to the child's second birthday, and consists of parent education, enhancement of support systems, linkages to commu-nity services, and emphasis on maternal and family strengths. The model is thought to reduce risk in part by interrupting potentially harmful gene-by-environment interactions (Olds, 2007). For example, reduction in alcohol and tobacco use associated with NFP is likely to have the greatest bene-fit for children genetically vulnerable to the effects of these agents. Three randomized trials examine the efficacy of this program with three dif-ferent samples: 1) Elmira, which involved predominantly White families; 2) Memphis, with primarily African American families, and 3) Denver, which included a large number of Latinos (see Olds, 2007, for a summary of these studies). Trials involved randomization to NFP or comparison services.

Combined results from the studies demonstrate intervention effects across a number of domains including prenatal health behaviors (reduc-tion in cigarette smoking and improvement in diet), pregnancy and delivery health-related outcomes (kidney, yeast infection, and hypertension), sen-sitive and competent parental care, fewer maltreatment reports, longer intervals between first and second children, and better long-term child outcomes (Eckenrode et al., 2001; Olds, Henderson, Kitzman et al., 1998; Olds, Henderson, Eckenrode et al., 1998). Longitudinal data collected fif-teen years after the program ended showed that NFP families were less likely to show negative outcomes in response to subsequent life stressors. The authors suggest that the program helped enhance self-efficacy and cop-ing skills that served to buffer families long after intervention ended (Izzo et al., 2005). Cost effectiveness analyses estimate a savings of $17,000 per family.

Child-Parent Psychotherapy

Relationship-based treatment makes flexible use of a variety of intervention modalities, including exploration of the child's and the parents' subjective experience, interpretation, developmental guidance, emotional support, modeling, concrete assistance with problems of living, and crisis intervention to provide parents and child with an all-encompassing therapeutic experience. Intervenors make it explicit that they are interested in the totality of the family's life and most particularly in the child's place within the family. Although originating in psychoanalytic theory and strongly influenced by attachment theory, this approach also incorporates intervention strategies that are consistent with social learning and cognitive behavioral approaches, including reframing, normalizing, and practicing new behavioral responses to challenging situations (Fraiberg, 1980; Lieberman, 1991, 2004; Lieberman, Silverman, & Pawl, 2000; Lieberman & Van Horn, 2008).

A number of relationship-based therapies have evolved since the initial development of infant-parent psychotherapy by Fraiberg (1980). These therapies include Toddler-Parent Psychotherapy (TPP; Lieberman, 1992), Preschool-Parent Psychotherapy (Toth et al., 2002), and Child-Parent Psychotherapy (CPP; Lieberman & Van Horn, 2005). Although they vary based on the child's age and developmental needs, they share the guiding goal of supporting and strengthening the parent-child relationship as a vehicle for establishing and protecting the child's mental health. In addition, they incorporate a focus on the parent's relational history and on contextual factors that affect the parent-child relationship and the child's developmental trajectory (e.g., culture, socioeconomic and immigration-related stressors, and trauma). Given their common core, Lieberman (2004) advocates the use of the term "Child-Parent Psychotherapy" when describing these related treatment models. Five randomized trials support the efficacy of CPP with infants, toddlers, and preschoolers. These trials, all of which involved one year of treatment, are described next.

Two trials involved infants. The sample in the first study consisted of anxiously attached Latino infants and their mothers (Lieberman, Weston, & Pawl, 1991). Mothers were Spanish speaking and had recently immigrated from Mexico or Central America. Dyads were considered to be at-risk due to infant attachment classification, lower SES, lack of acculturation and family supports, and high incidence of maternal depression, anxiety, and trauma exposure. Treatment began when the child was twelve months and ended when the child was two years old. Posttreatment data showed significant intervention effects. Treatment group toddlers scored lower than comparison group toddlers in avoidance, resistance, and anger and higher in partnership in solving disagreements with mother. Intervention mothers had higher scores than comparison group mothers in empathy and interactiveness with their children. In addition, at posttest, mothers and toddlers in the intervention group did not differ from a comparison group of securely attached dyads on any of the measures.

The second study examined the efficacy of intervention with infants considered at-risk because they had been maltreated or were living in maltreating families (Cicchetti, Rogosch, & Toth, 2006). The majority (74.1 percent) were reported to be ethnic minorities. Children were randomized to CPP, a Psychoeducational Parenting Intervention (PPI), or Community Standard (CS). Results indicated significant changes in attachment classification for CPP and PPI groups compared to CS. In the CPP group, the rate of secure attachment changed from 3.1 percent to 60.7 percent, whereas it remained stable for the CS group: 32.7 percent at intake and 38.6 percent at exit. Similar change was found for the PPI group.

One randomized trial involved toddlers. Cicchetti and colleagues examined CPP's ability to ameliorate the negative effects of maternal depression (Cicchetti, Toth, & Rogosch, 1999; Cicchetti, Rogosch, & Toth, 2000; Toth, Rogosch, Cicchetti, & Manly, 2006). Participants included mothers of toddlers who had experienced major depression at some time since their child's birth. Mothers were predominantly White (92.4 percent) and not economically disadvantaged; 73.4 percent were ranked in the two highest socioeconomic levels according to Hollingshead's four-factor index. Dyads were randomly assigned to CPP or comparison. The data showed significant intervention effects for cognitive and relational functioning. At posttest, when children were aproximately thirty-six months old, CPP children showed significantly greater improvements in attachment security and cognitive abilities than children who did not receive the intervention.

Two randomized trials examined CPP's efficacy with preschool-aged children. Toth et al. (2002) examined the efficacy of CPP in altering maltreated preschoolers' representations of their mothers and themselves. They randomly assigned low-income, maltreated preschoolers, of whom 76.2 percent were reportedly ethnic minorities, to CPP, a psychoeducation home visitation program (PHV), or CS. Multiple findings suggested that CPP was more effective than PHV and CS in improving representations of self and caregivers. CPP children showed greater declines in maladaptive maternal attributions and negative self-representations, and improvements in relationship expectations. Lieberman, Van Horn, and Ghosh Ippen (2005) randomly assigned children aged three to five years old to CPP or case management plus community treatment (CM-CT). Children were ethnically diverse (37 percent mixed ethnicity, 28 percent Latino, 14.5 percent African American, 10.5 percent White, 7 percent Asian, and 2 percent of other ethnicity). They were considered to be at-risk due to exposure to interparental domestic violence and other types of traumas, including physical abuse (49 percent), community violence (46.7 percent), and sexual abuse (14.4 percent). In addition, families were predominantly low income, with 41 percent having incomes below the federal poverty limit, and mothers had experienced multiple chronic traumas. The data revealed significant CPP intervention effects despite the fact that many families experienced additional stressors and traumas over the course of treatment. At posttreatment, CPP children showed significantly greater reductions in total behavior problems and traumatic stress symptoms. CPP mothers showed significantly greater reductions in avoidant symptomatology. Follow-up data showed that improvements in children's behavior problems and maternal symptoms continued six months after treatment ends (Lieberman, Ghosh Ippen, & Van Horn, 2006).

The results detailed here are particularly significant given that children in these studies faced multiple significant risks including poverty, maternal psychopathology, and interpersonal trauma. Together, the studies show that a core focus on the parent-child relationship and important contextual factors that affect this relationship is an efficacious way to improve the mental health of young children. They demonstrate that interventions of this type not only reduce symptoms but also increase protective factors associated with a healthy developmental trajectory.

Public Policy and Intervention in Infancy and Early Childhood

The models described earlier deliver intervention in the contexts where children and parents live (schools and home). They target multiple risk factors and enhance protective factors. Research demonstrates clearly that intervention is feasible, efficacious, and cost effective and results in significant positive changes in young children's developmental trajectories. Intensive interventions such as CPP can result in change even in the presence of maternal depression and maternal and child trauma exposure. The data also show that there are many pathways to mental health, so there are many ways to intervene to safeguard the mental health of young children.

However, it is not enough to have effective intervention models. We need societal-level mechanisms to enable these models and others like them to be implemented on a regular basis so they reach those children at-risk. Studies on adverse childhood experiences (www.acestudy.org) demonstrate that risks such as childhood physical, sexual, or emotional abuse, parental substance abuse, and interparental domestic violence are related to a host of negative physical and mental health outcomes including depression, suicide attempts, smoking, and alcoholism (Chapman et al., 2004; Dube et al., 2003). Early childhood intervention holds the promise of preventing some of these negative outcomes. As noted by Knitzer (2007), "new policy priorities are clearly in order." Among many other things, we need policies and funds that support high-quality early education programs, intensive preventive and intervention services, training of early childhood mental health professionals to build capacity for early childhood mental health, and screening to identify children at risk. In addition, we need a shift in policies that permits, encourages, and funds intervention for caregivers of vulnerable children. As we look at the next millennium, knowing what the field has learned about early childhood mental health and the importance of intervening early to prevent long-term biological, emotional, cognitive, and social impairments, we must intervene not only in the lives of the families with whom we work, but in the systems in which we exist.

REFERENCES

Ainsworth, M. D. S., Blehar, M. C., Waters, E., & Wall, S. (1978). *Patterns of Attachment: A Psychological Study of the Strange Situation.* Hillsdale, NJ: Erlbaum.

Anda, R. F., Felitti, V. J., Bremner, J. D., Walker, J. D., Whitfield, C., Perry, B. D., Dube, S. R., & Giles W. H. (2006). The enduring effects of abuse and related adverse experiences in childhood: A convergence of evidence from neurobiology and epidemiology. *European Archives of Psychiatry and Clinical Neuroscience,* 256, 174–86.

Bennett, N. G., Li, J., Song, Y., & Yang, K. (1999). *Young Children in Poverty: A Statistical Update.* New York: National Center for Children in Poverty, Columbia University Mailman School of Public Health.

Bianchi, S. M., & Raley, S. (2005). Time allocation in working families. In S. M. Bianchi, L. M., Casper, & R. B. King (Eds). *Work, Family, Health, and Well-Being.* Mahwah: NJ: Lawrence Erlbaum and Associates.

Bogat, G. A., DeJonghe, E., Levendosky, A. A., Davidson, W. S., & von Eye, A. (2006). Trauma symptoms among infants exposed to intimate partner violence. *Child Abuse and Neglect,* 30(2), 109–25.

Campbell, F. A., Ramey, C. T., Pungello, E., Sparling, J., & Miller-Johnson, S. (2002). Early childhood education: Young adult outcomes from the Abecedarian Project. *Applied Developmental Science,* 6(1), 42–57.

Caspi, A., Sugden, K., Moffitt, E. T. E., et al. (2003). Influence of life stress on depression: Moderation by a polymorphism in the 5-HTT gene. *Science*, 301(5631), 386–9.

Chapman, D. P., Whitfield, C. L., Felitti, V. J., Dube, S. R., Edwards, V. J., & Anda, R. F. (2004). Adverse childhood experiences and the risk of depressive disorders in adulthood. *Journal of Affective Disorders*, 82, 217–25.

Childtrends Data Bank. (2007). Children in poverty. Downloaded June 15, 2007, from http://www.childtrendsdatabank.org/pdf/4_PDF.pdf.

Cicchetti, D., & Lynch, M. (1993). Toward an ecological/transactional model of community violence and child maltreatment: Consequences for children's development. *Psychiatry: Interpersonal and Biological Processes*, 56(1), 96–118.

Cicchetti, D., Rogosch, F. A., & Toth, S. L. (2000). The efficacy of toddler-parent psychotherapy for fostering cognitive development in offspring of depressed mothers. *Journal of Abnormal Child Psychology*, 28, 135–48.

Cicchetti, D., Rogosch, F. A., & Toth, S. L. (2006). Fostering secure attachment in infants in maltreating families through preventive interventions. *Development and Psychopathology*, 18, 623–50.

Cicchetti, D., & Sroufe, L. A. (2000). Editorial: The times, they've been a'changing. Special Issue: Reflecting on the past and planning for the future of developmental psychopathology. *Development and Psychopathology*, 12(3), 255–64.

Cicchetti, D., Toth, S., & Maughan, A. (2000). An ecological-transactional model of child maltreatment. In A. J. Sameroff, M. Lewis, & S. M. Miller (Eds.), *Handbook of Developmental Psychopathology* (2nd ed., pp. 689–722). Dordrecht, The Netherlands: Kluwer Academic.

Cicchetti D., Toth S. L., Rogosch F. A. (1999). The efficacy of toddler-parent psychotherapy to increase attachment security in offspring of depressed mothers. *Attachment and Human Development* 1, 34–66.

Cummings, M., & Davies, P. T. (1994). Maternal depression and child development. *Journal of Child Psychology and Psychiatry*, 35, 73–112.

Downey, G., & Coyne, J. C. (1990). Children of depressed parents. *Psychological Bulletin*, 108, 50–76.

Dube, S. R., Felitti, V. J., Dong, M., Giles, W. H., & Anda, R. F. (2003). The impact of adverse childhood experiences on health problems: Evidence from four birth cohorts dating back to 1900. *Preventive Medicine*, 37, 268–77.

Eckenrode, J., Zielinski, D., Smith, E., Marcynyszyn, L., Henderson, C., Kitzman, H., Cole, R., Powers, J., & Olds, D. (2001). Child maltreatment and the early onset of problem behaviors: Can a program of nurse home visitation break the link? *Development and Psychopathology*, 13(4), 873–90.

Fraiberg, S. (1980). *Clinical Studies in Infant Mental Health.* New York: Basic Books.

Galinsky, E., Howes, C., Kontos, S., & Shinn, M. (1994). *The Study of Children in Family Child Care and Relative Care.* New York: Families and Work Institute.

Goldberg, S. (1990). Attachment in infants at risk: Theory, research, and practice. *Infants and Young Children*, 2, 11–20.

Greenough, W. T. (1991). Experience as a component of normal development: Evolutionary considerations. *Developmental Psychology*, 27, 14–17.

Greenough, W. T., & Black, J. E. (1992). Induction of brain structure by experience: Substrates for cognitive development. In M. R. Gunnar & C. A. Nielson

(Eds.), *Developmental Behavior Neuroscience*, Vol. 24 (pp. 155–200). Hillsdale, NJ: Erlbaum.

Greenspan, S. I., & Wieder, S. (1993). Regulatory disorders. In C. H. Zeanah (Ed.), *Handbook of Infant Mental Health* (pp. 280–90). New York: Guilford Press.

Harris, W. W., Lieberman, A. F., & Marans, S. (2007). In the best interests of society. *Journal of Child Psychology and Psychiatry*, 48(3/4), 392–411.

Helburn, S. W. (Ed.). (1995). *Cost, Quality, and Child Outcomes in Child Care Centers, Technical Report*. Denver, CO: Department of Economics, Center for Research in Economic and Social Policy, University of Colorado at Denver.

Howes, C., Rodning, C., Galluzzo, D., & Myers, I. (1988). Attachment and childcare: Relationships with mother and caregiver. *Early Childhood Research Quarterly*, 3, 403–16.

Hrdy, S. B. (1999). *Mother nature: A history of mothers, infants, and natural selection.* New York: Pantheon Books.

Izzo, C. V., Eckenrode, J. J., Smith, E. G., Henderson, C. R., Cole, R., Kitzman, H., & Olds, D. L. (2005). Reducing the impact of uncontrollable stressful life events through a program of nurse home visitation for new parents. *Prevention Science*, 6(4), 269–74.

Johnson, D. L., & Walker, T. (1991). A follow-up evaluation of the Houston Parent-Child Development Center: School performance. *Journal of Early Intervention*, 15(3), 226–36.

Kagan, J. (1981). *The Second Year of Life.* Cambridge: Harvard University Press.

Kaufman, J., Yang, B. Z., Douglas-Palumberi, H. et al. (2006). Brain-derived neurotropic factor-5-HTTLPR gene interaction and environmental modifiers of depression in children. *Biological Psychiatry*, 59, 958–65.

Keating, D. P., & Simonton, S. Z. (2008). Health effects of human development policies. In J. House, R. Schoeni, A. Pollack, & G. Kaplan (Eds.), *Making Americans Healthier*. New York: Russell Sage.

Knitzer, J. (2000). Early childhood mental health programs. In J. P. Shonkoff & S. J. Meisels, (Eds.), *Handbook of Early Childhood Intervention, Second Edition* (pp. 906–56). New York: Cambridge University Press.

Knitzer, J. (2007). Putting knowledge into policy: Toward an infant-toddler policy agenda. *Infant Mental Health Journal*, 28(2), 237–45.

Lamb, M. E. (1998). Nonparental child care: Context, quality, correlates. In W. Damon, I. E. Sigel, & K. A. Renninger (Eds.), *Handbook of Child Psychology, Volume 4: Child Psychology in Practice, 5th Edition* (pp. 73–134). New York: John Wiley & Sons, Inc.

Lennon, M. C., Aber, J. L., & Blum, B. B. (1998). *Program, Research, and Policy Implications of Evaluations of Teenage Parent Programs*. New York: Research Forum on Children, Families, and the New Federation.

Lieberman, A. F. (1991). Attachment theory and infant-parent psychotherapy: Some conceptual, clinical, and research considerations. In Cicchetti, D., Toth, S. L. (Eds.), *Rochester Symposium on Developmental Psychopathology, Vol. 3: Models and Integrations* (pp. 261–87). Rochester, NY: University of Rochester Press.

Lieberman, A. F. (1992). Infant-parent psychotherapy with toddlers. *Development and Psychopathology*, 4, 559–574.

Lieberman, A. F. (1993). *The Emotional Life of the Toddler.* New York. Free Press.

Lieberman, A. F. (2004). Traumatic stress and quality of attachment: Reality and internalization in disorders of infant mental health. *Infant Mental Health Journal,* 25, 336–51.

Lieberman, A. F., Ghosh Ippen, C., & Van Horn, P. (2006). Child-Parent Psychotherapy: 6-month follow-up of a randomized controlled trial. *Journal of the American Academy of Child and Adolescent Psychiatry,* 45, 913–18.

Lieberman, A. F., Silverman, R., & Pawl, J. H. (2000). Infant-Parent Psychotherapy: Core concepts and current approaches. In C. H. Zeanah (Ed.), *Handbook of Infant Mental Health, Second Edition* (pp. 472–84). New York: Guilford.

Lieberman, A. F., & Van Horn, P. (2005). *Don't Hit My Mommy: A Manual for Child-Parent Psychotherapy with Young Witnesses of Family Violence.* Washington, DC: Zero to Three Press.

Lieberman, A. F., & Van Horn, P. (2008). *Psychotherapy with Infants and Young Children: Repairing the Effects of Stress and Trauma on Early Attachment.* New York: Guilford.

Lieberman, A. F., Van Horn, P. J., & Ghosh Ippen, C. (2005). Toward evidence-based treatment: Child-Parent Psychotherapy with preschoolers exposed to marital violence. *Journal of the American Academy of Child and Adolescent Psychiatry,* 44, 1241–8.

Lieberman, A. F., Weston, D. R. & Pawl, J. H. (1991). Preventive intervention and outcome with anxiously attached dyads. *Child Development,* 62, 199–209.

McDonald, R., Jouriles, E. N., Ramisetty-Mikler, S., Caetano, R., & Green, C. E. (2006). Estimating the number of American children living in partner-violent families. *Journal of Family Psychology,* 20(1), 137–42.

Meisels, S. J., & Shonkoff, J. P. (2000) Early childhood intervention: A continuing evolution. In J. P. Shonkoff & S. J. Meisels (Eds.), *Handbook of Early Childhood Intervention, Second Edition* (pp. 3–34). New York: Cambridge University Press.

NICHD Early Child Care Network. (1996). Characteristics of infant child care: Factors contributing to positive caregiving. *Early Childhood Research Quarterly,* 11(3), 269–306.

Nores, M., Belfield, C. R., Barnett, W. S., & Schweinhart, L. (2005). Updating the economic impacts of the High/Scope Perry Preschool Program. *Education Evaluation and Policy Analysis,* 27(3), 245–61.

Olds, D. L. (2007). Preventing crime with prenatal and infancy support of parents: The Nurse-Family Partnership. *Victims & Offenders,* 2, 205–25.

Olds, D., Henderson, C. Jr., Eckenrode, J., Pettit, L. M., Kitzman, H., Cole, B., Robinson, J., & Powers, J. (1998). Reducing risks for antisocial behavior with a program of prenatal and early childhood home visitation. *Journal of Community Psychology,* 26(1), 65–83.

Olds, D., Henderson, C. Jr., Kitzman, H., Eckenrode, J., Cole, R., & Tatelbaum, R. (1998). The promise of home visitation: Results of two randomized trials. *Journal of Community Psychology,* 26(1), 5–21.

Olson, K., & Pavetti, L. (1996). *Personal and Family Challenges to the Successful Transition from Welfare to Work.* Washington, DC: The Urban Institute.

Ou, S., & Reynolds, A. J. (2006). Early childhood interventions and educational attainment: Age 22 findings from the Chicago Longitudinal Study. *Journal of Education for Students Placed at Risk,* 11(2), 175–98.

Parker, F. L., Piotrkowski, C. S., Horn, W., & Greene, S. (1995). The challenge for Head Start: Realizing its vision as a two-generation program. In I. Sigel & S. Smill (Eds.), *Advances in applied developmental psychology, Vol. 9: Two-generation programs for families in poverty* (pp. 135–59). New Jersey: Ablex.

Pawl, J. H., & St. John, M. (1998). *How You Are Is as Important as What You Do.* Washington, DC: Zero to Three/National Center for Clinical Infant Programs.

Reynolds, A. J., Ou, S., & Topitzes, J. W. (2004). Path of effects of early childhood intervention on educational attainment and delinquency: A confirmatory analysis of the Chicago Child-Parent Centers. *Child Development*, 75(5), 1299–1328.

Rutter, M., Moffit, T. E., & Caspi, A. (2006). Gene-environment interplay and psychopathology: Multiple varieties but real effects. *Journal of Child Psychology and Psychiatry*, 47, 226–61.

Rutter, M., & Sroufe, A. (2000). Developmental psychopathology: Concepts and challenges. *Development and Psychopathology*, 12, 265–96.

Sameroff, A. J. (2000). Developmental systems and psychopathology. *Development and Psychopathology*, 12, 297–312.

Sameroff, A. J., Bartko, W. T., Baldwin, A., Baldwin, C., & Seifer, R. (1998). Family and social influences on the development of child competence. In M. Lewis & C. Feiring (Eds.), *Families, Risk, and Competence* (pp. 161–85). Mahwah, NJ: Erlbaum.

Sameroff, A. J., & Emde, R. N. (1989). *Relationship Disturbances in Early Childhood: A Developmental Approach.* New York: Basic Books.

Sameroff, A. J. & Fiese, B. H. (2000). Transactional regulation: The developmental ecology of early intervention. In J. P. Shonkoff & S. J. Meisels (Eds.), *Handbook of Early Childhood Intervention* (pp. 135–59). Cambridge, UK: Cambridge University Press.

Sameroff, A. J., Seifer, R., & Zax, M. (1982). Early development of children at risk for emotional disorder. *Monographs of the Society for Research in Child Development*, 47(7), Serial No. 199.

Scheeringa, M. S., & Zeanah, C. H. (2001). A relational perspective on PTSD in early childhood. *Journal of Traumatic Stress*, 14(4), 799–815.

Scheeringa, M. S., Zeanah, C. H., Myers, L., & Putnam, F. W. (2005). Predictive validity in a prospective follow-up of PTSD in preschool children. *Journal of the American Academy of Child and Adolescent Psychiatry*, 44(9), 899–906.

Schor, J. (1992). *The Overworked American: The Unexpected Decline of Leisure.* New York: Basic Books.

Schweinhart, L. J. (2002). How the High/Scope Perry Preschool Study grew: A researcher's tale. Phi Delta Kappa Center for Evaluation, Development, and Research, No. 32. Downloaded June 15, 2007, from http://www.highscope.org/Content.asp?ContentId=232.

Schweinhart, L. J. (2007). Crime prevention by the High/Scope Perry Preschool Program. *Victims and Offenders*, 2, 141–60.

Shonkoff, J. P., & Phillips, D. A. (2000). *From Neurons to Neighborhoods: The Science of Early Childhood Development.* Washington, DC: National Academy Press.

Sroufe, L. A. (1979). Socioemotional development. In J. D. Osofsky (Ed.), *Handbook of Infant Development* (pp. 462–515). New York: John Wiley & Sons.

Sroufe, L. A., Carlson, E. A., Levy, A. K., & Egeland, B. (1999). Implications of attachment theory for developmental psychopathology. *Development and Psychopathology* 11, 1–13.

Sroufe, L. A., Carlson, E., & Schulman, S. (1993). Individuals in relationships: Development from infancy through adolescence. In D. C. Funder, R. D. Parke, C. A. Tomlinson-Keasey, & K. Widaman K. (Eds.) *Studying Lives through Time: Personality and Development* (pp. 315–42). Washington, DC: American Psychological Association.

Sroufe, L. A., Egeland, B., Carlson, E., & Collins, W. A. (2005). *The Development of the Person: Minnesota Study of Risk and Adaptation from Birth to Adulthood.* New York: Guilford Press.

Stern, D. N. (1995). *The Motherhood Constellation: A Unified View of Parent-Infant Psychotherapy.* New York: Basic Books.

Thomas, A., Chess, S., & Birch, H. (1968). *Temperament and Behavior Disorders in Children.* New York: Brunner/Mazel.

Thompson, R. A. (1999a). Early attachment and later development. In J. Cassidy & P. R. Shaver (Eds.), *Handbook of Attachment: Theory, Research, and Clinical Applications* (pp. 265–86). New York: Guilford.

Thompson, R. A. (1999b). The individual child: Temperament, emotion, self, and personality. In M. H. Bornstein & M. E. Lamb (Eds.), *Developmental Psychology: An Advanced Textbook, Fourth Edition* (pp. 377–409). Mahwah, NJ: Lawrence Erlbaum Associates, Publishers.

Tiernari, P., Wynne, L. C., Sorri, A. et al. (2004). Genotype-environment interaction in schizophrenia-spectrum disorder: Long-term follow-up of Finnish adoptees. *British Journal of Psychiatry*, 184, 216–22.

Toth, S. L., Maughan, A., Manly J. T., Spagnola, M., & Cicchetti, D. (2002). The relative efficacy of two interventions in altering maltreated preschool children's representational models: Implications for attachment theory. *Developmental Psychopathology*, 14, 877–908.

Toth, S. L., Rogosch, F. A., Cicchetti, D., & Manly, J. T. (2006). The efficacy of Toddler-Parent Psychotherapy to reorganize attachment in young offspring of mothers with major depressive disorder: A randomized trial reorganizes attachment in the young offspring of mothers with major depressive disorder. *Journal of Consulting & Clinical Psychology*, 74(6), 1006–16.

U.S. Department of Health and Human Services (2007). Child maltreatment 2005. Retrieved June 1, 2007, from http://www.acf.hhs.gov/programs/cb/pubs/cm05/cm05.pdf.

van den Boom, D. C. (1994). The influence of temperament and mothering on attachment and exploration: An experimental manipulation of sensitive responsiveness among lower-class mothers with irritable infants. *Child Development*, 65, 1457–77.

van den Boom, D. C. (1995). Do first-year intervention effects endure? Follow-up during toddlerhood of a sample of Dutch irritable infants. *Child Development*, 66, 1798–1816.

van Ijzendoorn, M. H., Juffer, F., & Duyvesteyn, M. G. C. (1995). Breaking the intergenerationl cycle of insecure attachment: A review of the effects of attachment-based intervention on maternal sensitivity and infant security. *Journal of Child Psychology and Psychiatry and Allied Disciplines*, 36(2), 225–48.

Winnicott, D. W. (1965). The theory of the parent-infant relationship. In D. W. Winnicott (Ed.), *The Maturational Processes and the Facilitating Environment* (pp. 37–55). New York: International Universities Press.

Wolff, E. N. (1996). *The Economic Status of Parents in Postwar America.* Paper prepared for the Task Force on Parent Empowerment., September 20, 1996, p. 9.

Zeanah, C. H., Boris, N. W., & Larrieu, J. A. (1997). Infant development and developmental risk: A review of the past 10 years. *Journal of the American Academy of Child and Adolescent Psychiatry*, 36, 165–78.

Zero to Three. (2004). *Diagnostic Classification of Mental Health and Developmental Disorders of Infancy and Early Childhood.* Revised. Arlington, VA.

9

Bringing a Population Health Perspective to Early Biodevelopment: An Emerging Approach

CLYDE HERTZMAN

This chapter describes the rationale for population-based monitoring of early child development (ECD); outlines our achievements to date in producing such ECD data for a large population (the kindergarten children of British Columbia, Canada); and shows how this work can support biodevelopmental research regarding the extent to which social gradients in brain and biological development are explained by experience "getting under the skin."

GRADIENTS IN EARLY CHILD DEVELOPMENT

Inequalities in child development emerge in a systematic fashion over the first five years of life, according to well-recognized factors: family income, parental education, parenting style, neighborhood safety and cohesion, neighborhood socioeconomic status, and access to quality childcare and other developmental opportunities. Three broad domains of ECD are of special relevance: physical, social/emotional, and language/cognitive. By age 5, as one goes from the families with the lowest to highest incomes, least to most parental education, and least to most nurturing and interactive parenting style, the average quality of early child experiences gradually increases and produces a gradient in child development (Hertzman et al., 2002). For example, data from the National Longitudinal Survey of Children and Youth from the late 1990s (Willms, 2002) showed a gradient in the risk of receptive language delay, increasing *gradually* from the children in the highest income decile of Canadian families (5.2 percent delayed) to the poorest decile (approximately 26 percent delayed).

By kindergarten age, development has been influenced by factors at all levels of society. These can be aggregated into three groups, from most proximal to most distal from the child: the family, the neighborhood or local

217

community, and the broader social/economic/political environment. At the level of the family, the qualities of stimulation, support, and nurturance in intimate circumstances contribute the most (Keating & Hertzman, 1999). These qualities, in turn, are influenced by the resources that families have to devote to child raising (represented by income); to their style of parenting; and to their tendency to provide a rich and responsive environment for language (often, but not always, associated with parental levels of formal education – Hart & Risley, 1995), and for play.

At the level of the neighborhood, children growing up in safe areas where the community is cohesive in relation to children – where it mobilizes resources formally (creates programs) and informally (treats its children like they belong there) – are less likely to be vulnerable in their development than children from similar family backgrounds living in unsafe and noncohesive neighborhoods. Children who have stable neighborhood environments during their early years also tend to be less vulnerable than those who are constantly changing their place of residence. Similarly, children from family backgrounds that contain multiple developmental risk factors tend to do better growing up in mixed socioeconomic neighborhoods than in poor ghetto areas (Kohen et al., 2002).

Finally, at the level of society, access to quality programs matters (Schweinhart, 2004; Ramey & Ramey, 2005). This includes the full range of childcare, family support, and family-strengthening programs; public health programs for high-risk children; and vision, hearing, dental, and broader social safety net functions such as parental leave and housing programs. Thus, the state of child development in any society is an "emergent property" of a complex of factors, many of them modifiable, at the intimate, civic, and societal level, that influence each child in unique combinations (Organisation for Economic Cooperation and Development, 2006).

DEVELOPMENTAL TRAJECTORIES

Exposure to both beneficial and adverse circumstances over the life course varies for each individual and constitutes a unique "life exposure trajectory," which will manifest as different expressions of health, well-being, behavior, and learning skills. Studies that collect data from the earliest stages of life and follow individuals over time provide the best lens on the relative importance of and interaction among these exposure-to-expression relationships. The possible long-term relationships cluster into three generic models that have been labeled *latency, cumulative,* and *pathway. Latency* refers to relationships between an exposure at one point in the life course and the probability of health expressions years or decades later, irrespective of intervening experience. *Cumulative* refers to multiple exposures over the life course whose effects on health combine. These may either be multiple exposures to a single recurrent factor (e.g., chronic poverty or persistent smoking) or a series of exposures to different factors. Finally, the term *pathways* represents dependent sequences of exposures in which exposure at one stage of the life course influences the probability of other exposures later in the life course, as well as associated expressions.

Latent, pathway, and cumulative phenomena are best understood from a human development perspective, beginning with the premise that human development, in general, and early child development, in particular, are the principal "agents" mediating the relationship among life course, human experience, and health outcome (Hertzman et al., 2001; Keating, this volume; Keating & Hertzman, 1999; National Research Council Institute of Medicine, 2000). This is in contrast to a more traditional epidemiological approach that would seek to explain discrete outcomes across the life course according to similarly discrete, prior risk factors. The developmental perspective accommodates complex forms of causation much more easily than a risk factor approach, and leaves much greater room for consideration of the underlying biological processes at work. In particular, the developmental perspective will accommodate the prospect that the three domains mentioned earlier (physical, social/emotional, and language/cognitive) can be considered, first, outcomes of the quotidian processes of daily living and, over time, determinants of subsequent health outcomes – that socioeconomic gradients in health, well-being, learning, and behavior, emerging as the life course unfolds, begin life as socioeconomic gradients in physical, social/emotional, and language/cognitive development.

To give an example, physical health in the early years is thought to be predictive of lifelong health, and illnesses such as coronary heart disease and elevated blood pressure have been statistically associated with health status in early life. Similarly, infants who are born at term but are small for their gestational age may be at increased risk for adult-onset diabetes, high blood pressure, and heart disease several decades later (Barker, 1995). Yet our research on the 1958 British Birth Cohort has suggested that the long-term consequences of gestational effects such as low birthweight are difficult to separate from early circumstances and cognitive trajectories (Jefferis et al., 2002). Taking math achievement in grade school, we have shown that high socioeconomic status (SES) children, on average, start school with above-average skills, and their advantage grows from age seven to sixteen. Low birthweight reduces, but does not eliminate, the advantage of growing up in a high-SES family. Conversely, low-SES children start school, on average, with below-average math skills, and this disadvantage increases from age 7 to 16 years. Low birthweight among children growing up in a low-SES family exacerbates this disadvantage. When these effects are combined, we have shown that low-birthweight children from privileged backgrounds still have a developmental advantage over normal-birthweight children from underprivileged backgrounds. Thus, socioeconomic disadvantage and gestational factors have both declared their developmental impacts early in the life course and parallel SES gradients in the expression of diabetes, high blood pressure, and heart disease later in life. The implication is that these disease processes are better studied by following developmental trajectories, rather than (for example) by separating low birthweight out, privileging it as a physical health antecedent, and studying it as a discrete risk factor.

A STRATEGIC PERSPECTIVE

The Human Early Learning Partnership of British Columbia has set itself the task of creating developmental trajectories for *all* BC children, with the goal of transforming the province into a population-based developmental laboratory. Given the difficulties of creating developmental trajectories on individuals, attempting population-wide trajectories might seem appallingly inefficient compared to strategic, random samples. Yet there are both political and scientific reasons for going the population-based route. On the political side, it turns out that officials are not interested in what we know about "randomly sampled" children. They are interested in what we know about *their* children – those who fall under their jurisdiction. Although political dynamics are not the subject of this chapter, it is important to note that *the more children we study, the greater the political and funding support there is for our work.*

The scientific argument is much less facile and more profound. If our long-term purpose is to understand how experience gets under the skin, it is necessary for those aspects of early experience that actually influence development to differ qualitatively among study subjects. As we will show, having knowledge of the developmental status of the total population can provide that starting point: allowing one to identify both those populations of children whose development is as we would predict it to be, based on the existing literature on the determinants of child development, as well as those who are doing systematically better or worse than predicted. In contrast, beginning with strategic samples relies on the assumption that known determinants of child development will have their predicted effects regardless of other aspects of context. If it is true that the experiences that get under the skin are the quotidian, then this is a risky assumption. Indeed, the most useful opportunities for understanding biodevelopment may come in contexts where quotidian experience is, somehow, pronounced enough to either buffer or exacerbate the impact of recognized determinants of child development.

MEASURING EARLY CHILD DEVELOPMENT AT THE LEVEL OF THE POPULATION

In 2004, under the auspices of the Human Early Learning Partnership, the province of British Columbia, Canada, became one of the world's first places to complete a population-based assessment of ECD. Assessments were done during the transition year from preschool to school, known in the United States as kindergarten, which begins when the children are in their fifth year. ECD was measured using the five scales of the Early Development Instrument (EDI): physical well-being; social competence; emotional maturity; language and cognitive development; and communication and general knowledge – that map onto the three broad domains of ECD whose importance has been discussed earlier. Early Development Indicator work began in BC in the Vancouver school district in February 2000, and all school districts had completed at least one population-based EDI evaluation by 2004 comprising data from approximately 44,000 kindergarten children from all walks of life across the province (Kershaw et al., 2006). We estimate that between 90 and 100 percent of kindergarten-age children were included from each geographic school district.

The EDI provides information that can be interpreted both backward and forward in time. The primary direction of interpretation for the purposes of ECD is backward. That is, the results of the EDI are interpreted to represent the outcome of the cumulative early experience that children in a given geographic area have had from birth to kindergarten entry. Variations in EDI outcomes by area are taken to represent average differences in the qualities of stimulation, support, and nurturance that children in those areas have experienced. The EDI can also be interpreted prospectively, in that the results frame the challenges that families, schools, communities, and governments will face in supporting their children's development from kindergarten onward (Janus & Offord, 2000).

The EDI requires kindergarten teachers to fill out a detailed checklist about each child in their class according to the five scales mentioned earlier. Each child's EDI assessment is analyzed so that the child receives a score between 0 and 10 for each scale. A score of 10 means that the kindergarten child is doing all the things she or he should be doing, all of the time, in relation to the given scale, whereas a score of 0 means she or he is not doing any of them at any time. These scores are never applied to individual children for educational or administrative purposes. Rather, they are used exclusively in a cumulative form – across neighborhoods, schools, language groups, and so on – for population-based analyses.

Our work has focused primarily on levels of childhood vulnerability within different domains of child development. For each EDI scale there is a score, somewhere between 0 and 10, that serves as a "vulnerability threshold." Children who fall below that score are said to be vulnerable in that aspect of their development. The appropriate interpretation of vulnerability is that the child is, on average, more likely to be limited in his or her development on the identified EDI scale than a child who receives scores above the cutoff. (Details of how the vulnerability cutoffs were established are beyond the scope of this chapter. For further details, please see the EDI Handbook at www.offordcentre.com/readiness/pubs/publications.html. For an overview of EDI measurement issues, see Keating, 2007). Because the EDI is not an individual diagnostic, EDI vulnerability cutoffs are particularly useful for inferences about groups of children. In other words, it is a meaningful use of the EDI to say something like:

> 20 percent of children in neighborhood A are vulnerable in their physical development, whereas only 5 percent are vulnerable in neighborhood B.

The approach taken in BC involves mapping child development according to the neighborhood of residence of the child, rather than the census unit, school catchment, or the school attended. Mapping provides a visual summary of ECD trends in the interests of making complex data meaningful to broad audiences. Maps depict information about the many intersecting environments in which families live and young children grow, including socioeconomic, natural, cultural, programmatic, and policy environments as they interact in and across neighborhood, community, regional, and provincial geographies. In this regard, maps invite observers to contemplate a broad understanding of early development that transcends the boundaries of any single policy envelope – such as education, health, childcare, welfare, or justice – to see how the interrelations between all of these areas influence children before they reach age 6 (see Figs. 9.1–9.3).

Figure 9.1 illustrates how the fifty-nine geographic school districts on all provincial maps are coded according to "quintiles" of vulnerability (that is, from the top to the bottom fifth). That is, the twelve (59/5 approximately equals 12) school districts with the smallest proportion of children

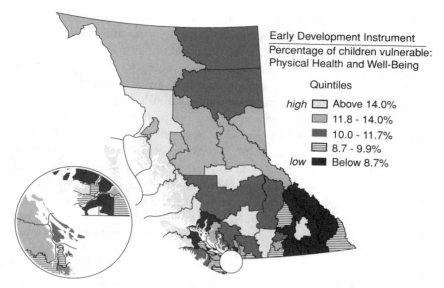

Figure 9.1. Proportion of students vulnerable on the physical health and well-being subscale of the EDI.

vulnerable on their physical development are shown as solid black ("low" in the figure key); followed by the second group in horizontal lines; the middle group in dark gray; the second highest vulnerability group in medium gray; and the most vulnerable group ("high" in the figure key) in light gray.

Children are accounted for in the geographic school district where they live, regardless of whether they go to public, independent, or Aboriginal school or if they cross school district boundaries to go to school. This is done because our primary interest is in the retrospective look at the 0- to 5-year period, and most of the determinants of ECD are found at the level of the family and neighborhood. Moreover, family types differ by neighborhood, as does the level of access to quality programs and services. Therefore, the most important population aggregation of the EDI is the neighborhood, embedded within the geographic school district.

The literature about the effects of neighborhoods on child development most frequently relies on data that are reported using census boundaries or other administrative units of analysis (Burton & Jarrett, 2000). The convenience of census or other survey boundaries comes with costs, however, including the fact that census boundaries often do not match local perceptions of neighborhood divisions. In response, we have worked closely with communities to benefit from local knowledge in determining neighborhood boundaries that more accurately reflect the daily experience of very young children. Local ECD coalition representatives were invited to draw lines on maps of their area to signal the presence of perceived divides in their community. Although some local coalitions opted to maintain the census or another existing boundary system, others opted for dramatically different breakdowns than those employed for survey data collection.

The initiative taken by local coalitions to establish neighborhood boundaries that reflect community perceptions resulted in the identification of 469 neighborhoods across the province. Our mapping team then digitized the local maps and built them into a province-wide file. The team shared this file with Statistics Canada and contracted the organization to perform a special run of BC census data that disaggregated information by the 469 neighborhoods, instead of the more traditional census boundaries.

Sample illustration – Physical Health and Well-Being Scale. The Physical Health and Well-Being scale of the EDI measures children's fine and gross motor development, levels of energy, daily preparedness for school (do they arrive late or hungry), washroom independence, and established handedness. Children who start school with age-appropriate motor skills, adequate energy levels from proper sleep and nutritional intake, and demonstrate age-appropriate independence can take full advantage of learning opportunities offered by school and enjoy engagement in social groups. Conversely, children who enter school more vulnerable in terms of their physical health are less favorably positioned to benefit from the social and educational opportunities that kindergarten provides. The typical profile of a child who falls below the physical vulnerability EDI cutoff is one who displays average or poor motor skills (both fine and gross), who is sometimes tired or hungry, usually clumsy, with flagging energy levels, and average overall physical development.

Variation in the physical vulnerability rates across BC's fifty-nine geographic school districts are shown in Figure 9.1, as described earlier. The rate of physical vulnerability among kindergarten children ranges from a low of 4.5 percent to a high of 35.1 percent across the fifty-nine districts. Figures 9.2 and 9.3 show neighborhood variation in the proportion of vulnerable children within two distinct groups of school districts. In virtually all cases, the range of variation in vulnerability within a school district, or a small group of school districts, is equal to or exceeds the variation seen between school districts across the province (Fig. 9.1). Why is this? The simple answer is that when one averages across whole geographic school districts, most of the important factors that influence child development are averaged together and their influence is lost. Instead, differences in family, community, and service access characteristics that influence child

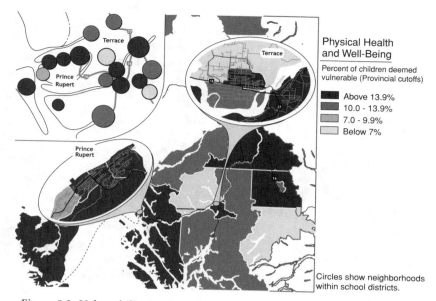

Figure 9.2. Vulnerability in physical health and well-being on the North Coast.

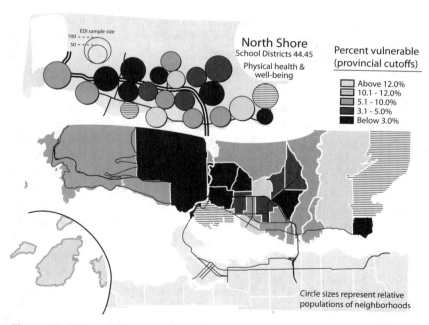

Figure 9.3. Vulnerability in physical health and well-being on the North Shore.

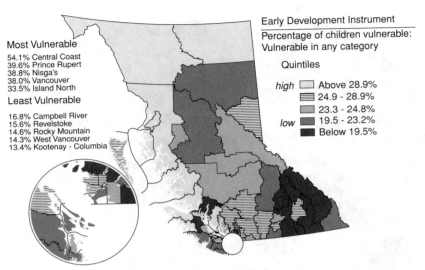

Figure 9.4. Proportion of students vulnerable on 1+ subscales of the EDI.

development reveal themselves when analysis is done at the level of the neighborhood.

Large neighborhood variations are not only seen in school districts that are moderate to high vulnerability. They are seen in low vulnerability school districts, too. Take, for example, Figure 9.3, which shows North Vancouver and West Vancouver school districts (collectively called the "North Shore"). On Figure 9.1, it can be seen that both are low vulnerability school districts, in the inset magnified area in the lower left of the figure. Yet, when broken down by neighborhood, Figure 9.3 reveals a greater than tenfold neighborhood variation in vulnerability. Although six neighborhoods are shown as solid black (less than 3.0 percent rate of physical health vulnerability), four are light gray (greater than a 12.0 percent rate), and the highest rate is more than 19 percent. That is, even in a low vulnerability school district there are high vulnerability neighborhoods by provincial standards. What, then, distinguishes a relatively low vulnerability school district from a relatively high vulnerability school district? The answer is that, in the high vulnerability school districts, a larger proportion of neighborhoods within the district show high vulnerability compared with the proportion of such neighborhoods in the low vulnerability school districts. Yet both groups of school districts will have both low and high vulnerability neighborhoods.

Vulnerability on one or more scales of the EDI: Figure 9.4 presents the share of children in each geographic school district who are vulnerable on *one or more* scales of the EDI. The rate of vulnerability on at least one scale varies from a low of 13.4 percent to a high of 54.1 percent, a fourfold difference among school districts. The districts that report the five lowest vulnerability rates are Kootenay-Columbia, West Vancouver, Rocky Mountain, Revelstoke, and Campbell River. Among these districts only one, West Vancouver, is highly affluent. The other four are stable resource communities that do not have wealthy or highly educated populations. The existence of such communities, which support better ECD than would be predicted on the bases of SES alone, validates our strategy of studying the whole population and, also, creates opportunities for studying how experience gets under the skin that will be described later in this chapter.

"ANCHORING" DEVELOPMENTAL TRAJECTORIES

The provincially standardized method for assessing student progress through school in BC is the Foundation Skills Assessment (FSA) program. Each year students at Grades 4 and 7 complete a series of tests on reading, arithmetic, and other academic skills that are then categorized as meeting expectations, exceeding expectations, or failing to meet expectations. It is also possible to identify those children who do not show up to school on the day that the FSAs are written or who begin the tests but do not complete them. Our plan is to link EDI forward in time to FSA scores on an individual basis, and backward in time to birth and health records, thus creating a developmental trajectory for each BC child who has done the EDI, starting with birth experience (birthweight, etc.), proceeding to early health status (inferred through medical and hospital services utilization), to state of development in kindergarten (using the EDI), to progress through school (using the FSAs). From these data, we will be able to identify more or less successful developmental trajectories, which can then be reaggregated by neighborhood or school in order to properly "nest" individuals within the context of interest. As of this writing, we have matched the children's unique Personal Education Number (PEN) to 94 percent of the EDI assessments carried out between 2000 and 2005 (making it possible to relate EDI to FSA on an individual basis); we have matched 97 percent of the Personal Education Numbers to a Personal Health Number (making it possible to merge health, education, and developmental information); and we have created a physical infrastructure and ethical framework that will allow us to link birth, physician and hospital services, EDI, and FSA records together on a person-specific, population-based but anonymous basis.

Table 9.1. *The influence of EDI vulnerability on school success*

EDI	Grade 4 FSA	
# of vulnerabilities	% Writing but failing to meet expectations	% Not passing
Numeracy		
0	7.5	12.3
1	11.8	22.2
2–3	18.7	33.8
4–5	27.5	55.6
Reading		
0	13.6	17.8
1	26.7	33.9
2–3	29.5	43.1
4–5	48.4	68.3

To date we have created EDI to FSA linkages for those children (approximately 6,800) who completed the EDI in either 2000 or 2001 and, thus, had completed their Grade 4 FSAs in 2004 or 2005. Table 9.1 shows some preliminary data from that linkage. The left-hand column of the table shows the number of EDI scales that each child was vulnerable on when they were in kindergarten (from 0 to 5 scales of the EDI). The middle column shows the proportion of those children who wrote and "failed to meet expectations" on each of the numeracy and reading tests of the FSA. The final column includes those who "failed to meet expectations," but also those who did not complete the FSA or who did not show up at school on the day that it was written; in other words, all who did not pass for one reason or another.

The patterns in Table 9.1 are similar for both numeracy and reading. In each case the proportion of children "failing to meet expectations" or "not passing" increases in a stepwise fashion with the number of vulnerabilities on the EDI. In other words, increasing developmental vulnerability at kindergarten is associated with school success regardless of the domain(s) in which vulnerability is found. Moreover, the effect size is large, with the proportions changing 3.5- to 4.5-fold from zero EDI vulnerabilities to four to five EDI vulnerabilities.

As we would expect, the proportion of children "not passing" is always higher than the proportion "failing to meet expectations." However, the table shows a further pattern. As one goes from children with zero to four to five EDI vulnerabilities, the gap between the proportions "failing to meet expectations" and "not passing" gets wider and wider. This means that as the number of EDI vulnerabilities in kindergarten rises, not only are the children less likely to "meet expectations" in Grade 4 but they are also less likely to attend and participate fully at that age.

SOCIOECONOMIC STATUS AND EDI VULNERABILITY

There is considerable evidence that the geography of opportunity has a significant statistical impact on a child's development irrespective of the SES of the child's household (Brooks-Gunn et al., 1993; Kohen et al., 2002; Boyle & Lipman, 2002; Curtis et al., 2004). Although statistically significant, however, the neighborhood effect is typically reported to be modest, accounting for between 5 and 10 percent of the variance in child outcomes (Burton & Jarrett, 2000) or less (Kohen et al., 2002). The approach that we have taken to the relationship between SES and *neighborhood* rates of vulnerability differs in an important way from most previous studies. The usual approach, taken by sample surveys, is to examine the influence of SES on individual children living within households that are, in turn, rooted within different neighborhoods grouped together according to common characteristics. In other words, in sample surveys the "neighborhood effect" is often based not on comparisons among children clustered in actual geographic neighborhoods, but among synthetic clusters of children from different neighborhoods that have some statistical characteristics (usually SES) in common. Our population-based approach avoids this problem by making comparisons among actual geographic neighborhoods, allowing the impact of both observed and unobserved neighborhood factors to express themselves – the broader social dynamics and institutions through which the citizenry organizes itself economically, culturally, socially, and so on. These broader community conditions and practices create an environment for "social care" that may well matter when it comes to child development (Kershaw et al., 2006). Our approach allows these elements of variance to be expressed.

Jencks and Mayer (1990) help to unpack the social care thesis by considering the multiple ways in which neighborhood settings may affect early development. They focused on five patterns of influence:

1. *Neighborhood resources*: Child outcomes are related to the level of resources available in communities, especially publicly provided or subsidized resources like community centers, parks, and childcare.

Table 9.2. *Variation in EDI vulnerability by BC neighborhood "explained" by neighborhood SES*

EDI scale	Proportion of variance explained
Physical health and well-being	33.8%
Social competence	20.9%
Emotional maturity	23.4%
Language and cognitive	27.2%
Communication skills and general knowledge	46.9%
One or more EDI vulnerabilities	42.7%

2. *Collective socialization*: Child outcomes relate to the social ties between community residents that facilitate the collective monitoring of children relative to shared neighborhood norms and practices, as well as positive role modeling. Neighborhood characteristics, such as poverty, residential instability, lone parenthood, and ethnic diversity, support or hinder the formation of this sort of neighborhood social organization.

3. *Contagion*: Child outcomes are influenced by the power of neighborhood relations, especially with peers, to spread problem behavior.

4. *Competition*: Child outcomes reflect competition between neighbors for scarce resources.

5. *Relative deprivation*: Child outcomes are influenced by how children and their families evaluate their own circumstances relative to neighbors and peers.

Contagion and collective socialization models suggest that affluent and/or family-oriented neighborhoods convey benefits to children, especially those from low-income families. By contrast, the competition and relative deprivation models imply that children from less privileged homes will struggle in affluent community contexts because they may not be able to keep up with classmates or they may suffer lower self-esteem if they compare themselves to others.

Starting with roughly 100 census variables we produced best-fit regression models for the five EDI scales, as well as "one or more vulnerability," that depict the relationship between SES and vulnerability rates across the 469 BC neighborhoods. In each case, seven or fewer census variables entered the model. Table 9.2 shows the proportion of neighborhood variation in EDI vulnerability accounted for by these models.

Readers should recall that this explanatory power is considerably higher than that attributed to the combined influence of neighborhood and family SES effects in traditional analytic studies, because the focus here is on ecological correlations for which neighborhood rates of vulnerability are the unit of analysis – not individual children. The six best-fit models collectively identify nineteen *neighborhood*-SES indicators to be significant predictors of the share of children at risk of developmental delays in BC as measured by EDI scores. These cover a range of constructs, such as neighborhood income characteristics, immigration and ethnic mix, level of childcare provided by men, occupational characteristics of men and women, residential transiency, and proportion of Aboriginal population. Despite access to roughly one thousand variables from the 2001 Census, we expect that the explanatory power of the SES-EDI models would be higher had we had access to additional information, including a child's period of residence in the neighborhood, because some studies report that children are only significantly influenced by neighborhood income if they have lived there for at least 3 years (e.g., Turley & Lopez, 2003); crime rates; the concentration of persistent poverty and/or unemployment, say over 5 to 6 years, in contrast to low income cutoff (LICO) and employment measures in the census, which only allow us to consider these issues at one point in time; depth of poverty; concentration of household high income; the general physical and social surroundings of the neighborhood, including such things as traffic; garbage/litter; people loitering or congregating; persons arguing, shouting, or otherwise behaving in a threatening manner; frequency of intoxicated people visible on the street; and the general condition of most nearby buildings; and the level of neighborhood cohesion, as indicated by the extent to which neighbors get together to deal with problems collectively; there are adults to whom children can look up; whether neighbors demonstrate a willingness to help one another; and the availability of neighbors who can be counted on to watch that children are safe.

EDI-SES COMPARISONS, COMMUNITY RESILIENCE, AND BIODEVELOPMENT

The work described in this chapter has allowed us to characterize every neighborhood in the province according to both ECD vulnerability and SES. We have taken advantage of these data to establish the relationship between each neighborhood (or school district) and the census characteristics that best explain variations in vulnerability to determine which neighborhoods (school districts) are doing 'as predicted" in terms of ECD;

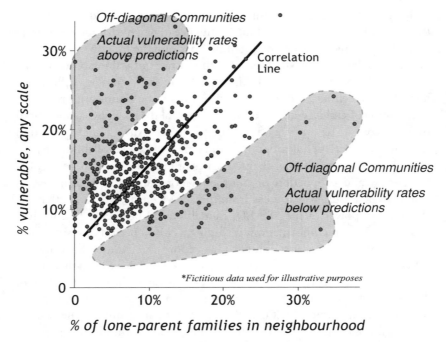

Figure 9.5. Conceptualization of off-diagonal communities.

which ones are doing better than expected ("positive off-diagonal communities"), and which ones are doing worse than expected ("negative off-diagonal communities"). This idea is depicted graphically in Figure 9.5, an approach that transforms the province into a massive sampling frame for the study of both community resiliency and biodevelopment.

Figure 9.6 presents the provincial map of vulnerability on the Emotional Maturity scale, modified for the purposes of this discussion. First, three-slice pie charts have been added, which are assigned to school districts and neighborhoods to summarize how communities compared to others in the province in terms of the three socioeconomic characteristics that correlate most strongly with the EDI scale under consideration. The pie slices are coded individually (Figure 9.6, middle right) to illustrate whether the community is more or less advantaged in terms of the socioeconomic variable at issue. Gray slices depict when the community is relatively advantaged in terms of the socioeconomic characteristic; horizontal lines show that the district is midrange; and solid black draws attention to districts that face the greatest obstacle in terms of the socioeconomic condition under

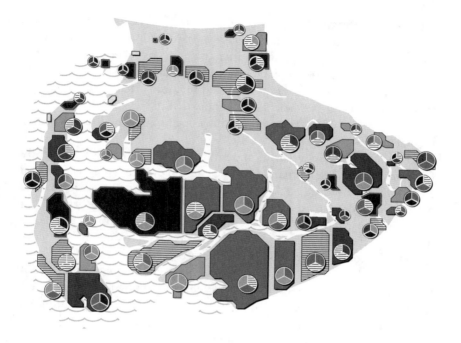

Emotional Maturity

Percent Vulnerable

- above 11.45 % *(11 districts)*
- 9.79 - 11.45 % *(11 districts)*
- 8.49 - 9.78 % *(14 districts)*
- 6.81 - 8.48 % *(10 districts)*
- below 6.80 % *(10 districts)*

Socioeconomic Status

Socioeconomic Variables

Employment Rate, Males with children

% Lone-parent Families

% Males, Management

Most Challenged

↕

Most Advantaged

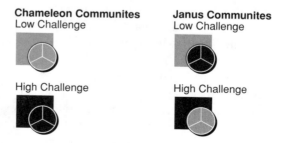

Chameleon Communites
Low Challenge

High Challenge

Janus Communites
Low Challenge

High Challenge

Figure 9.6. Illustrating "off-diagonal" school districts.

consideration. Second, each school district retains its approximate original shape, but is re-sized (top of Figure 9.6) according to the numbers of children who live there. Although there is some abstraction of shape, the resulting display corrects for BC's uneven population distribution while still retaining the visual cues of shape and proximity that are often crucial to map interpretation. Readers should note that the size of school districts indicates the size of the population under age 6. The size does *not* reflect their actual geographic size.

The relationship between SES and child development can be read directly from maps with SES pie charts (as shown in top of Figure 9.6). *Chameleon communities* (those that take on the pattern of their SES variables; see bottom of Figure 9.6) show the relationship most directly. A gray pie on an identical background shows communities that are relatively privileged in terms of the selected socioeconomic characteristics and are also enjoying less childhood vulnerability. This pattern is what we would expect to find. Solid black-on-solid black patterns illustrate a similar story about the relationship between SES and vulnerability, although one that moves in the opposite direction: the most socially disadvantaged communities with the highest rates of vulnerability on emotional maturity on the EDI. Communities that struggle with more disadvantaged social and economic circumstances typically suffer higher rates of vulnerability.

Some of the most interesting communities are those that do the opposite of the chameleon: they stand out for having EDI vulnerabilities that are distinctly different from their SES pie charts (bottom right of Figure 9.6). In such communities, children's development is less closely associated with socioeconomic characteristics than we would predict. For example, some school districts (such as Quesnet, in the upper middle of the Figure 9.6 map) have a pie chart that is entirely solid black, indicating high socioeconomic vulnerability; yet their vulnerability on emotional maturity is relatively low, portrayed as dotted gray in the background. These areas on the map are good-news stories because they represent communities that are overcoming socioeconomic circumstances that would generally result in higher rates of vulnerability among kindergarten children. *From the standpoint of community resilience,* does it have something to do with the community resources and assets that citizens enjoy, the level of trust in the community, the level of collective socialization or cohesion, or something else entirely? *From the standpoint of biodevelopment,* are the intimate environments that directly shape young children's brains and biologies somehow protected from the rigors of the broader socioeconomic environment? Contrasting communities can also have EDI outcomes that are particularly worrisome.

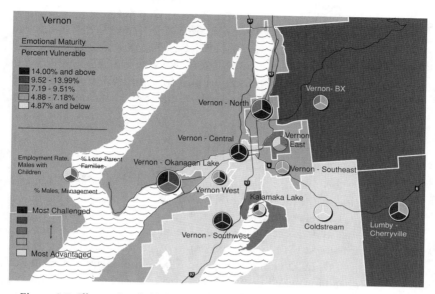

Figure 9.7. Illustrating "off-diagonal" neighborhoods within a school district.

Communities that enjoy relatively favorable SES pie charts may report midrange or low-range vulnerability rates. This combination paints the picture of a community that has vulnerability rates that fall below expectations based solely on social and economic conditions in the community.

Elucidating population-level variability in SES-ECD associations is a key precondition to understanding how social circumstances drive biological variability in health, well-being, learning, and behavior across the life course. Illumination of both on- and off-diagonals – why some poor children thrive and some advantaged children founder – is an essential step in understanding social partitioning of these outcomes.

Figure 9.7 illustrates the EDI-SES relationship for Emotional Maturity at the neighborhood level within a single school district, the district of Vernon. As a district, Vernon enjoys low challenge EDI results on the emotional scale. It appears to be achieving these results by overcoming significant SES barriers that otherwise associate with greater levels of vulnerability. The southwest, west, and central parts of the city stand out provincially for having vulnerability levels well below what would be expected based on their SES. These neighborhoods are joined by Vernon Southeast and Coldstream, which exemplify the low challenge *chameleon* pattern of SES pies that accompany low vulnerability rates and high neighborhood average scores. Within Vernon, relatively high rates of vulnerability are evident in the eastern communities of BX and Lumby-Cherryville. BX stands out because of its relatively advantaged social and economic characteristics. Vernon residents have reason to question why positive SES conditions in this neighborhood are *not* translating into low vulnerability rates when their community is generally able to reduce the negative influence of more difficult SES indicators.

FROM POPULATION-BASED DATA TO BIODEVELOPMENT

We have long been aware that human experience is capable of "getting under the skin," as the metaphor goes, and influencing brain and biological development. However, a big unanswered question is: to what extent? Our population-based data system is a crucial resource here, because it creates the opportunity to study biodevelopment in neighborhoods characterized according to both ECD vulnerability and SES – in principle, allowing us to better understand whether, and how, early experience gets under the skin to influence gene expression, brain, and biological development *to an extent that can explain population-based differences in ECD and subsequent developmental trajectories.* By coordinating research projects according to on- and off-diagonal status, and whenever possible with the same children, we have the best chance to understand the different familial and social settings in which individual children play, participate, and learn; and illuminate how, and to what extent, children's psychobiological characteristics are associated with their experiences of these settings.

At the present time, we are at the stage of refining our understanding of the mechanisms through which social settings get under the skin; posing research questions that query how to operationalize psychobiological measures at the community level among young children; and, finally, how to take account of epigenetic processes. Our long-term goal is an integrative approach that extends from genetic and molecular levels to neural systems, and includes functional outcomes that reflect both basic processes and the environmental context in which children live.

WHERE TO LOOK FOR BIODEVELOPMENTAL PROCESSES THAT INFLUENCE HUMAN DEVELOPMENT

Biological embedding. One key starting point is the concept of "biological embedding" of early experience (Hertzman, 2000). Biological embedding occurs when

- experience gets under the skin and alters human biodevelopment,
- systematic differences in experience in different social environments lead to different biodevelopmental states, and
- the differences are stable and long term, and influence health, well-being, learning, and/or behavior over the life course.

As an example of biological embedding, consider the rat model of the life course of the hypothalamic-pituitary-adrenal (HPA) axis (see also Boyce, this volume; Gunnar & Loman, this volume). The HPA axis is highly relevant because of its role in our perception of and response to stressful circumstances. HPA stimulation leads to the secretion of the hormone cortisol, which, in turn, has widespread metabolic effects on various organ systems. These effects are adaptive in the acute stress response phase, focusing the body's energy on the immediate task at hand and reducing metabolic processes that do not contribute to the immediate response. However, over the long term, cortisol can damage these same organ systems as a result of high exposure (McEwen, 1998).

In rats, the apparently minimal intervention of removal from the cage in early life can bring about a cascade of events that permanently condition the way the HPA axis functions over the remainder of the life course (Sapolsky, 1992). This conditioning effect can only be created by intervention during a limited and specific period in early life, which suggests that it depends on appropriate stimulation during a highly circumscribed window of opportunity in brain and biological development. Most important, once the HPA axis has been conditioned, the effects appear to be lifelong. In one study, rat-pup handling reduced total lifetime exposure of corticosterone (the rat equivalent of cortisol in humans) to the brain. Chronic overexposure to corticosterone, in turn, endangered selected neurons in the brain's hippocampus, such that the rate of loss of hippocampal neurons was reduced in the handled rats over their whole life span. Because of cognitive functions' sensitivity to relatively small degrees of hippocampal damage, the handled rats, by 24 months of age (elderly by rat standards), had been spared some of the cognitive deterioration typical of aging. Rats not handled as pups showed a progressive deterioration in their memory, cognitive processing, and learning performance with age; in contrast, much less deterioration occurred in aged rats handled in infancy (Meaney, 2001). Furthermore, female offspring adopted the high-licking and suckling behavior when they became mothers. In other words, a pattern of intergenerational transmission was taking place, such that the more adaptive nurturing pattern was passed to those who previously had been "predisposed" to the nonadaptive pattern (Weaver et al., 2006).

By comparing the development of rat pups that were frequently versus infrequently licked and suckled by their mothers, it was possible to elucidate a mechanism by which systematic differences in the function of the HPA axis emerge (Meaney, 2001). He showed that licking and suckling initiate a biochemical cascade that leads to long-term *alterations in the expression of genes* influencing the development not only of response patterns in the HPA axis but also of higher-order executive functions in the brain (Weaver et al., 2004; Szyf, 2003a). In colloquial terms, Meaney speaks about a "life is going to suck" pattern, wherein the less-licked and suckled rats end up with a more reactive HPA axis (good for fight or flight, but bad for sustaining learning and memory functions over time) and a reduction of neuron-to-neuron synapses in the cerebral cortex. It is a model wherein high-quality early nurturance leads, *through mediation of gene expression,* to a more tranquil HPA axis, greater capacity for complex learning, and reduced age-related declines in learning and memory capacity. In other words, the early nurturant environment has been *biologically embedded.*

Does the HPA axis have a similar life story in human society that it has among laboratory rats? Evidence on this point has been much slower to accumulate. However, several lines of inquiry converge with the evidence just presented and will be briefly mentioned here. Early maternal-child attachment affects both the HPA axis function and behavior, such that poorly attached toddlers have both more reactive HPA axes and less adaptive behavioral responses in social conflict situations (Gunnar & Nelson, 1994). An extreme example of this is the plight of Romanian orphans (Gunnar et al., 2001). Other investigators have identified systematic social class differences in basal cortisol levels among both primary and secondary school children (Lupien et al., 2001). In the 1958 British Birth Cohort, we have demonstrated systematic differences in cortical patterns at age 45 in relation to growth and adiposity (Power et al., 2006), lifetime SES (Li, Power, Kelly, Kirschbaum, & Hertzman, 2007), cognitive development (Power, Li, & Hertzman, 2008), and social-emotional development (Power et al., 2010).

In Eastern Europe during the last twenty years of the Soviet era, there was a mounting health crisis such that heart disease mortality rates for middle-aged men were similar in Sweden and Lithuania in the mid-1960s, but by the mid-1990s they were fourfold higher in Lithuania than in Sweden. One hypothesis was that the average life experience for people in places like Lithuania was like that of low socioeconomic groups in the West.

Accordingly, Kristenson and colleagues (1998) studied 50-year-old men from one city in Lithuania and one city in Sweden, and examined their cortisol response pathways, using a challenge protocol. She found that the pattern of salivary cortisol before, during, and after these standardized stressors was such that the Lithuanian men had a higher baseline cortisol, and then a blunted response to the stressor (what she described as a "burnout" pattern). The Swedish men had lower baseline levels, a sharp response to the stressor, regardless of which one, and then a quicker return to baseline levels; ultimately the Swedes' cortisol levels came down further. This demonstrated that *population* differences in HPA axis function could emerge from systematic differences in the conditions of life experienced by populations living under different political regimes.

Candidate systems: Where is the best place to start a search for biological embedding in human populations? Keating (2009) has proposed that there are certain "candidate systems" that serve as transducers between society and human biology. That is, among all the ways that experience can get under the skin, certain biological pathways better conform to what we understand about humans as a social species than others. To be sure, there are many ways for experience to get under the skin. Inhalation of toxic fumes is one way. A massive transfer of kinetic energy from a moving automobile is another. However, these are processes that are either discrete and/or distinct from daily life. Candidate systems for biological embedding must meet four basic criteria:

- The system is influenced by quotidian experiences (often early in life), such that differential qualities of experience have the capacity to lead to a differently functioning system.
- The system responds to quotidian experiences throughout the life course.
- The system, if dysfunctional, has the biological capacity to influence health, well-being, learning, and/or behavior.
- Differential functioning of the system across the life course, to the extent that outcomes are affected, derives from the quotidian.

There are four "candidate systems" that meet these criteria and are worthy of mention here: the HPA axis and its accompanying secretion of cortisol; the autonomic nervous system (ANS) in association with epinephrine and norepinephrine; the development of memory, attention, and other executive functions in the prefrontal cortex; and the systems of social affiliation involving the primitive amygdala and locus coeruleus with accompanying higher order cerebral connections, mediated by serotonin and other hormones (Keating, 2009).

Epigenomics: The epigenome is responsible for regulation of the genome. It consists of chromatin and its modifications (Strahl & Allis, 2000; Jenuwein & Allis, 2001) as well as modification of DNA by the chemical process of methylation of cytosine rings found at the dinucleotide sequence CG (Razin & Szyf, 1984). In contrast to the genome, which is identical in different cell types and through life, the epigenome is dynamic and varies between cell type and temporally during life. It is known to vary because of developmental, physiological, environmental, and pathological signals and helps to confer cell type and temporal identities of gene expression programs (Szyf, 2003b, 2009).

What distinguishes DNA methylation in vertebrate genomes is that not all CGs are methylated in any given cell type (Razin & Szyf, 1984). DNA methylation is laid down by enzymes and different CGs are methylated in different cell types, generating cell-type-specific patterns of methylation (Razin & Szyf, 1984). It was hypothesized more than two and a half decades ago that DNA methylation silences genes (Razin & Riggs, 1980), and recent whole-genome mapping of the DNA methylation landscape confirmed the association of DNA methylation with inactive genes. DNA methylation silences gene expression by either inhibiting the binding of transcription factors to their recognition elements in DNA (Comb & Goodman, 1990) or by precipitating inactive chromatin (Nan et al., 1997; Jones et al., 1998). Aberrant DNA methylation would thus have a consequence similar to an inactivating mutation, whereas aberrant hypomethylation will have a consequence similar to an activating mutation.

To date, no one has specifically examined whether social adversity has an impact on the epigenomic landscape. Convergent findings from diverse programs of molecular, neuroscientific, and population health research suggest that socioeconomic disparities in health, well-being, learning, and behavior could be attributable, at least in part, to early experiences that epigenomically regulate the timed expression of key genes. Our hypothesis is that human experience leaves its mark on the epigenome, thus changing gene expression programming in ways compatible with the concept of biological embedding. In other words, systematic differences in human experience will lead to systematic differences in methylation patterns (and possibly other epigenomic phenomena) in loci important for health, well-being, learning, and behavior, such as the candidate systems described earlier. Because of the leading role of the brain as the principal organ of human development, we have labeled this research thrust "Social Disparities in Epigenetic Regulation of Neurodevelopment."

As a hypothesis-generating exercise, we are studying DNA methylation in a subset of individuals ($n = 40$) from the 1958 British Birth Cohort selected according to extremes of socioeconomic adversity in early life (as well as exposure to maternal smoking in pregnancy and evidence of child maltreatment) and in adulthood, for whom extensive genotyping is available. Extensive genotyping is important in order to analyze the relationship between epigenotype and genotype at loci where we find differences in methylation. To generate a comprehensive base of hypotheses as to how socioeconomic adversity is associated with epigenetic marks (and/or vice versa) we are currently doing mapping of approximately twenty thousand promoter and regulatory regions in the genome. Taking advantage of an extensive and continually expanding resource of bioinformatics databases and analysis tools for probing gene function, we will generate hypotheses regarding the biological plausibility and possible mechanisms that would relate social adversity to specific gene expression and potentially to health, well-being, learning, and behavior.

This hypothesis-generating exercise will, in turn, inform the "Social Disparities in Epigenetic Regulation of Neurodevelopment" research in BC, which will address the question of how early childhood environments work together with epigenetic vulnerabilities to generate socioeconomically partitioned developmental and health outcomes in early life. Two long-term goals of this program of research are to elucidate interactions among social environmental, genomic, and neurobiological factors on child development and the genesis of disease; and to examine linkages between social disparities in children's early experiences and the development and calibration of stress-responsive neural circuitry. The sampling frame for this research is the provincial grid of EDI-SES according to neighborhood, described earlier. After two complete rounds of EDI work in the province (2000–4 and 2004–7) we are now in a position to identify the "stable" on- and off-diagonal communities in terms of children's development. This, in turn, will allow us to sample from localities where ECD differs in important ways and where broad socioeconomic conditions are either transmitted to children or buffered/exacerbated by local and family circumstances. In this way we hope to build an analytical bridge connecting early experience, biodevelopment, locality, and population patterns of human well-being.

REFERENCES

Barker D. J. (1995). The Wellcome Foundation Lecture 1994. The fetal origins of adult disease. *Proceedings of the Royal Society of London: B Biological Science*, 262, 37–43.

Boyle, M., & Lipman, E. (2002). Do places matter? Socioeconomic disadvantage and behavioural problems of children in Canada. *Journal of Consulting and Clinical Psychology*, 70(2), 378–89.

Brooks-Gunn, J., Duncan, G., Klebanov, P., & Sealand, N. (1993). Do neighborhoods influence child and adolescent development? *American Journal of Sociology*, 99(2), 353–95.

Burton, L., & Jarrett, R. (2000). In the mix, yet on the margins: The place of families in urban neighborhood and child development research. *Journal of Marriage and the Family*, 62, 1114–35.

Comb, M., & Goodman, H. M. (1990). CpG methylation inhibits proenkephalin gene expression and binding of the transcriptionfactor AP-2. *Nucleic Acids Res*, 18, 3975–82.

Curtis, Lori J., Dooley, M., & Phipps, S. (2004). Child well-being and neighborhood quality: Evidence from the Canadian National Longitudinal Survey of Children and Youth. *Social Science and Medicine*, 58, 1917–27.

Gunnar, M. R., Morison, S. J., Chisholm, K., & Schuder, M. (2001). Salivary cortisols in children adopted from Romanian orphanages. *Development and Psychopathology.* 13, 611–28.

Gunnar, M. R., & Nelson, C. A. (1994). Event-related potentials in year-old infants: Relations with emotionality and cortisol. *Child Development*, 65, 80–94.

Hart, B., & Risley, T. R. (1995). *Meaningful differences in the everyday experience of young American children.* Baltimore: Paul H Brooks Publishing Co.

Hertzman, C. (2000). The biological embedding of early experience and its effects on health in adulthood. *Annals of the New York Academy of Sciences*, 896, 85–95.

Hertzman, C., McLean, S., Kohen, D., Dunn, J., & Evans, T. (2002). *Early Development in Vancouver: Report of the Community Asset Mapping Project (CAMP).* Human Early Learning Partnership [cited. Available from http://www.earlylearning.ubc.ca/vancouvermaps.pdf.]

Hertzman, C., Power, C., Matthews, S., & Manor, O. (2001). Using an interactive framework of society and life course to explain self-rated health in early adulthood. *Social Science and Medicine*, 53, 1575–85.

Janus, M., & Offord, D. R. (2000). Reporting on readiness to learn in Canada ISUMA. *Canadian Journal of Policy Research*, 1, 71–5.

Jefferis, B., Power, C., & Hertzman, C. (2002). Birth weight, childhood socioeconomic environment, and cognitive development in the 1958 birth cohort. *British Medical Journal*, 325, 305–11.

Jencks, C., & Mayer, S. (1990). The social consequences of growing up in a poor neighborhood. In L. Lynn & M. McGeary (Eds.), *Inner-city poverty in the United States.* Washington, DC: National Academy Press.

Jenuwein, T., & Allis, C. D. (2001). Translating the histone code. *Science*, 293, 1074–80.

Jones, P. L. et al. (1998). Methylated DNA and MeCP2 recruit histone deacetylase to repress transcription. *Nat Genet*, 19, 187–91.

Keating, D. P. (2007). Formative evaluation of the Early Development Instrument: Progress and prospects. *Early Education and Development*, 18(3), 561–70.

Keating, D. P. (2009). Social interactions in human development: Pathways to health and capabilities. In P. Hall & M. Lamont (Eds.), *Successful societies:*

Institutions, cultural repertoires, and population health. New York: Cambridge University Press.

Keating, P., & Hertzman, C. (1999). *Developmental health and the wealth of nations: Social, biological, and educational dynamics.* New York: Guildford Press.

Kershaw, P., Irwin, L., Trafford, K., & Hertzman, C. (2006). *British Columbia Atlas of Child Development.* Vancouver: UBC Press.

Kohen, D., Brooks-Gunn, J., Leventhal, T., & Hertzman, C. (2002). Neighborhood income and physical and social disorder in Canada: Associations with young children's competencies. *Child Development,* 73(6), 1844–60.

Kristenson, M., Orth-Goméer, K., Kucinskienë, Z., Bergdahl, B., Calkauskas, H., Balinkyniene, I., et al. (1998). Attenuated cortisol response to a standardized stress test in Lithuanian versus Swedish men: The LiVicordia study. *International Journal of Behavioral Medicine,* 5(1), 17–30.

Li, L., Power, C., Kelly, S., & Kirschbaum, C., & Hertzman, C. (2007). Lifetime socioeconomic position and cortisol secretion patterns in mid-life. *Psychoneuroendocrinology,* 32(7), 824–33.

Lupien, S. J., King, S., Meaney, M. J., & McEwen, B. S. (2001). Can poverty get under your skin? Basal cortisol levels and cognitive function in children from low and high socioeconomic status. *Development and Psychopathology,* 13, 653–76.

McEwen, B. (1998). Protective and damaging effects of stress mediators. *New England Journal of Medicine,* 338, 171–9.

Meaney, M. J. (2001). Maternal care, gene expression, and the transmission of individual differences in stress reactivity across generations. *Annual Review of Neuroscience,* 24, 1161–92.

Nan, X., Campoy, F. J., & Bird, A. (1997). MeCP2 is a transcriptional repressor with abundant binding sites in genomic chromatin. *Cell,* 88, 471–81.

National Research Council Institute of Medicine. (2000). *From neurons to neighborhoods: The science of early childhood development.* Washington, DC: National Academy Press.

Organisation for Economic Cooperation and Development. (2006). *Starting strong II. Early childhood education and care.* Paris, France. ISBN 92–64–03545–1.

Power, C., Li, L., Atherton, K., & Hertzman, C. (2010). Psychological health throughout life and adult cortisol patterns at age 45y. *Psychoneuroenocrinology,* doi:10.1016/j.psyneuen.2010.06.010

Power, C., Li, L., & Hertzman, C. (2006). Patterns of growth from birth to adulthood, adiposity and cortisol in mid-adulthood. *Journal of Clinical Endocrinology & Metabolism,* 91, 4264–70.

Power, C., Li, L., & Hertzman, C. (2008). Cognitive development and cortisol patterns in mid-life: Findings from a British birth cohort. *Psychoneuroendorcinology,* 33(4), 530–9.

Ramey, C. T., & Ramey, S. L. (2005). *Early learning and school readiness: Can early intervention make a difference?* Washington, DC: Georgetown University Center on Health and Education.

Razin, A., & Riggs, A. D. (1980). DNA methylation and gene function. *Science,* 210, 604–10.

Razin, A., & Szyf, M. (1984). DNA methylation patterns. Formation and function. *Biochim Biophys Acta,* 782, 331–42.

Sapolsky, R. M. (1992). *Stress, the aging brain, and the mechanisms of neuron death.* Cambridge, MA: MIT Press.

Schweinhart, L. J. (2004). *The High/Scope Perry Preschool Study through Age 40. Summary, conclusions, and frequently asked questions.* Ypsilanti, MI: High/Scope Educational Research Foundation.

Strahl, B. D., & Allis, C. D. (2000).The language of covalent histone modifications. *Nature*, 403, 41–5.

Szyf, M. (2003a). DNA methylation enzymology. In *Encyclopedia of the human genome.* New York: Macmillan Publishers, Nature Publishing Group.

Szyf, M. (2003b). Targeting DNA methylation in cancer. *Ageing Res Rev*, 2, 299–328.

Szyf. M. (2009). Early life, the epigenome, and human health. *Acta Paediatrica*, 98, 1082–4.

Turley, R., & Lopez, T. (2003). When do neighborhoods matter? The role of race and neighborhood peers. *Social Science Research*, 32(1), 61–79.

Weaver, I. C. G., Cervoni, F. A., Champagne, F. A., Alessio, A. C. D., Sharma, S., Seckl, J. R., Dymov, S., Szyf, M., & Meaney, M. J. (2004). Epigenetic programming by maternal behavior. *Nature Neuroscience*, 7, 847–54.

Weaver, I. C., Meaney, M. J., & Szyf, M. (2006). Maternal care effects on the hippocampal transcriptome and anxiety-mediated behaviors in the offspring that are reversible in adulthood. *Proceedings of the National Academy of Sciences*, 103, 3480–5.

Willms, J. D. (Ed.). (2002). *Vulnerable children.* Edmonton, AB: University of Alberta Press.

10

Society and Early Child Development:
Developmental Health Disparities in
the Nature-and-Nurture Paradigm

DANIEL P. KEATING

The long-standing nature versus nurture debate has generated many controversies and was in turn fueled by them. The question of whether differences among individuals should be attributed to their genes or their environment, their inheritance or their circumstances, has been – and remains – central to how policy makers and the public alike think about major issues concerning social arrangements. It has become so fundamental to public discourse that it is hard to see beyond this legacy and grasp the full implications of the emerging nature-and-nurture paradigm captured in the preceding chapters of this volume. On the evidence presented in this volume, and continuing to accumulate rapidly in research centers around the world as developmental scientists use new tools and technologies to explore the new paradigm, the impact of this shift in thinking will be profound.

The greatest impact may well be in an area that has always been among the most controversial in the long nature-nurture standoff: What accounts for individual differences in competence and health? There is overwhelming evidence that individual differences in the full range of developmental outcomes – cognitive, behavioral, and social competence, along with physical and mental health, an aggregate for which we have used the term "developmental health" (Keating & Hertzman, 1999) – are not random but rather patterned by socioeconomic circumstances (Keating, 2009b).

The social patterning of health outcomes recurs so regularly across different historical periods, despite shifts in the intervening variables, that Link and Phelan (2005) have described it as a "fundamental cause." The association of socioeconomic status (SES, or the alternate term often used in social epidemiology, socioeconomic position, SEP) with developmental outcomes has been similarly well known throughout the history of developmental science, and has been the focus of research across a wide range of child health and development outcomes from externalizing behaviors

245

to neurodevelopment to obesity (Bornstein & Bradley, 2003; Bornstein & Sawyer, 2006; Corwyn & Bradley, 2005; Hardaway & McLoyd, 2009; McLoyd, 1998; Noble, McCandliss, & Farah, 2007; Simonton, 2008; Yancey & Kumanyika, 2007).

Arguments about the sources of these social disparities have often been framed for public discourse in terms of the nature-nurture polarity. Those arguing from nature have pointed to behavior genetic findings of substantial heritability of a range of characteristics, and more recently, using newer technologies, have identified specific genetic variability associated with those outcomes. Those arguing from nurture have pointed to the strong regularities of social patterns as well as to the indeterminacy of genetic influences – specific allelic variations do not always lead to identical phenotypes.

The societal salience of these theoretical dispositions is hard to overstate. They have been tied closely to policy choices reflective of competing political and moral philosophies. If one holds the view that nature dominates, then the wisdom of substantial social investments aimed at enhancing the outcomes of those at the disadvantaged end of the social distribution can be seriously challenged. Sometimes the challenge is explicit and controversial, as in Jensen's (1969) early claim that compensatory preschool education had failed – a claim that four decades of subsequent research has shown to be not only premature but also misguided, with substantial evidence of both academic and other benefits, including health (Keating & Simonton, 2008). More often, it remains an implicit barrier to significant investments in intervention, prevention, and social policy initiatives, based on an unstated or even unconscious assumption that they are unlikely to be effective.

If instead one holds the view that nurture rules, then such investments are seen as necessary, a perspective that is frequently accompanied by a belief that universal programs will provide equivalent benefits to all. Establishing the right policies and practice should yield equality of outcomes if one gets the social engineering right. Although these descriptions are caricatures to some extent, they are frequently reflected in public debates about social investments, even as these debates become increasingly rare in scientific discourse.

Taking note of the increasing sterility of the nature-nurture duality – whether characterized as a stalemate or an oscillating pendulum in policy discourse – is by now commonplace. The findings reported in this volume provide clear evidence that the new nature-and-nurture paradigm is already functioning vigorously in developmental science, with solid results, strong conceptual frameworks, and limitless prospects.

What is less clear is how this new paradigm can be effectively translated to public and policy debates, to rejuvenate them in a productive fashion. One obstacle to effective translation is that the story behind this new paradigm is inherently more complex – ranking contingencies and interactions as paramount over main effects – and thus the route to action is substantially more complicated. Implications from main effects are easier to grasp – "Don't waste money on a lost cause" versus "Any problem can be solved for everyone with the right program" – even though they are misguided or even potentially harmful. To take the full implications of the new interactive and contingent paradigm on board, we need to build a causal model that captures these emerging complexities, but one that is at the same time coherent in its specification of the causal links.

A second challenge to overcome is that subjecting such a causal model to empirical tests will require intensive interdisciplinarity. The processes through which gene-environment interactions function, such as neural sculpting or gene methylation, do not make their effects known in a sufficiently direct way to guide thinking about action. The links from gene-environment interactions at the micro level to developmental health at the population level are numerous, and each of them requires strong empirical and theoretical specification. This level of specification is needed in order to identify the most promising avenues from theory to practice, and to identify the causal links most in need of empirical testing.

We have been refining and testing key elements of a proposed model, and attempting to identify the causal links in such a model (Boyce & Keating, 2004; Keating, 2009b; Keating & Hertzman, 1999). Drawing on recent research advances across a wide range of traditional and emergent disciplines from population health to epigenomics, we can begin to specify with greater precision the developmental sources of ubiquitous social disparities, and to identify potentially productive ways to mitigate or resolve the historically implacable problems that arise because of such disparities.

The primary goals of this chapter are (1) to identify new insights that have come from disparate disciplines as they have begun to bring the nature-and-nurture paradigm to bear on the issue of social disparities; (2) to propose and describe a conceptual framework based on an interdisciplinary integration across these disparate disciplines; and (3) to draw out some major implications for policy and practice that arise from the proposed model. This conceptual framework aims to organize both new and long-standing evidence that speak to the issue of social disparities, and to articulate how developmental mechanisms function together in a system that yields the robustly observed empirical patterns of social disparities.

In the first part of this chapter, an overview of the conceptual model is presented as a working hypothesis that seeks to specify the causal links between variations in social circumstances and a wide range of developmental and health outcomes. The model proposes that these causal links are rooted in the differing developmental histories of individuals, including variations in social interactions, material resources, and physical exposures. It also proposes that these experience and exposure differences reciprocally influence and are influenced by the child's behavior and biology over the course of developmental history, including both genetic starting points and the impact of experience on core behavioral and biological functioning. Understanding how experiential differences "get under the skin," through mechanisms such as neural sculpting and gene-environment interactions, constitutes a critical link in the causal chain. These nature-and-nurture dynamics operate across the course of development, but based on substantial evidence, among which are the findings presented elsewhere in this volume, a foundational role of early child development is clear. In the second section, a targeted overview of the current state of research on the major links in the model illustrates the existing empirical support and identifies high-priority future directions for research. The final section draws out some major implications of this working model for policy and practice as we seek to address the pressing issue of social disparities.

A DEVELOPMENTAL MODEL OF SOCIAL DISPARITIES

The major elements of the proposed model are depicted in Figure 10.1. As an effort to provide a working hypothesis of how differences in social circumstances become instantiated as social disparities in a wide range of health and developmental outcomes, the model focuses on testable elements. This section begins with an overview of the major causal links, identifies the major constraints on the types of hypotheses that hold promise for providing the necessary empirical tests of the model, and ends by taking note of the scope and limitations of the model.

Describing the Developmental Mediator Model of Social Disparities

The proposed developmental mediator model for explaining social disparities in developmental health is depicted in Figure 10.1, and the key components are described in this section. In general, the model includes both experiential and child mediators, which are reciprocally related to each other. A mediator model is one in which an established empirical

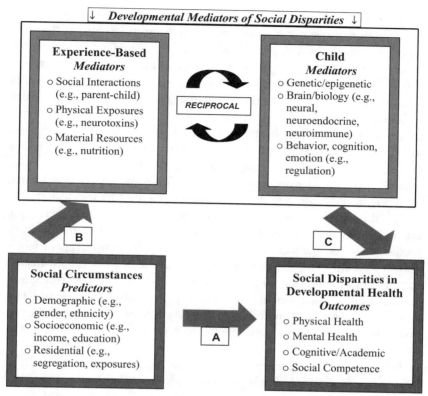

Figure 10.1. Causal model of social disparities in developmental health.

relationship is hypothesized to be explained by one or more mediators that intervene between the predictor(s) and the outcome(s). Mediation may account for this relationship in full or in part, and may involve one or a set of mediators.

There are multiple ways to test such a claim. Specific pathways require testing through co-variance structure modeling, including structural equation modeling (SEM) or multilevel/hierarchical linear modeling (MLM/HLM); quasi-experimental intervention and prevention studies; and/or investigations using formal experimental designs. The comprehensive model here serves as a conceptual framework to describe the hypothetical connections, within which specific investigations can be generated. The strongest tests of the causal chain will require a convergence of evidence across multiple studies based on different methodologies (Kraemer, Stice, Kazdin, Offord, & Kupfer, 2001; Rutter, this volume).

A common difficulty of comprehensive causal models is that the number of interactions grows exponentially with merely linear increases to the set of potential causal links. This paradox is not easily resolved. The approach used here is to focus on the potential of the conceptual model to generate research that illuminates key elements of the causal chain, identifying high-priority areas where investigations of the causal links will provide significant gains for understanding how social circumstances generate social disparities in developmental health. The next section thus focuses on existing evidence to identify nature-and-nurture subsystems that are primary candidates for explaining large portions of developmental health disparities that arise from social circumstances, mediated by developmental mechanisms. First, however, an explanation of the key features of Figure 10.1 is in order.

Social disparities predict health outcomes (Arrow A): This restates the starting observation of this chapter, that disparities in social circumstances predict observed inequalities in health and development. The variables known to matter empirically as predictors (in the box labeled *Predictors*) include not only income and social status (SES/SEP) but also education, race/ethnicity, gender, neighborhood or community of residence, and parents' marital status. Moreover, these factors interact with each other in both amplifying and compensating fashions (Hardaway & McLoyd, 2009; Keating, 2009b).

It is important to note that the shape of the relationship captured in Arrow A can be described as a gradient, such that lower levels of SES/SEP are uniformly associated with lower levels of virtually any measured developmental health outcome, and that this is a monotonic and most frequently linear relationship. There are some deviations from strict linearity – for example, there may be an asymptote on the effect of high income at the upper end, yielding a curvilinear (quadratic) relationship – but the general effect is one that can be largely captured as a "social gradient" (Keating, 2009b).

Conceptually, there are three general strategies that have been used in trying to account for the social gradient in developmental health. The first is to explore in greater depth the co-variance structure of the relationship depicted as Arrow A in Figure 10.1. Which specific predictors matter more than others? How do they interact? Do they relate similarly or differently to different outcomes? And so on. Research based on these questions, which is briefly reviewed here, has been highly informative about the nature of the relationship between variations in social circumstances and social disparities in outcomes. In one sense, they set the constraints for any proposed causal account. Analyses at this level alone, though, may not be able to support robust causal accounts, in part owing to structural limitations of the necessarily large databases, especially the signal-to-noise ratio of survey indicators, and in part because the causal links in social disparities are embedded in developmental mechanisms – which is the principal focus of the model presented here.

A second highly productive approach has been to move to a more macro level, comparing populations and societies to each other on the nature of the gradient and on other characteristics of those societies. One important distinction here is that between the developed and the developing world. In the developing world, the gradient effect plays a role, but in international comparisons of such countries, it is often the fundamental material differences that play the largest role (Engle, Black, Behrman, deMello, Gertler et al., 2007). In contrast, in the developed world that resides at the "flat end of the health/wealth curve" (depicted, for example, in Lynch, Smith, Harper, Hillemeier et al., 2004), a key finding is that the slope of the social gradient – that is, how steep the outcome differences are from the top through the middle to the lower end of the SES distribution – is routinely associated with an average level of societal performance on numerous developmental health indicators (Keating, 2009a, 2009b). These and related findings, also summarized later, provide further empirical constraints on the types of causal accounts that can work to explain the gradient effect in the absence of major differences in access to essential material resources.

A third approach, and the primary focus of the model proposed in this chapter, is to probe at the more micro level in an effort to "unpack" Arrow A by understanding the underlying developmental mechanisms. The reciprocally related experience-based and child mediators shown in Figure 10.1 depict the proposed mechanisms through which circumstances are translated into outcomes. The strong empirical findings derived from the first two strategies – detailed co-variance analyses and population comparative analyses – provide significant constraints on the nature of the candidate mediator systems that can successfully unpack Arrow A by providing an adequate causal account. The evidence reviewed here regarding Arrow A specifies these constraints in more detail.

Note also that the outcomes (in the box labeled *Outcomes* in Figure 10.1) encompass the full range of developmental health. One area of special interest is the degree to which different developmental health outcomes share common developmental pathways. The considerable overlap among developmental health outcomes suggests such a possibility (Essex, Boyce, Goldstein, Armstrong, Kraemer, & Kupfer, 2002), although this is not sufficient evidence for common mechanisms or pathways. Yet if there are common

mechanisms and pathways linking social disparities with multiple outcomes, the potential implications for societal approaches are substantial.
For example, do countries with steep gradients and poorer health outcomes
also have steep gradients and worse literacy (Keating, 2009a)? If there is a
generalized country effect of steeper/worse and flatter/better across a number of indicators, the search for underlying mechanisms would be different
than if the findings revealed substantial trade-offs among different outcomes. A systematic exploration of common versus distinct pathways is a
key goal for future research.

*Relationship of differing circumstances to variation in experience-based
and developmental mediators (Arrow B):* This link is important because
the child does not experience social addresses, demographic location, or
SES/SEP. Rather, the child experiences particular qualities of social interactions, types of physical exposures, and levels of material resources (depicted
in the box labeled Experience-Based *Mediators*). These interact reciprocally
with a range of child factors, from genetic to biological to behavioral. Over
time, these interactions become "biologically embedded" in the organism
(Keating & Hertzman, 1999).

Within the set of experience-based mediators, the specific category of
social interactions includes experiences that range from more micro level
and immediate (close, regular, face-to-face, intimate), through meso level
(representing the nature of one's interactions in groups, from kindergarten
to the workplace), to more macro level (representing the broad range of
social networks and community membership).

Social disparities in circumstances are structured differently across different societies, and are generated through institutional arrangements,
cultural patterns, and generalized social capital. These social disparities are thus constituted differently across societies, through the interface of historically conditioned institutional and cultural structures and
processes (Hall & Lamont, 2009). Social disparities then affect subsequent health and capabilities via social interactions, physical exposures,
and material resources – in conjunction with how that variability is
embedded in and responded to by individuals, both behaviorally and
biologically.

*Experience-based mediators and child mediators reciprocally affect each
other over the course of development.* This reciprocal relationship (shown
in Figure 10.1 as the pair of arrows labeled *Reciprocal*) comprises the core
of the proposed mediator model. How developmental processes and functions are formed and expressed is the central question of developmental
science and is a core concern of the preceding chapters in this volume.

Recent research in developmental neuroscience has enabled a unique new look into this particular "black box." We are only beginning to grasp the important processes here, such as the sensitivity of neural systems to early developmental experiences, the iterative relationship between brain and behavior over the course of development, and the life-course impact of early development (Boyce & Ellis, 2005; Boyce & Keating, 2004; Gottlieb & Willoughby, 2006; Keating & Hertzman, 1999; Worthman & Kuzara, 2005). However, our current understanding and evidence are sufficient to propose with some confidence what the broad themes of this reciprocal relationship will be, particularly in light of the constraint that the downstream consequences for developmental health must be a key part of the story (as shown in Arrow C, described next).

The set of child *Mediators* includes the genetic starting points and epigenetic changes to gene regulation; the biological embedding of experience in neural, neuroendocrine, and neuroimmune functions; and the behavioral, cognitive, and socioemotional patterns that emerge over the course of development. Again, the main thrust of this model, and the focus of much of the developmental science reviewed in this volume, is on the reciprocal relationship between these two sets of mediators.

Developmental processes and functions affect downstream outcomes of developmental health (Arrow C). This completes the causal chain of the model depicted in Figure 10.1. Specifically, the contention is that developmental processes and functions are expressed in a wide range of outcomes, contingent on subsequent life-course circumstances. This relationship places further constraints on the nature of possible explanations of the social gradient in developmental health. To summarize, the test of mediation is that the causal chain shown as Arrows B and C (encompassing as well the reciprocal relationship between experience-based and child mediators) reduces or eliminates the observed direct effect of social disparities on developmental health outcomes shown as Arrow A, and in so doing provides a basis for causal inferences that have important implications for both practice and policy.

Scope and Limitations of the Model

The model depicted in Figure 10.1 seeks to provide a plausible set of working hypotheses to explain how variations in social circumstances are translated, over the course of development, into social disparities in a wide range of developmental health outcomes. It is situated in an even broader context, of course, not all of which can be readily accommodated in such a model without compromising its coherence. Two key limitations are noted here, to forestall misunderstanding about the scope of the model.

The social contexts of social circumstances. On its face, the model seems to suggest that the social patterning identified as social "circumstances" reflects stand-alone, objective realities. This is of course not the case. Gender, for example, is a biological reality, but its social meaning is clearly not co-extant with the biology. How gender functions as a social reality is a different question. Similar caveats apply to all the sources of variation in the *Predictors* noted or implied in the model.

More substantively, we need to understand how the broader social ecology has an impact on the meaning of those social circumstances, as well as possible direct effects on how those social circumstances function as causal links to eventual social disparities in *Outcomes*. At this broader level, we need to understand how societal structures and practices serve to create the context within which disparities are generated.

Put another way, social circumstances are embedded in social and market policy regimes. A particularly promising perspective for understanding this type of embedding focuses on the interaction of institutions and cultural repertoires. These societal-level interactions shape how social circumstances are distributed across populations in different societies – and how those variations are given meaning by the members of the society (Hall & Lamont, 2009).

Two brief examples illustrate the importance of this societal variability, which is a contributing source to health and development that lies beyond the scope of the mediator model shown in Figure 10.1. Lamont (2009) has studied "everyday destigmatizing strategies," which are part of a cultural tool kit (Swidler, 1986) employed by groups marginalized owing to social class, ethnicity, country of origin, and so on. The significant point here is that stigmatization, as a social process closely affiliated with social circumstances, does not operate as a direct cause on social disparities in outcomes. How stigmatization functions in different societies and what strategies the stigmatized groups and/or individuals employ in response to them are essential elements in a causal chain. These functions and strategies are clearly embedded in the institutional/cultural nexus, but are not a part of the model depicted here. One implication is that, although the mediator model is proposed as a general model, how it operates in a particular setting will depend on the details operating at the societal level.

The second example illustrates this with a focus on how "collective" identity, as a societal variable, interacts with and influences individual identity, health, and behavior. In a time-series investigation, Chandler,

Lalonde, Sokol, and Hallett (2003) reported decreased prevalence of youth suicide in Canadian first nations' tribes that had achieved greater collective autonomy compared with other tribes that had not sought such autonomy. This is an example of collective identity as both a protective factor (in the positive case of communities with more firmly established collective identities) and a risk factor (in the negative case of communities that had not done so). The potential responses to identity threat arising from stigma are numerous. Awareness of them, and of their risks and benefits, may well be important mediators of the link between societal-level variables and developmental health.

Static versus dynamic systems. Another limitation of the model depicted in Figure 10.1 is that although it incorporates a dynamic system reciprocity between experience-based and child mediators, it does not illustrate the larger dynamic system within which the overall model should be understood to function. This limitation in scope is intentional to maximize the coherence of the proposed mediator model and its potential value as a working hypothesis.

There are of course a number of levels in the broader dynamic system within which this mediator model operates. The societal ecology impacts described in the preceding section are one such extension. Another is that there is a transgenerational system that serves to reproduce disparities in *Outcomes* at Generation 1 as variations in *Predictors* in Generation 2. As the processes depicted in Figure 10.1 iterate across generations, the reproduction of differential health and competence creates, in part, the differences in opportunities for succeeding generations (Hardaway & McLoyd, 2009). This could be depicted as a reciprocal (rather than unidirectional) Arrow A in the figure, but at considerable cost to coherence.

However, the practical implications of this transgenerational dynamic system should not be underestimated. Enhancing population developmental health is very likely to have recurring benefits for a society that attains this by creating a more virtuous cycle for future generations – and the opposite possibility of a vicious cycle that perpetuates poorer health and competence for a population or for subpopulation groups ignored at society's peril (Hardaway & McLoyd, 2009). International comparative data, some of which is described next, illustrate both possibilities.

Leading candidate systems. Given the complexity of the model in Figure 10.1, it may be useful to anticipate the leading developmental systems that will emerge after a review of the evidence for the direct and indirect effects shown in the figure. Three leading potential mediator systems that appear to meet all the constraints and fit the conceptual model are a sense of control, a sense of connection, and the emergence of consciousness. It is important to note that these and other systems function in a complex interdependence, leading toward integration as human capabilities, which vary among individuals (Keating, 2009b). Furthermore, these candidate systems do not represent an exhaustive list, but rather those for which the evidence is strongest that they function as developmental mediators of social disparities in overall health and competence. A comprehensive review of each of these literatures is beyond the scope of this chapter, but the broad outlines of current research convey how they may function as mediators of the social gradient in health.

EVIDENCE FOR THE DEVELOPMENTAL MODEL
OF SOCIAL DISPARITIES

Social Circumstances to Social Disparities in Outcomes:
Evidence for Arrow A

This summary suggests why there has been broad interest in the social gradient in health for both science and policy. The focus of this attention has increasingly been on explaining the social gradient in health, and there are many candidates. To date, most of these have been investigated at the societal and population level, with respect to measurable differences that may account for the gradient effect. These include, among others, income inequality, differential access to health care, differential health behaviors, the psychosocial impact of status differences, direct effects of discrimination, and the level of social cohesion or social capital in the society. Somewhat less attention, and much of this is more recent, has been given to the possible developmental mechanisms through which differential experiences across individuals, linked to the disparities in circumstances, get under the skin (Boyce & Keating, 2004; Hertzman & Power, 2006; Keating & Hertzman,1999; Worthman & Kuzara, 2005). As noted earlier, the number of potential interactions among the sources of social disparities, in conjunction with the number of interactions among the proposed social mechanisms, generates an exponentially expanding model that may prove intractable to empirical test. Considering how a given set of social inputs actually makes a difference in the underlying biological and developmental mechanisms potentially offers a simplifying model that can point toward a resolution of what matters at the social level. That story is taken up in the section that follows the discussion of the gradient effect.

    In addition to the importance of knowing how the social gradient functions in its own right, such knowledge is helpful in that it sharply constrains the characteristics that candidate developmental mediator systems would need to have. For simplicity, the major evidence on the social gradient, first of health and then on developmental health more broadly, is summarized here in terms of six broad propositions. The relative certainty of the empirical observations varies across propositions. Propositions 1 through 4 are very well supported and often replicated; propositions 5 and 6 have substantial evidence, sufficient to inspire considerable confidence, but it should be noted that some of the strongest databases to test these propositions have come on line in recent years, and continue to come on line as this is written. Thus, a major research agenda is to further probe their empirical robustness. They are best viewed as strong working hypotheses, but hypotheses nonetheless. Future research with newly available databases, especially utilizing analytic advances that maximize the extraction of information from them (including structural modeling, MLM, and multiple imputation to minimize the effect of sample attrition across time), will add dramatically to our understanding in the near future. This represents a crucial research agenda that needs to be informed by our best understanding of potential social and developmental determinants.

1. *Social and Economic Effects on Health Are Pervasive.*
As noted earlier, there is robust evidence to support the conclusion that SES is strongly associated with a wide range of health outcomes. The direct effect of income differences between individuals is strong, in the range of 6.5 to 7 years of life expectancy when averaging across race and gender, and more than 16 years of life expectancy when comparing the most advantaged to the most disadvantaged groups in U.S. national samples, as calculated from Williams (2005, table 8.3). Similar income effects in national samples are found for morbidity, self-reported health, and depressive symptoms (Fiscella & Franks, 2000). Recent systematic reviews have found this income-to-health effect everywhere it has been investigated, and this effect persists despite secular trends that have shifted the underlying mechanisms (Link & Phelan, 2005; Lynch et al., 2004; Marmot, 2004). A similar picture emerges for the general relationship between education and health: the effects of education are similar to those for income, with large effect sizes based on national samples (Marmot, 2004; Ross & Wu, 1995; Schnittker, 2004). For example, the average Cohen's d across two U.S. national samples reported by Ross and Wu (1995, tables 1 and 2) were .76 for self-reported health and .73 for self-reported physical functioning, an effect size typically regarded as large.

Indeed, it is this pervasiveness of the relationship between social disparities and health outcomes across societies and across history that led

Link and Phelan (2005) to characterize SES as a "fundamental cause" of health inequalities. A further claim of their model is that it should be regarded as a fundamental cause because even as intervening mechanisms have been addressed (through public health interventions to assure clean living conditions or the advent of a new medical technology, for example), the SES-health relationship recurs through new (or newly revealed) intervening mechanisms.

2. The Relationship between Disparities and Health Can Be Described as a Gradient.

As noted earlier, the slope of the gradient is virtually always monotonic (that is, continuously ascending or descending) and approximately linear, without major thresholds or breaks. This indicates that the gradient is not a function of one segment of the population. For example, if the social gradient were entirely a function of poverty, we would expect to see a break or discontinuity in the line representing the gradient between poverty and nonpoverty segments of the population. Yet the far more frequent observation is that the gradient effect obtains throughout the population: those in the lower middle class have better health than those in poverty, but worse health than those in the middle class, who in turn have worse health than those in the upper middle and professional classes.

The precise shape of the relationship remains a matter of empirical scrutiny. For the income-to-health relationship, some curvilinearity has been observed, such that at the higher end of the income distribution, the benefit of having more income is less than that at the lower end of the distribution (Lynch et al., 2004; Wolfson, Kaplan, Lynch, Ross, & Backlund, 1999). For some purposes, this is a distinction that does not make much difference to the major points of the gradient account – no one has reported clear, replicable thresholds or discontinuities in the gradient – but it does matter for determining the precise contribution of income inequality (as opposed to income per se) to the explanation of the gradient, described in more detail later.

In this sense, *slope* is a useful oversimplification – for curvilinear relationships we would include a quadratic regression estimate (which is also a linear estimate, albeit curvilinear) in addition to absolute slope – but for the purposes of this discussion, slope is adequate as a general term. To examine this with greater precision, a test of the slope/mean relationship for curvilinear relationships would have two slope predictors, the first- and second-order beta weights in a regression model. A detailed description of mathematical models to test a range of hypotheses about gradients has been provided by Willms (2003).

This linearity of the gradient effect is an important observation, both as a clue to the underlying processes that drive it (in that they imply more general population health processes, rather than specific deprivation accounts) and as a social policy consideration (in that programs aimed exclusively at improving health at the lowest end of the gradient will not address all, or perhaps even the majority, of population health problems).

3. *There Are Multiple Social Disparities That Make a Difference in Health.* Although most of the research attention has focused on the roles of income (and of income inequality) and education as they affect health, it is clear that there are a substantial number of other sources of social disparities that similarly have an impact on health outcomes. Race and ethnicity clearly have a strong relationship with health outcomes, independent of income and education, but also as they interact with them. In terms of years of life expectancy beyond age 25 in the United States, for example, the average income differentials tend to be larger, at a little more than 8 years, and black-white differentials at a bit more than half of that, around 4.4 years. Gender is another source of disparity in health outcomes, with women having about a 6.5-year life expectancy advantage over men. The extreme values, as these multiple sources are considered together, are even larger. From age 25 forward, the average life expectancy in the United States for low-income African American males is 66.6; for higher income white females, the figure is 82.8, a cumulative difference of 16.2 years, as calculated from Williams (2005, table 8.3). As a comparison, this is the difference between the average life expectancy in Indonesia versus Japan (World Development Indicators, 2010).

Comparisons among multiple sources of disparities have given rise to a number of continuing controversies in the field. Deaton and Lubotsky (2003), for example, have argued that the focus on income as a source of health inequality has sometimes failed to account adequately for the impact of race differences on that pattern. Another set of comparisons of sources of social disparities in health are those between income and education. Clearly both matter, but one can ask which matters more. This is often difficult to disentangle because the shape of the gradient changes depending on which lens one is looking through. Some recent reviews have suggested that the education effect may be primary, especially given that it continues to have a significant effect even at high levels of income (Deaton, 2002; Schnittker, 2004).

In addition to these sources of social disparities in health, there are a number of other social factors that have an impact on the health outcomes of individuals, including variations in the level of economic or racial segregation, the level of public expenditures within a jurisdiction, marital status, and others. Confusion and controversy arise from how these multiple sources of social disparity are conceptualized, in particular whether they should be viewed as characteristics of persons or of populations, and whether they should be viewed as demographic circumstances versus social mechanisms for the transmission of inequality.

It is useful to note at this point, though, that much unnecessary confusion (as opposed to actual scientific disagreement) arises from shifting between a population and an individual perspective. One classic example is the use of a (presumed) counterfactual, noting the excellent health outcomes of particular individuals despite major socioeconomic disadvantages. This is of course a false counterfactual because the social gradient of health is a population observation rather than an individual prediction – lower SES individuals often live to very old age, and some higher SES individuals suffer premature deaths – but it is sometimes difficult to overcome a cognitive bias or habit of mind that arises from thinking "automatically" in terms of either individuals or populations. A comprehensive, systemic explanation will need to function adequately at both levels, requiring a flexible approach that can move readily from the individual to the population and back again. Further development of research methods and analytic strategies that integrate population data (such as survey indicators) and underlying mechanisms (such as developmental processes) will be essential to this approach to understanding the sources of social disparities.

There is one final point that we return to next. Having recognized the contribution of multiple sources of the social gradient in health, and also that these sources may amplify or mask each other, the question of how to identify the core mechanisms of social disparities becomes considerably more difficult. As potential interactions multiply in an exponential fashion, the ability to identify the variables that "control" the generation of health disparities is dramatically compromised. In such a circumstance, it may be helpful to look at the question through a different lens, such as through the model depicted in Figure 10.1.

4. *The Gradient Effect Is Found Also in Developmental Outcomes.*
The social gradient in health, including research on the effects of steeper versus flatter gradients, has until recently focused on explicit health outcomes, most often mortality but also on general morbidity and self-rated health, as well as on specific diseases. More recently, this work has been extended to look at a broader range of outcomes, especially developmental outcomes such as literacy, mathematics achievement, mental health, and

emotional and behavior problems. Where this work has been extended to these developmental outcomes, the same gradient pattern emerges. Thus, for the full range of outcomes from physical and mental health to competence and coping, which can be combined into an omnibus indicator that we have termed "developmental health," there emerges a familiar portrait: social position is related to developmental health, and societies with sharper social status differences appear to have lower developmental health overall (Keating & Hertzman, 1999b; Willms, 1999, 2003).

These convergent findings serve to sharpen the search for explanations of the social gradient effect. Potential explanations need to account for the social gradient in health, the comparative advantage of flatter gradients, and how both of those effects are observed not only in physical health but also in a range of developmental outcomes.

5. *The Effects of Disparities Are Observed throughout the Life Course.*

The next key element of the social gradient, in addition to its pervasiveness and the convergence of its patterns, is observed in longitudinal data sets, where the prospective impact of early developmental experiences can be directly observed. Where such evidence is available, it appears that SES of the family of origin reveals a social gradient similar in pattern and magnitude to that obtained when the individual's SES is the independent variable (Boyce & Keating, 2004; Hertzman & Power, 2006; Power & Hertzman, 1999). Although the generalizability of this finding is constrained by the availability of appropriate longitudinal data sets, the evidence to date is generally supportive (Hertzman, 2001).

The implications of these provisional but key findings are profound, in that they suggest that the SES gradient is both portable (as individuals, on average, carry that legacy into their adult lives) and enduring (as life-course health effects). Previously we described the consequences in this way: " . . . failure to attend to childhood circumstances may create, at the population level, a range of societal burdens that are hard to subsequently shift. Conversely, investments in early development may generate enduring societal opportunities" (Boyce & Keating, 2004, 419). Drawing on a range of work (Hertzman, 1999; Hertzman & Power, 2006; Kuh & Ben-Shlomo, 2004), we described several hypothetical patterns of influence through which childhood circumstances may affect health outcomes over the life course, and for both scientific and policy purposes, it is helpful to attempt a conceptual disaggregation of such patterns.

One pattern progresses through stability of socioeconomic circumstances from childhood to later life, with an accumulation of risk from a range of sources. In other words, individuals who grow up in more stressful circumstances tend to experience those same circumstances throughout their development. The implication of this cumulative pattern is that broad societal differences in the quality of the social and physical environment play a major role in developmental health outcomes, through the overall degree of social partitioning and the SES-related patterns that emerge in early life (Hatch, 2005; Luecken & Lemery, 2004).

A second set of childhood influences may be thought of as pathway effects, constituted by chains of risk or protection. If we construe the previous set as representing the cumulative effects of contexts over time, this next set focuses, in contrast, on the ways in which early circumstances constrain or enable trajectories of health and development. For example, educational attainment plays a substantial role in both subsequent health and social status. Beyond simple continuity, the early acquisition of competencies, skills, and dispositions likely affects the pathways leading toward future health, well-being, and developmental attainments. Such associations are likely attributable to sequences of linked exposures, in which early risk factors increase the likelihood of subsequent exposures, which in turn augment the probability of encountering others. In a recent review of the literature on the health effects of human development investments, we found that the evidence for direct effects was less than that for indirect effects operating through subsequent education and employment (Keating & Simonton, 2008). This is perhaps attributable in part to the duration so far of the most relevant preventive intervention studies, but it emphasizes the importance of understanding the full set of effects arising from differential experiences throughout the life course.

Another indicator of such pathway or linked effects is the phenomenon of resilience, in which significant early disadvantage is surmounted. The developmental trajectories of many children growing up in adverse or suboptimal circumstances belie the expected declines in physical and/or mental health that are known to be associated with such circumstances. Observations of anomalously good outcomes emerging from impoverished or unsupportive social settings are nonnormative, but far from rare. In such accounts, resilience appears often derived from the establishment of chains of protection – individual characteristics and forms of individual support that predispose toward the pursuit of health-protective developmental trajectories (Masten, 2001).

A third type of influence can be thought of as critical period, sensitive period, or latent effects. Even after removing the effects from other, later sources – adult SES, differential developmental pathways, and so on – there is often a nontrivial impact of childhood circumstances on life-course health outcomes. For example, the early instantiation of a dys-regulated stress-response system (described later in the chapter) may affect developmental trajectories through success in selected environments (path-way effects), but in addition, may create a health risk that will become manifest only at a later stage in the life course as stressors accumulate and/or grow more intense. Note, in this example, that the early experien-tial calibration of stress-responsive neural circuitry might alternately affect later health via risk accumulation (cumulative, long-term costs of repeated activation of biological circuits), pathway effects (dysregulated reactivity biasing developmental trajectories toward risk induction or away from risk protection), or critical period effects (early exposure in, for example, infancy alone results in biological response profiles that jeopardize adult health and adaptation) – and it may reflect the simultaneous operation of all three types of effects.

Thus, consideration of the characteristics of the social gradient effect (pervasive, convergent, and enduring) and patterns of transmission (cumu-lative, pathway, and critical period) supports the hypothesis of "biological embedding," a process "whereby systematic differences in psychosocial/material circumstances, from conception onward, embed themselves in human biology such that the characteristics of gradients in developmental health can be accounted for" (Keating & Hertzman, 1999, 11). Hertzman and Frank (2006) systematically reviewed this construct and the current evi-dence regarding biological embedding. Entertaining the biological embed-ding hypothesis therefore entails a detailed consideration of the biological and behavioral mechanisms linking child development and later develop-mental health gradients.

6. *The Steepness of the Gradient Predicts Average Health across Societies.*
As comparisons among societies on the gradient effect have accumu-lated, another important characteristic of the social gradient has emerged. Although linear gradients are found in virtually all populations, the steep-ness of the gradient can vary substantially from society to society, country to country, state to state, city to city, or community to community. In addition, the steepness of the gradient – the degree of difference in a devel-opmental health outcome from the top to the bottom of the within-group distribution – is mathematically independent of the average outcome for a given jurisdiction (country, state, city, community). In other words, it is possible to have steep gradients associated with higher or lower aver-age outcomes for the population as a whole. Yet the evidence consistently

points to an association of steeper gradients with overall lower outcomes, and flatter gradients with overall higher developmental health outcomes (Keating, 2009a, 2009b) – thus the importance of a policy goal of trying to "flatten up" by both raising and leveling socioeconomic gradients (Willms, 2003).

This association of steeper gradients with worse outcomes, and flatter gradients with better outcomes, often yields a pattern that has been described as "fan convergent" (Willms, 1999, 2003). The differences in health outcomes among those at the top of the social gradient in different countries or communities are typically much smaller than the differences at the bottom or middle of the gradient. In comparative terms, this means there would be a substantial net benefit to nearly all segments of the population in "flattening up" the gradient, with little evident risk to those already enjoying comparatively better outcomes.

Evidence for Developmental Mediation of Social Disparities: Arrows B and C

Given the characteristics of the social gradient outlined in the earlier overview, we can ask, "What are the characteristics of candidate developmental systems that can yield these empirical patterns? What candidate developmental systems operating in early childhood are most likely to provide causal links between disparities in social circumstances and disparities in developmental health outcomes?" The range of possibilities is vast, although knowledge of the characteristics of the social gradient sharpens the focus to those influences that (a) are reliably socially patterned; (b) begin early in life; (c) have pervasive effects across a number of outcomes; (d) are portable by the organism; and (e) have long-term downstream consequences. A complete review of the scientific literatures that have dealt with different aspects of this topic is beyond the scope of this chapter, which will instead highlight major categories of research that are likely to serve as developmental mediators as depicted in Figure 10.1. Two classes of mediators that may have broad systemic effects are considered first: physical exposures and material resources. Exposures to environmental toxins, for example, have a range of known negative effects, although these are often systemic rather than specific to particular features of developmental health. Similarly, inadequate material resources, such as in nutritional deficiencies, are likely to have broad effects on the organism. Both sets of effects are multifaceted, complex, interactive, and systemic.

Social interactions that are associated with differential patterns of early development are considered after the overview of physical exposures and material resources. In both cases – physical and social mediators – the evidence for both Arrows B and C is considered jointly, that is, the links from social circumstances through the mediators to the longer term consequences for developmental health.

Physical exposures and material resources as developmental mediators. In a recent overview of the impact of environmental exposures to metals, Hu, Shine, and Wright (2007, 161–2) identified their importance:

> Why are children most vulnerable to neurotoxins? During fetal life and early childhood, neurons must undertake migration, synaptogenesis, selective cell loss, myelination, and a process of selective synaptic pruning before development is complete. Even minor inhibitory or excitatory signals imposed by neurotoxicants at early stages of central nervous system development therefore can cause alterations to subsequent processes.

Negative effects of exposures to metals as neurotoxins early in life have been frequently observed, with the most extensive research on lead exposure, but also for arsenic, cadmium, and manganese. Moreover, there are reasons to believe that mixtures of these metals, which have been less frequently researched but that occur with great frequency at known toxic sites (such as the U.S. "Superfund" sites), may interact to increase the negative impact on neurodevelopment (Hu et al., 2007; Wang & Fowler, 2008). Some specific pathways between exposures and brain/behavior patterns have been identified for some neurotoxicants, but there is also the likelihood of broader or more general toxicity, especially where metal mixtures are concerned.

Thus, there is ample evidence to conclude that differences in exposures to neurotoxins have significant negative effects on the developing child, especially in neurodevelopment and neurobehavior. There can be little doubt that such exposures satisfy the condition specified by Arrow C in Figure 10.1, namely that excess exposures lead to decreases in developmental health.

What then of the other condition for mediation? Are variations in exposures socially patterned by socioeconomic, demographic, or community factors? There has been a vigorous debate over a number of years regarding the social distribution of environmental hazards, with paradoxical findings emerging based on level of analysis and which social disparity variable is the focus (SES, race, residential location). In a recent review of this literature, Brulle and Pellow (2006, 117) argue that a broader view of the evidence leads to the conclusion that the distribution of environmental hazards is clearly socially patterned:

> [R]egarding the literature on environmental inequality, although the race versus class debate has produced exceptional methodological advances in the study of environmental racism/inequality, it has missed the larger picture. The distribution of environmental harm does involve, and has always involved, both race and class. The social production of environmental inequality cannot be understood through a singularly focused framework that emphasizes one form or [of] inequality to the exclusion of others.

Taken together, the social patterning of environmental hazard and the demonstrated risk to developmental health arising from exposure to such hazards indicate clearly that differential exposures are a likely developmental mediator of social disparities. At this point, however, there is insufficient evidence to estimate how substantial a mediating role these differential exposures to neurotoxicants may play at the population level.

It is important in this regard to note the differences between attributable and relative risk. *Attributable risk* addresses the question of how much of the population variance can be attributed to a specific cause or set of causes. *Relative risk* addresses the level of increased risk that is observed among those who have experienced the risk versus those who have not. Both are important. Even if differential exposures account for only a small amount of disparities occurring in the population – and at this point it is difficult to estimate the proportion of population variance accounted for by differences in exposures – it remains important to understand this mediating pathway with respect to relative risk. If the risk of an adverse outcome is high for those who are exposed, as it seems to be, and if such exposures are socially conditioned, as they seem to be, then this type of developmental mediation is socially important, without regard to how much of the total population variance it can explain.

It may be the case that the population-level impact of differential exposures is larger that previously assumed. There is a substantial though recent literature that has begun to identify an additional broad mechanism, beyond the impact on the developing neural system, through which environmental effects may operate. There is increasingly robust evidence of epigenetic effects arising from a wide variety of environmental exposures:

> The epigenome consists of DNA methylation marks and histone modifications involved in controlling gene expression. It is accurately reproduced during mitosis and can be inherited transgenerationally. The

innate plasticity of the epigenome also enables it to be reprogrammed by nutritional, chemical, and physical factors. (Dolinoy & Jirtle, 2008, 4)

As Boyce has noted in this volume, it is not only physical exposures but also differences in early life social interactions that produce epigenetic modifications to gene expression, identified through the effects of differential maternal care in rodent models studied by Meaney, Szyf, and colleagues (Meaney, Szyf, & Seckl, 2007; Szyf, Weaver, Champagne, Diorio, & Meaney, 2005).

Thus, developmental differences in both physical exposures and social experiences have significant biological effects on the developing organism, through two broad mechanisms: direct effects on cell and system functioning, and indirect effects on gene expression. The fact that the epigenetic effects may have transgenerational consequences heightens their probable impact on population developmental health (Dolinoy & Jirtle, 2008; Meaney, 2001; Morgan, Sutherland, Martin, & Whitelaw, 1999; Reik, 2007; Wright & Baccarelli, 2007).

There are a number of sources of material variation in addition to differential exposures to harmful substances. Some of these are more direct, such as nutrition (Engle et al., 2007; Wachs, 2009), and others are more indirect, such as the ability of families to provide access to developmental opportunities, which are to some extent a function of the family's material resources (Hardaway & McLoyd, 2009). As noted earlier, the differences between the developed and the developing world are more generally attributable to material deprivation (Engle et al., 2007).

One of the continuing tensions among researchers who study social disparities is cast in a different light given the biological effects, both cellular and epigenetic, of both physical and social experiences. This is the tension between those who emphasize the impact of material differences and those who focus on social and cultural impacts on the social gradient (Keating, 2009b). Within the developed world, it is likely that both material and social differences influence the severity of the social gradient and the overall performance of different societies. In fact, there is a strong probability that material resources and social affordances interact to generate the shape of the social gradient.

Differential nutrition provides a good example of this. In contrast to the impact of undernutrition in the developing world (Engle et al., 2007), the increasing prevalence of obesity in the United States and elsewhere has become a major concern. The behavioral determinants of obesity – fruit and vegetable consumption versus fast-food and low-nutrient-density consumption, physical activity and outdoor time versus television watching, differential sleep disruption – are clearly socially conditioned, but as noted by Yancey and Kumanyika (2007, S172),

> When we ask **why** the environments of minority and low-income children are relatively less conducive to healthy eating and physical activity, we confront the all-too-familiar reality that people who are socially and politically disadvantaged with respect to the larger social structure are, in fact, socially and politically disadvantaged in many respects. . . . [F]ood availability, food advertising, school policies, recreational facilities, and opportunities for safe, affordable physical activity . . . are not exempt from the forces of . . . stratification. [Emphasis in original.]

In addition, it is likely that there are significant interactions among material contributors to the social gradient. Among the compensatory possibilities for neurotoxicant exposure, for example, is enhanced nutrition to counteract or mitigate some of the physiologic effects of a high loading of metals (Hu et al., 2007).

One of the most understudied approaches to understanding the developmental mediators of social disparities is the likelihood of interactions among differences in exposures, material resources, developmental opportunities, and social interactions (the last of which is the focus of the following section). Interactions among risk factors may play a disproportionate role, owing not only to cumulative risks but also because they may well amplify each other's negative impact. These interactions, however, are understudied for a good reason: the lack of sufficiently large, representative samples in which multiple developmental mediators can be examined. The promise of the U.S. National Children's Study is to address this research gap specifically, allowing the examination of multiple interactions with sufficient power to statistically resolve some of these pressing questions (Landrigan, Trasande, Thorpe, Gwynn et al., 2006).

Social interactions as developmental mediators. That causal links exist from differences in social circumstances to differences in physical conditions during development, to biological effects on the organism, and, in turn, to later life outcomes of developmental health, fits easily within everyday assumptions about how the world works. It has required more effort to establish that variations in social interactions during early life follow a similar path of causal links, including the biological embedding of such experiences in the organism, with similar downstream links to developmental health outcomes. The evidence assembled throughout this volume and summarized in a number of reviews of the literature (e.g., Boyce &

Keating, 2004; Keating, 2009b; Keating & Hertzman, 1999) establishes the viability of such causal pathways. The reciprocal effects from social interactions to brain-behavior functioning depicted in Figure 10.1 are the focus of a burgeoning literature, with increasing attention to the nature of those reciprocal relationships, the gene-environment interactions that generate them, and the epigenetic mechanisms that extend their reach across the life course and transgenerationally.

A conceptual overview of this large research literature for the purpose of elucidating the causal links in Figure 10.1 involving social interactions can focus on three major candidate systems (Keating, 2009b): a sense of connection, rooted largely in the serotonergic (and possibly oxytonergic) system (Barr, this volume); a sense of control, rooted largely in the stress response system (Boyce, this volume; Gunnar & Loman, this volume); and an emergent conscious system, rooted largely in the prefrontal cortex and related systems as they become functionally linked over the complex course of brain development (Nelson, this volume). These and other systems become integrated into a set of capabilities that are rooted in the social and cultural context within which development proceeds. The goal of this necessarily brief overview is to establish that these are high-potential candidate systems to serve as developmental mediators of social disparities in developmental health (Keating, 2009b).

A SENSE OF CONNECTION: SOCIAL RELATIONSHIPS AND THE SOCIAL REWARD SYSTEM. The first and developmentally earliest candidate developmental mediator system can be characterized as a sense of social connection. The serotonergic system plays a key role in generating positive emotions associated with social relationships, and is the target of a major class of antidepressant pharmaceuticals: selective serotonin reuptake inhibitors (SSRIs), such as fluoxetine (Prozac) and related formulations. Dysregulation of this system, arising as an interaction between genetic factors and negative early developmental experiences, has been associated with substance abuse and with heightened aggression in both humans and nonhuman primates, as well as contributing to persistent vulnerability to depressive disorders (Caspi, Sugden, Moffitt, Taylor et al., 2003; Suomi, 1999, 2000).

The social interaction differences that affect the development of this fundamental social connection (and the functioning of the serotonergic system) are rooted in the early transactions between caregiver (typically the mother) and child (Barr, this volume). Both animal and human evidence point toward disruptions in this core relationship as one primary source of variability in this brain-behavior system. It is also important to note, however, that there is genetic variation in the serotonergic system (in the serotonin transporter gene, 5-HTTLPR). In both animal and human studies, this plays out as a gene-environment interaction such that genetic risk is activated only in circumstances of disrupted parent-child relationships. This has been studied experimentally through peer-rearing manipulations in rhesus macaques (Suomi, 2000) and, in humans, through longitudinal investigations that describe a developmental history of abuse/neglect (Caspi et al., 2003). A recent meta-analysis (Risch, Herrell, Lehner, Liang et al., 2009) has raised questions about the likelihood of the generalizability of gene-environment interactions in human studies, although Rutter (this volume; Rutter, Thapar, & Pickles, 2009) has argued that this critique is problematic in that it does not specify the pathways being tested.

The observed consequences of a disrupted serotonergic system are manifold, including elevated risks for depression, aggression, and substance abuse (Barr, Newman, Lindell, Shannon et al., 2004; Bennett, Lesch, Heils, Long et al., 2002; Champoux, Bennett, Shannon, Higley et al., 2002; Pihl & Benkelfat, 2005; Suomi, 2000). Each of these is linked in turn with adult morbidity and early mortality. In addition, effective functioning of this system seems to act as an antagonist and/or compensator for disruptions to the stress response system, and thus disruptions in its function make it less likely that otherwise valuable coping strategies will be available or effective, such as the protective factors found in human and nonhuman primate studies of having kin and/or social support networks.

This system involving social connection may well be significantly influenced by a related neuroendocrine system involving oxytocin and vasopressin (Barr, this volume; Carter, 2003; Febo, Numan, & Ferris, 2005). This system has received considerable recent attention as a potential basis for basic trust and social bonding among individuals in social situations, particularly in intimate relations (especially in maternal nursing and in pair-bonded sexual relations). This system provides a valuable illustration of research directions in the investigation of prospective candidate systems linking social disparities in circumstances to important outcomes in developmental health, through the operation of fundamental developmental mechanisms. In contrast with the hypothalamic-pituitary-adrenal (HPA) axis and the serotonergic systems, we are just beginning to learn how this system develops in humans, and what its consequences are for developmental health, although the prospects are that it may play a significant protective role.

Thus, there is evidence that this system displays the full set of Arrow B and C linkages in the mediator model shown in Figure 10.1: social interactions influence core developments in brain-behavior processes; the developmental functions influence future interactions and reflect the impact of the early developmental history; and there are downstream consequences for a number of areas of developmental health, including social competence and substance abuse, among others. So far, these effects invariably involve gene-environment interactions, highlighting the importance of understanding both genetic vulnerabilities and risks arising from a disrupted pattern of social interactions in early life.

A SENSE OF CONTROL: SOCIAL HIERARCHIES AND THE STRESS RESPONSE SYSTEM. The second candidate as a developmental mediator system focuses on the notion of psychosocial control that figures prominently in contemporary theories of social disparities and health (Marmot, 2004). The functioning of the stress response system (principally based in the limbic hypothalamic-pituitary-adrenal [L-HPA] axis) is a prime candidate for understanding the origins and impact of this psychosocial variable. The now commonplace methodology of the collection of cortisol (a central part of the L-HPA axis function) nonintrusively from saliva samples has greatly expanded the information about how the human stress response system reacts to challenges, and about its diurnal rhythms, both of which vary across individuals and can thus be related to behavioral and health characteristics.

Threats to status or dominance in social hierarchies activate a stress response, which can become dysregulated. In studies of nonhuman primates, Sapolsky (2005) found patterns of dysregulation in response to a variety of status threats. McEwen's (2005, 315) notions of allostasis and allostatic overload sought to capture the long-term consequence of continuing dysregulation of the stress response system:

> The term "allostasis" has been coined to clarify ambiguities associated with the word "stress." Allostasis refers to the adaptive processes that maintain homeostasis through the production of mediators such as adrenalin, cortisol, and other chemical messengers. These mediators of the stress response promote adaptation in the aftermath of acute stress, but they also contribute to allostatic overload, the wear and tear on the body and brain that result from being "stressed out" . . . result[ing] in many of the common diseases of modern life.

The precise physiological mechanisms through which these effects occur are a source of intense current investigation, including the study of other features of the biological stress response system, and our understanding of how the L-HPA system develops and functions has increased rapidly in the last decade or so (Boyce, this volume; Boyce & Ellis, 2005; Gunnar & Loman, this volume; Gunnar & Donzella, 2002; Hertzman & Frank, 2006; McEwen, 2005; Sapolsky, 2005). Cortisol is one of the glucocorticoids (GC), a class of adrenal steroid hormones secreted during stress that also includes hydrocortisone in primates and corticosterone in rodents. They are "double-edged" as they "help mediate adaptation to short-term physical stressors, yet are pathogenic [disease-producing] when secreted chronically" (Sapolsky, 2005, 651). It should also be noted that the L-HPA axis is not the only physiological system involved in stress response. Rather, it is a complexly regulated system involving the sympathoadrenomedullary (SAM) axis, which produces the catecholamines epinephrine and norepinephrine, and the neuroimmune system, which produces the key agents in host defense against infection and injury and has been shown to be affected by psychosocial factors (in the field of psychoneuroimmunology, PNI; Coe, 1999). In addition to the primary agents (HPA, SAM, and PNI), there are a number of other physiological systems that serve as antagonists and regulators, including serotonin and oxytocin (Carter, 2003; Suomi, 2000).

There is solid evidence that L-HPA function is shaped by early social experience in both animal and human models, described by Boyce (this volume) and by Gunnar and Loman (this volume), and summarized succinctly by Boyce and Ellis (2005, 278):

> [R]odent, nonhuman primate, and human research all point to a common conclusion: that both genetic and environmental factors contribute to the calibration of biological stress response systems over the course of early development. . . . Stress reactivity . . . appears to become "canalized" over time, revealing progressively greater resistance to change and diminishing plasticity . . . [F]indings suggest . . . that social contextual effects over the first 3 to 5 years of life may have particular potency in the calibration of stress responsive biological systems.

This calibration is affected by both normative variation in early developmental experiences, as in elevated cortisol levels among children in low-quality daycare settings compared with those in higher quality settings or at home (Geoffroy, Côté, Parent, & Seguin, 2006), and also dramatically by extreme variations of abuse and neglect (Boyce & Ellis, 2005; Gunnar & Donzella, 2002; Hertzman & Frank, 2006). The early developmental experiences in close relationships "prepare" the organism for the challenges it is likely to face in the average expected environment predicated on the conditions of that early environment. To avoid teleology in this formulation, it is important to note that neither the organism nor its parents consciously predict what the future will hold. Yet there appears to be an evolutionarily preserved variability in the stress response system that responds to the likely environment, based on current experience.

Taken together, the cumulative evidence on the stress response system, including the psychosocial impacts on its expression, the impacts of early development on its calibration, the social distribution of those impacts, and the health consequences of dysregulation in this system, supports the plausibility of this as a route for mediation of the social gradient in developmental health. Put more succinctly (Gunnar & Loman, this volume),

> Stress is a normal part of human existence, but overwhelming stress in the face of inadequate physical and emotional support appears to pose a risk for human development throughout life.

THE EMERGENCE OF CONSCIOUSNESS: IDENTITY AND MEANING. A third, later-developing system – both phylogenetically and ontogenetically – includes aspects of perceived purpose, hope, meaning, and identity. This system is uniquely human, evolving as a level of reflective consciousness that represents a coevolution of culture and mind. An interesting recent intersection of paleoanthropology and neuroscience has vastly increased our understanding of human speciation with respect to this coevolution of culture and mind (e.g., Donald, 2001; Dunbar, 1996, 2004). In short, the core idea is that the primary competitive advantage in the speciation of modern humans (*Homo sapiens*) was improved social coordination within larger groups, and eventually between groups. Based on a long buildup of various protocomponents, language, culture, and mind coevolved rapidly to support such social coordination. In fact, there is a strong argument that the evolution of fully modern humans (us) from early modern humans occurred in the context of a millennia-long competition with archaic humans (Neanderthals) in the Levant, a competition that we "lost" for the first thirty thousand years or so, but returned to "win" around forty to fifty thousand years before the present (BP) because of new adaptations that employed "flexible, symbolically reinforced alliance networks with wide situational variability in group size" (Shea, 2003, 183) – in other words, a reworked interface of mind, culture, and social structure. A similar conclusion based on distinct patterns between Neanderthals and modern humans emphasized the similarities in material extraction but dissimilarities in social networks: "[I]t is the development and maintenance of larger social networks, rather than technological innovations or increased hunting prowess, that distinguish modern humans from Neanderthals in the southern Caucusus" (Adler, Bar-Oz, Belfer-Cohen, & Bar-Yosef, 2006, 105). A newly emerging field of neuroecology focuses on the evolutionary mechanisms of human cognition (Sherry, 2006).

The biodevelopmental focus of this individually and socially coordinated activity is largely the prefrontal cortex of the brain (PFC), which undergoes substantial development during adolescence, including increased speed of transmission as well as the amount and complexity of connections to other brain systems, especially the cerebellum (Giedd, 2008; Giedd, Schmitt, & Neale, 2007; Keating, 2004). The implication of this is that the achievement of the specifically human level of consciousness involves an increased role of the PFC in the governance of brain and biological systems.

This third candidate mediator system focuses on developments in the second decade of life. The PFC is disproportionately larger (relative to body-size-corrected brain volume) in humans, both archaeologically and by comparison to contemporary nonhuman primates. The PFC achieves this differential size largely during the adolescent transition, and is connected in developmental time with increased speed and amount of connections to other brain systems. A second implication from the nature of PFC growth – proliferation of new synapses in early adolescence, and pruning of those during later adolescence – is that this growth is sensitive to developmental experience, in much the same way that early experience shapes other brain and biological systems (like the L-HPA and serotonergic systems).

Because research in this specifically human domain cannot be clearly guided by animal models, the specific nature of developmental shaping is less well known. Yet we do know that this is a period of increasing behavioral self-regulation, including advanced cognitive facilities of logic and emotion-attention regulation and the development of self-aware identity, during a period when greater autonomy of action increases the risks from behavioral misadventure (Steinberg, Dahl, Keating, Kupfer, Masten, & Pine, 2006). Similarly, precise connections from these acquisitions and outcomes in developmental health are more difficult to specify, given the complexity of the system, but there is an established link from related aspects such as hopelessness or lack of meaning to a range of negative health outcomes, as well as increased health risks from behavioral misadventure arising from the differential growth of brain systems subserving exploration and risk-taking versus judgment (Cauffman, Shulman, Steinberg, Claus, Banich et al., 2010; Steinberg, 2008; Steinberg et al., 2006).

Another reason to incorporate this candidate developmental mediator system into the model is that it is likely to be a mediator of the functioning of the stress response system during adulthood. Two elements stand out as the most influential in producing the stress response: cognitive uncertainty and the prospect of negative social evaluation (Dickerson & Kemeny, 2004). Both of these are significant functions of the self system, and thus the experience of stress, both psychosocially and physiologically, is substantially affected by the way in which prefrontal (conscious) functions have been shaped by developmental experiences. Given the multiplicity of ways in which individuals can cope with stressors, the ability to navigate alternate responses in a conscious fashion, utilizing available capabilities, may prove to be a powerfully organizing protective factor. Future research will be needed to more firmly establish this mediator pathway and its central mechanisms.

Summarizing the working hypothesis depicted by the model in Figure 10.1, the claim is that the social gradient is rooted in a causal system that links social circumstances to differential exposures, experiences, and social transactions, which interact reciprocally with biological and developmental mechanisms over the course of development, leading to the social patterning of downstream, life-course consequences for developmental health. If this account is essentially correct – and the overview of evidence on the developmental mediator model suggests that it is – there are important implications for how we think about policy, prevention, and intervention on the issue of social disparities. Most significantly, the developmental mediator model carries with it the recognition that social disparities can be successfully addressed, and that a major route lies through a better understanding of the underlying developmental mechanisms. As well, it potentially grounds social policy formation and evaluation in a new and firmer interdisciplinary terrain. Some of the major implications of this conceptual framework for societal decision making are discussed in the following section.

ALLEVIATING SOCIAL DISPARITIES: NATURE-AND-NURTURE IN POLICY AND PRACTICE

Successfully addressing the long-standing problem of social disparities will require an integration of research efforts addressed to population patterns and individual developmental mechanisms. One goal of the model presented here is to provide scientific support for an integrated approach to broad social policy, targeted prevention efforts, and specific interventions at the clinical level. The major topics included in this overview start with lessons at the population level, drawing on what has been learned from the emergent field of population health. Following that overview, those lessons are extended to take account of research on "unpacking" the social gradient.

Distribution of Resources: Income Inequality

There are two major goals of the population health approach: to understand health in terms of its population dynamics rather than as merely an aggregate of individual pathways; and to understand the determinants of population health, first at the social level and then as it affects individual health (Adler & Newman, 2002; Hertzman, 1999; Hertzman & Siddiqi, 2009). Disciplinary terminology in this area, as in many others, can hinder understanding. In the population health and social epidemiology literatures, *determinants* is the standard term to identify factors that have an independent influence on some outcomes. This language can be a red flag for developmental scientists, who associate the term with the notion of "determinism," which, in a variety of flavors (genetic determinism, environmental determinism, infant determinism), is anathema because it is taken to imply fixed factors whose expression is not contingent on other inputs or subsequent events – the core interests of developmentalists. Developmental scientists are more likely to use terms seen as more neutral, such as predictors, mediators, and moderators. If a strong theoretical claim is intended, then terms like mechanisms or developmental origins are employed, with a concomitant elevation of the evidentiary requirement. Most users of "determinants" do not have in mind any kind of determinism, but the term is a common source of misunderstanding.

A major research and policy interest has historically been to identify the potential sources of health inequalities. The most frequently studied mechanisms have been at a societal level, and the most studied social level mechanism has been the role of income inequality. There is strong logic to support this hypothesis. First, it seems clear that absolute physical deprivation, fundamental access to resources, is not likely to be a major source of health inequality in advanced industrial societies, on the grounds that historical trends have reduced the contribution of physiological inequalities (Fogel, 2003), and because the country comparisons of the relationship between average income and average health are close to zero among the industrialized nations. This has come to be known as the "flat end of the (country comparative) health/wealth curve" and is illustrated, for example, in Lynch et al. (2004, fig. 3).

For these reasons alone, attention might well turn to the *distributional* aspects of wealth or income as a different type of explanation for the social gradient in health. There is a potential for conflation of the *income inequality gradient* and the *social gradient in health* that should be noted and avoided. The social gradient that has been the focus of this chapter can be thought of as *outcome inequality*. One, among many, possible accounts of differences between countries (or states, or metropolitan areas) in the social gradient is the variation of *income inequality* across those jurisdictions.

It is easy to understand the policy interest in the possible contribution of income inequality to both average health and health inequalities. If it were the case that differences among countries in the level of income inequality accounted for a significant amount of the average health differences among them, this would add a powerful argument for advocates of income and wealth redistribution.

Given this, it is no surprise that this particular question has increasingly engaged the interests of economists as well as social epidemiologists (Beckfield, 2004; Deaton, 2002; Dunn, Burgess, & Ross, 2005; Dunn, Frohlich, Ross, Curtis, & Sanmartin, 2006; Ross, Wolfson, Kaplan, Dunn, Lynch, & Sanmartin, 2006). Given the stakes of this policy debate, it is also no surprise that it continues to rage. Several recent systematic reviews and meta-analyses raise a host of issues that will not be sorted out any time soon. Much of the debate arises from differential assumptions about what to include in the model, and what analytic strategy to pursue in isolating any effect of income inequality.

In a systematic review with a helpful historical overview, Lynch and colleagues (2004) concluded that income inequality does not explain international differences in population health, although it does appear to contribute to state-level differences in the United States. They note that there is ongoing controversy as to whether these differences are properly attributed to income inequality or to some other aspects of those regions – a classic hidden variable problem. Deaton and Lubotsky (2003) argued that the black-to-white ratio of residents explained the effects of income inequality, although that conclusion was subsequently contested by Ross and colleagues (2006, 215, fig. 8.7) who showed no such relationship in an analysis of U.S. metropolitan areas. Beckfield (2004) found no international effect of income inequality when using a fixed-effects model that enables an analysis of *changes* in income inequality as they are related to *changes* in average health. Wolfson and colleagues (1999) tested a claim that the income inequality relationship was a statistical artifact of the curvilinearity of the individual income to health relationship, and found that across U.S. states, the artifact could not explain all of the income inequality effect. Dunn, Burgess, and Ross (2005) reported significant effects of differential public expenditures across U.S. states on all-cause mortality, suggesting another possible "third" variable explanation. They found that the income inequality effect was only partly attenuated by the level of public expenditures. On balance, the evidence suggests that income inequality is a significant contributor to overall health outcomes, in some geographical regions and at some but not all levels of aggregation. What seems safer to conclude is that income inequality does not explain all health differences across societies.

Note, though, that even if income inequality made no contribution to overall health differences, this finding does not reduce in any way the reality of the social gradient in developmental health. It would mean that income inequality does not contribute to the explanation of average differences in health or to society-level health inequalities and would require that we look elsewhere for more satisfactory explanations.

Health Policy

Adler and Newman (2002) provided a cogent summary of a range of possible explanations for health inequalities that can be usefully divided into aspects that have been the direct focus of health policies of societies, and those that are viewed as influencing health but are not themselves health policies. Perhaps the most obvious direct target of the effect of health policy on the social gradient is access to health care. Clearly, if lower SES individuals do not have access to care, their health is likely to suffer. Yet as an overall contributor to the social gradient, differential access to health care is a relatively weak predictor of health outcomes. It is important to recall here the distinction between relative and attributable risk. In this case, if one has a treatable disease (i.e., an exposure), then lack of access to health care can dramatically increase relative risk; but differential access to health care has not been shown to be a large attributable risk, because the proportion of individuals at any given time who have the exposure (a health problem for which medical treatment is critical) is relatively small. This is more than a technical distinction because in policy discussions, one line of thinking leads toward viewing a particular risk as large, whereas another line of thinking leads one to see it as small. Both are accurate descriptions of reality; which one matters in any given case depends on the question being asked.

A second area of potential explanation for the social gradient in health is differences in the physical environment, especially through differential exposures to environmental toxins, as discussed earlier. This dates back, of course, to the origins of a formally defined field of public health. Because exposure to harmful biological and toxic industrial agents is not randomly distributed, but instead is partly conditioned by where one stands on SES, it is logical to look for differential exposures as a major way to explain the social gradient. Because there are many types of potential exposures, whose precise biological mechanisms and time courses are a major focus of current research, as they exert a health effect alone or in combination with other exposures, it is difficult to know how fully these risks explain the social gradient. Large, prospective studies with direct sampling of differential exposures, such as the U.S. National Children's Study, would advance our understanding considerably. Universal programs to reduce or eliminate known harmful exposures, or to protect against infectious exposures through vaccination, have great public health benefits. They do not necessarily reduce the social gradient in health if the residual exposures or lapses in protective measures are themselves conditioned on social position, but they nonetheless continue to enhance overall population health.

Health Behavior

A final factor from the research literature on health to be considered here is the impact of health behaviors on the social gradient. Virtually all health behaviors are correlated with social position – smoking, diet, exercise, to name a few – and they do contribute to the social gradient in health. Health-promotion efforts to alter these lifestyle characteristics and to improve health are often effective, at least partially, especially if they are linked to legal and financial initiatives, as has been the case recently with bans on smoking in public places and sharp increases in tobacco taxes. However, as a number of observers have noted, these health-promotion efforts may exacerbate the social gradient, because the health-improving behaviors may be taken up more readily and commonly in higher SES groups (Adler & Newman, 2002, 69).

Some proponents of a social determinants model of health regard the health-promotion movement with suspicion, owing not only to the (usually unintended) added benefit to the already advantaged but also to its potential to become a message that "blames the victim." One parody of a healthy lifestyle list ("Stop smoking, eat a balanced diet, get more exercise . . . ") arising from those who advocate a greater focus on social determinants begins with "Don't be born to poor parents, don't live in bad neighborhoods, obtain safe working conditions. . . . " To what extent health behaviors account for differences in health outcomes varies considerably, from a modest amount to nearly half of the differences, but there is a stronger consensus that health-promotion efforts that do not take the social gradient into account will be less likely to have an impact on the whole population (Adler & Newman, 2002; Williams, 2005).

In the seminal Whitehall studies of British civil servants working in their head office in London, Marmot (2004) and colleagues examined the patterning of the health outcomes across employment levels from clerical to top executives. This was intentionally designed to control for each of the factors named earlier: all the participants had identical public health insurance (at least in principle); all worked in the same physical environment and were at income levels that likely minimized risky exposures in their residential settings; and health behaviors were measured. The striking finding was that, despite these similarities, and even after controlling for health behavior differences, there were large health differences across nearly all types of disease, and these were related to social status – clerical at the bottom, top executives at the top. Moreover, the measurable variables still left unexplained more than half, perhaps up to three quarters, of the social gradient in health. On these grounds, Marmot (2004) focused on more psychosocial explanations, as a function of status in itself.

The findings that differences in material resources, health care access, healthy behaviors, and environmental exposures do not eliminate the social gradient have been taken by some observers to mean that they do not matter. It is worth emphasizing that they do matter, considerably. The amount of variance they explain in overall health and in the social gradient is far from trivial, even if much of the social gradient remains unexplained by these factors. In a recent summary of progress on addressing the social gradient, Williams noted (2005, 129),

> Evidence clearly shows that social disparities in health are large, pervasive across health measures, persistent over time, and costly to society. Moreover, interventions aimed at improving health that are not coupled with those that seek to reduce social disadvantage are unlikely to substantially reduce disparities.

If we view this in light of the set of six social gradient propositions noted earlier, the urgency of the current situation becomes even more apparent. If (a) the severity (steepness) of the social gradient in health is robustly related to the average performance of a society; (b) the effect is pervasive across not only health but also the full range of developmental health, including the level and distribution of individual capabilities in the society; (c) that these effects are mediated through a causal system that includes developmental and biological embedding; and (d) the effects have life-course and even transgenerational consequences, then the centrality of successfully addressing the social gradient for an agenda focused on social disparities is brought into sharp relief. Including the degree to which the distribution of capabilities throughout the population operates through these mediators, the breadth and depth of population competence sufficient to build the knowledge economies of the foreseeable future also come into play (Keating, 1998).

Distribution of Developmental Opportunities

If social and biodevelopmental mediators form the primary causal chain that gives rise to the social gradient, there are a number of implications for how we should think about social policy, and the institutional and cultural milieu within which policy functions. There is also a substantial research agenda implied by the mediator model described in this chapter, which will require a concerted interdisciplinary approach of significant duration.

First, the social gradient can be viewed as a "leading indicator" of population developmental health, both because of the connection between the gradient and the average performance of the society, and because the life-course consequences of what capabilities are being developed have such a long reach. Explicit attention to the distribution of opportunities for acquiring developmental health, in addition to average performance, is highlighted in this approach.

Second, the range of policies that affect how developmental and health opportunities are distributed is very wide. It includes not only health and education policies, but the full range of social arrangements that affect early developmental experiences. These of course affect not only health outcomes but also the development of capabilities in the more inclusive sense – which in turn have consequences for subsequent health outcomes (Keating & Simonton, 2008). Differences in overall public expenditures have been found to be a predictor of mortality (Dunn, Burgess, & Ross, 2005), suggesting that investments in human well-being may play a significant role at the societal level.

Third, it potentially focuses attention on those aspects of social arrangements where there is the greatest leverage for change, based on a clearer understanding of the underlying mechanisms, and also allows an analysis of the potential impact on the social gradient of developmental health arising from shifts in social policy or arrangements whose primary target is neither health nor development. As a by-product of this type of analysis, it is possible to learn more about what the action components of social policies and social arrangements actually are.

In addition to these broad implications, there are two specific dialogues worth noting about social policy research for which this mediator model and the supporting evidence have significant implications: the balance among universal, targeted, and clinical interventions, incorporated into a shift to evidence-based intervention, prevention, and policy; and the centrality of monitoring developmental health outcomes to assess both planned and unplanned change. The complex systems involved are likely not amenable to blueprints, but do respond to iterative design changes implemented with accurate feedback about effectiveness.

Evidence-Based Policy and Practice: Universal, Targeted, and Clinical Approaches

Across a variety of disciplines, there has been a strong movement in recent years toward an insistence on evidence-based approaches to intervention, prevention, and policy. At one level, this seems entirely noncontroversial. Most of us would worry about a medical treatment that had not been evaluated for efficacy and for the avoidance of harm. Yet this seemingly simple picture becomes more complicated quite rapidly. Waiting for treatments to be established in medicine's gold standard randomized control trial (RCT) methodology seems sensible, unless a loved one could benefit now, but not later, from an experimental therapy. From there, one can consider circumstances in which RCTs are not feasible, where it would be impossible for the nature of treatment to be blind to the therapist or patient. For example, RCTs for community-level interventions are not likely to be politically feasible. If the outcome measures are themselves controversial in some quarters, such as academic achievement tests, then the problems are magnified.

Yet, despite the problems, the advantages of evidence-based approaches to early child development outweigh the problems. Following fads, serially repeating failed efforts, and never developing a cumulative knowledge base about what works provoke excessive costs in time, money, and missed opportunities. Several issues deserve consideration. The first is to acknowledge that there is no single gold standard for establishing the evidence base for interventions across the board. Rigorous experimental (RCT but not only RCT), quasi-experimental, and time-series designs for "natural experiments" make it entirely feasible, with a bit of work, to separate intervention, prevention, and policy efforts into unsuccessful, promising, and proven categories (Keating & Simonton, 2008).

The second is that the search for more promising evidence-based approaches could be accelerated if it were more closely guided by validated mediator models. The argument of this chapter is that understanding the underlying mechanisms affords a purchase on more likely levers for change than is otherwise available. It is far easier to move toward evidence-based interventions if the underlying mechanisms are known. Finally, it is likely, based on the earlier cited research, that local context will matter a great deal (Smedley, Syme, & Committee on Capitalizing on Social Science and Behavioral Research to Improve the Public's Health, 2001).

Because different researchers see different targets of opportunity in trying to address identified problems in health generally and in the social gradient in developmental health specifically, because their respective disciplines equip them with different tools, and because resources for addressing problems are limited, competition over how best to proceed is guaranteed. Thus, it is important to devise a rational and nonideological way to evaluate competing approaches.

Universal approaches have the benefits of potentially widespread positive effects, equity in application, and potentially strong political support if the approach is or becomes popular. Targeted approaches focus on problems that even well-designed universal programs are not able to address because of differential access (psychosocial if not material), and they are often cost effective if the intervention is too expensive to be implemented universally. Clinical interventions have maximally specific application for individuals who are already displaying a problem in health and development, and have the benefit, if applied early enough, to prevent many more difficult downstream consequences. Yet they are often the most expensive, and can absorb resources that would have greater benefits if applied further upstream rather than after a problem has been diagnosed.

Given that a principled preference is not possible, it becomes an empirical question as to how best to allocate available resources. A well-developed model that incorporated estimated prevalence of disorders, estimated effect sizes for treatments, and accuracy of a screening regime was proposed by Offord, Kraemer, Kazdin, Jensen, Harrington, and Gardner (1999). Such an allocation model would be helpfully informed by understanding the relevant underlying mechanisms more precisely, including which programs have the strongest evidence base and what developmental timing maximizes the effect of an intervention. As well, a systematic way to assess the prevalence of specific issues on a local scale could potentially be used to maximize the effective allocation of resources (Offord et al., 1999). Note also that to the extent that policy, prevention, and intervention are systemically connected, as they surely are, a systemic approach to reducing social disparities is likely to have a larger payoff than disconnected approaches. Searching for such synergies is likely to become increasingly important, and would be facilitated by a better understanding of the developmental mediator system that underlies observed social disparities.

The focus of this chapter has been on understanding and using a developmental mediator model to address social disparities. It should be noted that there are a number of health problems in childhood that have not been found to be socially patterned, such as neurodevelopmental disorders like autistic spectrum disorder (ASD), likely because they have a strong genetic component. Although such disorders do not contribute to social disparities at the population level, it is likely that they can also be better understood through an understanding of their developmental mediators, including gene-environment interactions and epigenetic influences. Whatever the source of developmental diversity, optimizing developmental outcomes will be enhanced by an adaptive approach to education and treatment that places a high priority on dealing with such diversity developmentally as opposed to categorically (Keating, 1990, 2008).

Monitoring Developmental Health

A final implication of this approach is that there will be a need for continuous monitoring of developmental health across society if a cumulative evidence base on "what works" is to be built. Centrally designed blueprints disseminated to the periphery will not typically suffice owing to the complexity of local social contexts. There are a variety of "human development indicators" initiatives underway, at the United Nations, Organisation of Economic Cooperation and Development, World Bank, and elsewhere. These efforts would benefit by being expanded, localized, and better grounded in underlying mechanisms.

An expansion would encompass a broader range of developmental health outcomes to ensure that key areas are not being missed, and the selection of domains to sample could be more clearly guided by an understanding of the underlying social and developmental mediators (Keating, 2001). Reporting on the social gradient in developmental health, along with the population trends, would be relatively easy and would provide an important set of leading indicators.

Similarly, early development could be oversampled, and longitudinal indicators could become commonplace rather than rare. The ability to consider trajectories as well as static outcomes would provide a much sharper picture of progress. Finally, national-level indicators, which would be highly informative regarding broader social policy and institutional issues, could be supplemented by community-level indicators that would be invaluable guides for tracking the effectiveness of multiple community interventions and programs.

One of the clear messages from the available evidence is that there are multiple social arrangements that can support the development of capabilities, both individual and collective, but also that some are more conducive than others. Finding what works to support this feature of successful societies would seem well worth the investment (Keating, 2009b). Using a lens based on an understanding of the crucial social and developmental mediators may substantially improve our ability to discern what works to support development throughout the population, and thus reduce social disparities in a meaningful way.

REFERENCES

Adler, D. S., Bar-Oz, G., Belfer-Cohen, A., & Bar-Yosef, O. (2006). Ahead of the game – Middle and upper palaeolithic hunting behaviors in the southern Caucasus. *Current Anthropology* 47(1), 89–118.

Adler, N. E., & Newman, K. (2002). Socioeconomic disparities in health: Pathways and policies. *Health Affairs*, 21(2), 60–76.

Barr, C. S., Newman, T. K., Lindell, S., Shannon, C., Champoux, M., Lesch, K. P., Suomi, S. J., Goldman, D., & Higley, J. D. (2004). Interaction between serotonin transporter gene variation and rearing condition in alcohol preference and consumption in female primates. *General Psychiatry*, 61(11), 1146–52.

Beckfield, J. (2004). Does income inequality harm health? New cross-national evidence. *Journal of Health and Social Behavior*, 45(3), 231–48.

Bennett, A. J., Lesch, K. P., Heils, A., Long, J. C., Lorenz, J. G., Shoaf, S. E., Champoux, M., Suomi, S. J., Linnoila, M. V., & Higley, J. D. (2002). Early experience and serotonin transporter gene variation interact to influence primate CNS function. *Molecular Psychiatry*, 7, 118–22.

Bornstein, M., & Bradley, R. (2003). *Socioeconomic Status, Parenting, and Child Development*. Mahwah, NJ: Lawrence Erlbaum Associates Publishers.

Bornstein, M., & Sawyer, J. (2006). Family systems. In *Blackwell Handbook of Early Childhood Development* (pp. 381–98). Malden, MA: Blackwell Publishing.

Boyce, W. T., & Ellis, B. J. (2005). Biological sensitivity to context: I. An evolutionary-developmental theory of the origins and functions of stress reactivity. *Development and Psychopathology*, 17, 271–301.

Boyce, W. T., & Keating, D. P. (2004). Should we intervene to improve childhood circumstances? In D. Kuh & Y. Ben-Shlomo (Eds.), *A Life Course Approach to Chronic Disease Epidemiology.* Oxford: Oxford University Press.

Brulle, R. J., & Pellow, D. N. (2006). Environmental justice: Human health and environmental inequalities. *Annual Review of Public Health,* 27, 103–24.

Carter, C. S. (2003). Developmental consequences of oxytocin. *Physiology and Behavior,* 79(3), 383–97.

Caspi, A., Sugden, K., Moffitt, T. E., Taylor, A., Craig, I., Harrington, H. L., McClay, J., Mill, J., Martin, J., Braithwaite, A., & Poulton, R. (2003). Influence of life stress on depression: Moderation by a polymorphism in the 5-HTT gene. *Science* 301(5631), 386–9.

Cauffman, E., Shulman, E. P., Steinberg, L., Claus, E., Banich, M. T., Graham, S., & Woolard, J. (2010). Age differences in affective decision making as indexed by performance on the Iowa Gambling Task. *Developmental Psychology,* 46(1), 193–207.

Champoux, M., Bennett, A., Shannon, C., Higley, J. D., Lesch, K. P., & Suomi, S. J. (2002). Serotonin transporter gene polymorphism, differential early rearing, and behavior in rhesus monkey neonates. *Molecular Psychiatry,* 7,1058–63.

Chandler, M. J., Lalonde, C. E., Sokol, B. W., & Hallett, D. (2003). Personal persistence, identity development, and suicide: A study of Native and Nonnative North American adolescents. *Monographs of the Society for Research in Child Development,* 68(2), vii–130.

Coe, C. L. (1999). Psychosocial factors and psychoneuroimmunology. In D. P. Keating & C. Hertzman (Eds.), *Developmental Health and the Wealth of Nations.* New York: The Guilford Press.

Corwyn, R., & Bradley, R. (2005). Socioeconomic status and childhood externalizing behaviors: A structural equation framework. In *Sourcebook of Family Theory and Research* (pp. 469–92). Thousand Oaks, CA: Sage Publications, Inc.

Deaton, A. (2002). Policy implications of the gradient of health and wealth. *Health Affairs,* 21(2), 13–30.

Deaton, A., & Lubotsky, D. (2003). Mortality, inequality, and race in American cities and states. *Social Science and Medicine,* 56(6), 1139–53.

Dickerson, S. S., & Kemeny, M. E. (2004). Acute stressors and cortisol responses: A theoretical integration and synthesis of laboratory research. *Psychological Bulletin,* 130(3), 355–91.

Dolinoy, D. C., & Jirtle, R. L. (2008). Environmental epigenomics in human health and disease. *Environmental and Molecular Mutagenesis,* 49, 4–8.

Donald, M. (2001). *A Mind So Rare: The Evolution of Human Consciousness.* New York: Norton.

Dunbar, R. I. (1996). *Grooming, Gossip, and the Evolution of Language.* London: Faber.

Dunbar, R. I. (2004). *The Human Story.* London: Faber.

Dunn, J. R., Burgess, B., & Ross, N. A. (2005). Income distribution, public services expenditures, and all cause mortality in U.S. States. *Journal of Epidemiology and Community Health,* 59(9), 768–74.

Dunn, J. R., Frohlich, K. L., Ross, N. A., Curtis, L. J., & Sanmartin, C. (2006). Role of geography in inequalities in health and human development. In J. Heymann, C. Hertzman, M. L. Barer, & R. G. Evans (Eds.), *Healthier Societies: From Analysis to Action.* New York: Oxford University Press.

Engle, P. J., Black, M. M., Behrman, J. R., deMello, M. C., Gertler, P. J. et al. (2007). Strategies to avoid the loss of developmental potential in more than 200 million children in the developing world. *Lancet,* 369, 229–42.

Essex, M. J., Boyce, W. T., Goldstein, L. H., Armstrong, J. M., Kraemer, H. C., & Kupfer, D. J. (2002). The confluence of mental, physical, social, and academic difficulties in middle childhood: II. Developing the MacArthur Health & Behavior Questionnaire. *Journal of the American Academy of Child and Adolescent Psychiatry,* 41, 588–603.

Febo, M., Numan, M., & Ferris, C. F. (2005). Functional magnetic resonance imaging shows oxytocin activates brain regions associated with mother-pup bonding during suckling. *Journal of Neuroscience* 25(50), 11637–44.

Fiscella, K., & Franks, P. (2000). Quality, outcomes, and satisfaction. Individual income, income inequality, health, and mortality: What are the relationships? *Health Services Research,* 35(1), 307–15.

Fogel, R. (2003). Secular trends in physiological capital: Implications for equity in health care. In *National Bureau of Economic Research Working Paper,* 9771.

Geoffroy, M., Côté, S., Parent, S., & Séguin, J. (2006). Daycare attendance, stress, and mental health. *Canadian Journal of Psychiatry/La Revue canadienne de psychiatrie,* 51(9), 607–15.

Giedd, J. N. (2008). The teen brain: Insights from neuroimaging. *Journal of Adolescent Health,* 42, 335–42.

Giedd, J. N., Schmitt, J. E., & Neale, M. C. (2007). Structural brain magnetic resonance imaging of pediatric twins. *Human Brain Mapping,* 28(6), 474–81.

Gottlieb, G., & Willoughby, M. T. (2006). Probabilistic epigenesis of psychopathology. In D. Cicchetti & D. Cohen (Eds.), *Handbook of Developmental Psychopathology.* New Jersey: John Wiley & Sons, Inc.

Gunnar, M. R., & Donzella, B. (2002). Social regulation of the cortisol levels in early human development. *Psychoneuroendocrinology,* 27(1–2), 199–220.

Hall, P. A., & Lamont, M. (Eds.) (2009). *Successful Societies: Institutions, Cultural Repertoires, and Health.* New York: Cambridge University Press.

Hardaway, C., & McLoyd, V. (2009). Escaping poverty and securing middle-class status: How race and socioeconomic status shape mobility prospects for African Americans during the transition to adulthood. *Journal of Youth and Adolescence,* 38(2), 242–56.

Hatch, S. L. (2005). Conceptualizing and identifying cumulative adversity and protective resources: Implications for understanding health inequalities. *Journal of Gerontology (B).* 60, 130–4.

Hertzman, C. (1999). Population health and human development. In D. P. Keating & C. Hertzman (Eds.), *Developmental Health and the Wealth of Nations: Social, Biological, and Educational Dynamics.* New York: Guilford Press.

Hertzman, C. (2001). Health and human society. *American Scientist,* 89(6), 538–45.

Hertzman, C., & Frank, J. (2006). Biological pathways linking social environment, development, and health. In J. Heymann, C. Hertzman, M. L. Barer, & R. G. Evans (Eds.), *Healthier Societies: From Analysis to Action*. New York: Oxford University Press.

Hertzman, C., & Power, C. (2006). A life-course approach to health and human development. In J. Heymann, C. Hertzman, M. L. Barer, & R. G. Evans (Eds.), *Healthier Societies: From Analysis to Action*. New York: Oxford University Press.

Hertzman, C., & Siddiqi, A. (2009). Population health and the dynamics of collective development. In P. A. Hall & M. Lamont (Eds.), *Successful Societies: Institutions, Cultural Repertoires, and Health*. New York: Cambridge University Press.

Hu, H., Shine, J., & Wright, R. O. (2007). The challenge posed to children's health by mixtures of toxic waste: The Tar Creek superfund site as a case study. *Pediatric Clinics of North America*, 54, 155–75.

Jensen, A. (1969). How much can we boost IQ and scholastic achievement? *Harvard Educational Review*, 39(1), 1–123.

Keating, D. P. (1990). Charting pathways to the development of expertise. *Educational Psychologist*, 25, 243–67.

Keating, D. P. (1998). Human development in the learning society. In A. Hargreaves, A. Lieberman, M. Fullan, & D. Hopkins (Eds.), *International Handbook of Educational Change*. Dordrecht, The Netherlands: Kluwer.

Keating, D. P. (2001). Definition and selection of competencies from a human development perspective. In *Additional DeSeCo Expert Opinions*. Paris: Organisation for Economic Cooperation and Development.

Keating, D. P. (2004). Cognitive and brain development. In R. Lerner, & L. Steinberg (Eds.), *Handbook of Adolescent Psychology*. New York: Wiley & Sons.

Keating, D. P. (2008). Developmental science and giftedness: An integrated lifespan framework. In F. Horowitz, D. Matthews, & R. Subotnik (Eds.), *The Development of Giftedness and Talent across the Life Span*. Washington, DC: American Psychological Association.

Keating, D. P. (2009a). *National Differences in Developmental Health: Levels and Inequalities*. Presented at the Biennial Meeting of the Society for Research in Child Development, Denver, Colorado (April).

Keating, D. P. (2009b). Social interactions in human development: Pathways to health and capabilities. In P. A. Hall & M. Lamont (Eds.), *Successful Societies: Institutions, Cultural Repertoires, and Health*. New York: Cambridge University Press.

Keating, D. P., & Hertzman, C. (Eds.) (1999). *Developmental Health and the Wealth of Nations: Social, Biological, and Educational Dynamics*. New York: Guilford Press.

Keating, D. P., & Simonton, S. Z. (2008). Health effects of human development policies. In J. House, R. Schoeni, A. Pollack, & G. Kaplan (Eds.), *Making Americans Healthier*. New York: Russell Sage.

Kraemer, H. C., Stice, E., Kazdin, A., Offord, D., & Kupfer, D. (2001). How do risk factors work together? Mediators, moderators, and independent, overlapping, and proxy risk factors. *American Journal of Psychiatry*, 158(6), 848–56.

Kuh, D., & Ben-Shlomo, Y. (Eds.) (2004). *A Life Course Approach to Chronic Disease Epidemiology.* Oxford: Oxford University Press.

Lamont, M. (2009). Responses to racism, health, and social inclusion as a dimension of successful societies. In P. A. Hall & M. Lamont (Eds.), *Successful Societies: Institutions, Cultural Repertoires, and Health.* New York: Cambridge University Press.

Landrigan, P. J., Trasande, L., Thorpe, L. E., Gwynn, C., Lioy P. J., D'Alton, M. E., Lipkind, H. S., Swanson, J., Wadhwa, P. D., Clark, E. B., Rauh, V. A., Perera F. P., & Susser, E. (2006). The National Children's Study: A 21-year prospective study of 100,000 American children. *Pediatrics,* 118, 2173–86.

Link, B. G., & Phelan, J. C. (2005). Fundamental sources of health inequalities. In D. Mechanic, L. B. Rogut, & D. C. Colby (Eds.), *Policy Challenges in Modern Health Care.* Piscataway, NJ: Rutgers University Press.

Luecken, L. J., & Lemery, K. S. (2004). Early caregiving and physiological stress responses. *Clinical Psychology Review,* 24(2), 171–91.

Lynch, J. W., Smith, G. D., Harper, S., Hillemeier, M., Ross, N., Kaplan, G. A., & Wolfson, M. (2004). Is income inequality a determinant of population health? Part 1. A systematic review. *The Milbank Quarterly,* 82, 1–77.

Marmot, M. (2004). *The Status Syndrome: How Social Standing Affects Our Health and Longevity.* New York: Times Books.

Masten, A. (2001). Ordinary magic: Resilience processes in development. *American Psychologist,* 56, 227–38.

McEwen, B. S. (2005). Stressed or stressed out: What is the difference? *Journal of Psychiatry Neuroscience,* 30(5), 315–18.

McLoyd, V. (1998). Socioeconomic disadvantage and child development. *American Psychologist,* 53, 185–204.

Meaney, M. J. (2001). Maternal care, gene expression, and the transmission of individual differences in stress reactivity across generations. *Annual Review of Neuroscience,* 24, 1161–92.

Meaney, M. J., Szyf, M., & Seckl, J. R. (2007). Epigenetic mechanisms of perinatal programming of hypothalamic-pituitary-adrenalfunction and health. *Trends in Molecular Medicine,* 13(7), 269–77.

Morgan, H. D., Sutherland, H. G. E., Martin, D. I. K., & Whitelaw, E. (1999). Epigenetic inheritance at the agouti locus in the mouse. *Nature Genetics,* 23, 314–18.

Noble, K., McCandliss, B., & Farah, M. (2007). Socioeconomic gradients predict individual differences in neurocognitive abilities. *Developmental Science,* 10(4), 464–80.

Offord, D. R., Kraemer, H. C., Kazdin, A. E., Jensen, P. S., Harrington, R., & Gardner, J. S. (1999). Lowering the burden of suffering: Monitoring the benefits of clinical, targeted, and universal approaches. In D. P. Keating & C. Hertzman (Eds.), *Developmental Health and the Wealth of Nations: Social, Biological, and Educational Dynamics.* New York: Guilford Press.

Pihl, R. O., & Benkelfat, C. (2005). Neuromodulators in the development and expression of inhibition and aggression. In R. E. Tremblay, W. W. Hartup, & J. Archer (Eds.), *Developmental Origins of Aggression.* New York: The Guilford Press.

Power, C., & Hertzman, C. (1999). Health, well-being, and coping skills. In D. P. Keating & C. Hertzman (Eds.), *Developmental Health and the Wealth of Nations: Social, Biological, and Educational Dynamics*. New York: Guilford Press.

Reik, W. (2007). Stability and flexibility of epigenetic gene regulation in mammalian development. *Nature*, 447, 425–32.

Risch, N., Herrell, R., Lehner, T., Liang, K., Eaves, L., Hoh, J., Griem, A., Kovacs, M., Ott, J., & Merikangas, K. (2009). Interaction between the serotonin transporter gene (5-HTTLPR), stressful life events, and risk of depression: A meta-analysis. *JAMA: Journal of the American Medical Association*, 301(23), 2462–71.

Ross, C. E., & Wu, C. L. (1995). The links between education and health. *American Sociological Review*, 60(5), 719–45.

Ross, N., Wolfson, M., Kaplan, G. A., Dunn, J. R., Lynch, J., & Sanmartin, C. (2006). Income inequality as a determinant of health. In J. Heymann, C. Hertzman, M. L. Barer, & R. G. Evans (Eds.), *Healthier Societies: From Analysis to Action*. New York: Oxford University Press.

Rutter, M., Thapar, A. & Pickles, A. (2009). Gene-environment interactions: Biologically valid pathway or artifact? *Archives of General Psychiatry*, 66(12), 1287–9.

Sapolsky, R. M. (2005). The influence of social hierarchy on primate health. *Science*, 308(5722), 648–52.

Schnittker, J. (2004). Education and the changing shape of the income gradient in health. *Journal of Health and Social Behavior*, 45(3), 286–305.

Shea, J. J. (2003). Neanderthals, competition, and the origin of modern human behavior in the Levant. *Evolutionary Anthropology*, 12, 173–87.

Sherry, D. F. (2006). Neuroecology. *Annual Review of Psychology*, 57, 167–97.

Simonton, S. (2008). Social inequalities in childhood obesity. *Obesity in Childhood and Adolescence, Vol 1: Medical, Biological, and Social Issues* (pp. 61–91). Westport, CT: Praeger Publishers/Greenwood Publishing Group.

Smedley, B. D., Syme, S. L., & Committee on Capitalizing on Social Science and Behavioral Research to Improve the Public's Health. (2001). Promoting health: Intervention strategies from social and behavioral research. *American Journal of Health Promotion*, 15(3), 149–66.

Steinberg, L. (2008). A social neuroscience perspective on adolescent risk-taking. *Developmental Review*, 28(1), 78–106.

Steinberg, L. D., Dahl, R., Keating, D. P., Kupfer, D., Masten, A., & Pine, D. (2006). Adolescent psychopathology. In D. Cicchetti & D. Cohen (Eds.), *Handbook of Developmental Psychopathology*. New Jersey: John Wiley & Sons.

Suomi, S. J. (1999). Developmental trajectories, early experiences, and community consequences. In D. P. Keating & C. Hertzman (Eds.), *Developmental Health and the Wealth of Nations: Social, Biological, and Educational Dynamics*. New York: Guilford Press.

Suomi, S. J. (2000). A biobehavioral perspective on development psychopathology. In A. Sameroff, M. Lewis, & S. Miller (Eds.), *Handbook of Developmental Psychopathology, 2nd Edition*. New York: Kluwer Academic/Plenum Publishers.

Swidler, A. (1986). Culture in action: Symbols and strategies. *American Sociological Review*, 51(2), 273–86.

Szyf, M., Weaver, I. C., Champagne, F. A., Diorio, J., & Meaney, M. J. (2005). Maternal programming of steroid receptor expression and phenotype through DNA methylation in the rat. *Frontiers of Neuroendocrinology*, 26(3–4), 139–62.

Wachs, T. D. (2009). Models linking nutritional deficiencies to maternal and child mental health. *American Journal of Clinical Nutrition*, 89(3), 935S–939S.

Wang, G., & Fowler, B. A. (2008). Roles of biomarkers in evaluating interactions among mixtures of lead, cadmium, and arsenic. *Toxicology and Applied Pharmacology*, 233, 92–99.

Williams, D. R. (2005). Patterns and causes of disparities in health. In D. Mechanic, L. B. Rogut, & D. C. Colby (Eds.), *Policy Challenges in Modern Health Care*. Piscataway, NJ: Rutgers University Press.

Willms, J. D. 1999. The effects of families, schools, and communities. In D. P. Keating & C. Hertzman (Eds.), *Developmental Health and the Wealth of Nations: Social, Biological, and Educational Dynamics*. New York: Guilford Press.

Willms, J. D. (2003). *Ten Hypotheses about Socioeconomic Gradients and Community Differences in Children's Developmental Outcomes*. Quebec, Canada: Human Resources Development Canada.

Worthman, C. M., & Kuzara, J. (2005). Life history and the early origins of health differentials. *American Journal of Human Biology*, 17(1), 95–112.

Wolfson, M., Kaplan, G. A., Lynch, J., Ross, N., & Backlund, E. (1999). Relation between income inequality and mortality: Empirical demonstration. *British Medical Journal*, 319, 953–7.

Wright, R. O., & Baccarelli, A. (2007). Metals and neurotoxicology. *Journal of Nutrition*, 137, 2809–13.

Yancey, A., & Kumanyika, S. (2007). Bridging the gap: Understanding the structure of social inequities in childhood obesity. *American Journal of Preventive Medicine*, 33(4), S172–S174.

INDEX